ELECTROTECHNOLOGY VOLUME 2

APPLICATIONS IN MANUFACTURING

ELECTROTECHNOLOGY VOLUME 2

APPLICATIONS IN MANUFACTURING

ROBERT P. OUELLETTE
Associate Technical Director, Energy, Resources and the Environment
METREK Division of The MITRE Corporation
McLean, Virginia

FRED ELLERBUSCH
Technical Staff
METREK Division of The MITRE Corporation
McLean, Virginia

PAUL N. CHEREMISINOFF
Consulting Engineer
Closter, New Jersey

Editors

ANN ARBOR SCIENCE
PUBLISHERS INC
P.O. BOX 1425 • ANN ARBOR, MICH. 48106

230 Collingwood, P. O. Box 1425, Ann Arbor, Michigan 48106

Library of Congress Catalog Card No. 77-85093
ISBN 0-250-40207-6

Manufactured in the United States of America

FOREWORD AND ACKNOWLEDGMENTS

A series of studies surveying the status of electric technologies was initiated and prepared for Eléctricité de France by the METREK Division of MITRE Corp. Much of this work was originally conducted under the technical and administrative direction of Dr. Robert Ouellette. The editors would like to acknowledge the following individuals for the work in preparation of the original reports:

Marcel Barbier	Norman Lord
Charles Bliss	Grant Miller
B. Bovarnick	Robert Pikul
Oscar Farah	Rip Rice
W. E. Jacobsen	Ralph Roberts
Mary Harlow	John Vlahakis

Special acknowledgment is also made to Mr. J. Bouchet, Direction Générale, Service d'Etude et de Promotion de l'Action Commérciale, Eléctricité de France, for his encouragement to the editors of Volumes 1 and 2 of this series to reorganize and present this information. The objective and interest is to see rapid introduction of the technology using electricity and diffusing this information that is favorable to its development. The principles of operation and main applications of processes are discussed, and an assessment made of economic and energy advantages, and the potential market for techniques wherever possible.

Robert P. Ouellette
Fred Ellerbusch
Paul N. Cheremisinoff

CONTENTS

1
CORELESS INDUCTION HEATING*

INTRODUCTION

Coreless induction heating is being used for an increasingly wide variety of industrial applications requiring the heating or melting of conducting solid materials. Applications span a large spectrum of machine size requirements, including a 600 ton/hr forging application down to spot heating of small parts such as in the assembly of plastic medical syringes.

Estimates of the total market for heating equipment and the portion of this market attributed to coreless induction heating are given in Table 1-1, which shows that the largest potential market for induction heaters is in heat processing and metal treating, amounting to $25.6 million in 1974. Of this market, 13.5% of the equipment installed in 1974 was coreless induction equipment, which represented an increase from 11% since 1958 for the share of the market it held. Total constant 1974 dollar amounts of equipment installed by 1974 amounted to $400 million.

The largest use of coreless induction heating equipment is in molten metal melting, refining and holding. In 1974, the total market for this type of equipment was $72.9 million of which 76% was supplied by coreless induction equipment; this represents a dramatic increase from a market share of 19% in 1958. By 1976 over $600 million in coreless induction equipment had been installed.

Induction heating equipment has also been used increasingly for hot forming, forging, rolling, piercing and extruding. Table 1-1 indicates its market share increased from 13% in 1958 to 41% in 1974 in a total market of $17.8 million. By 1974 $100 million in induction heating equipment had been installed in this market.

The greatest potential market is for large-scale machines for heat processing and metal treating. It is anticipated that the market share will

---

*By: Grant Miller

Table 1-1. Summary of Market Data for Existing Coreless Induction Heating Applications

Actual Submarket	Total Market Size per Year ($ millions)			Induction Heating as a Percentage of Total Market Size			Total Cumulative Induction Heating Capacity Installed by 1974 ($ millions)
	1958	1965	1974	1958	1965	1974	
Molten metal, melting, refining and holding furnaces	22.9	58.9	72.9	19%	42%	76%	600
Hot-forming, forging, rolling, piercing and extruding	18.8	33.1	17.8	13%	11%	41%	100
Heat processing and metal treating	95.9	159	215.6	11%	10%	13.5%	400

increase appreciably in the near future. Bobart has indicated that consumption of induction heating equipment in this market is expected to increase (in constant 1974 dollars) from $35 million in 1974 to $55 million in 1980 and $85 million in 1985.[1]

Generally, the large-sized and capacity machines require the use of considerably automated monitoring, handling and protection devices, while small-scale induction heaters may be entirely manually controlled. Thus, costs of inductor heating equipment vary greatly depending on size; capacity machines require the use of considerably automated monitoring, handling and protection devices, while small-scale induction heaters may be entirely manually controlled. Nevertheless, equipment costs may be estimated based on the operating frequency of the equipment (Table 1-2). The equipment costs from Table 1-2 are comparable or just slightly more expensive than alternative heating method equipment.

Generally, equipment for large-scale heating applications utilizes the 60-cycle frequency for heating below Curie temperature and 180 cycles for heating above. Energy usage in kWh for heating steel is given in Table 1-2. For example, if 2 tons/hr of steel are to be heated to 200°F using 60-cycle power, the number of kWh required over a 5-hr period would be:

$$2 \text{ tons } (200 \text{ lb/ton}),\ 200^{\circ}\text{F } (5 \text{ hr}) (5.24 \times 10^{-4} \text{ kW}) = 2090 \text{ kWh of energy}$$

Table 1-2. Cost and Efficiency Data for Coreless Induction Heating Equipment[a]

Frequency Range	Estimated Efficiency	$/kW	Energy Usage for Heating Steel 1°F (Specific Heat = 1.16)
60-cycle	60-70%	60-70	5.24×10^{-4} kW/(lb/hr)/day
180-cycle	50-60%	90-110	6.18×10^{-4} kW/(lb/hr)/day
960-cycle	45-50%	105-120	6.80×10^{-4}
3000-cycle	45-50%	120-145	7.15×10^{-4}
10,000-cycle	45-50%	145-165	7.15×10^{-6}
450-cycle (RF)	40%	200-225	8.50×10^{-4}

[a]Costs include allowance for work-handling apparatus.

The primary advantages of coreless induction heating equipment include:

- dramatically reduced energy costs over fossil fuel alternatives;
- reliance on a secure and predictable energy supply (electricity);
- reduction in the scale and size of equipment providing savings of floor space; and
- large-scale iron heating, reduction of scale loss from 3% for fossil fuel furnace to 1% for an induction furnace.

For most applications, coreless induction heating is a capital-intensive technology, but due to its lower energy consumption it is economically advantageous over competing technologies, especially motor generators and fossil-fueled furnaces.

For some applications, newer, competing technologies are surfacing. For surface treatment of metals, for example, laser technology has recently been introduced and is cheaper than induction heating. Another alternative to induction heating, for nonconductive products, is microwave heating, and technology development on infrared heating is underway.

GENERAL PRINCIPLES

Coreless induction heating is being used for an increasing variety of industrial applications requiring heating or melting of conducting solid materials. Some current uses include:

1. Forging: heating or metal before shaping;
2. Melting: reducing metal or ores to a molten state;
3. Soldering: joining of two separate parts by heating, usually by the introduction of a soldering metal;
4. Annealing: heating to remove or prevent internal stress;
5. Tempering: heating and subsequent quenching of materials to produce the desired state of hardness and elasticity;

6. Bonding: joining of two separate parts through heating;
7. Shrink-Fitting: joining of two separate parts by expanding an outside part, positioning an inside part and doing shrinking upon cooling to provide a tight joint;
8. Coating: covering with a layer through heating and flow of material;
9. Crystal Growing: careful temperature control allows growth of large, pure and stable crystals; and
10. Sputtering: evaporative disposition of metal on a surface.

These applications span a large spectrum of temperature and machine size requirements. Generally each application is sufficiently specific to require some specialized design and integration effort, especially in small-scale applications such as bonding or shrink-fitting; it is less true of large-scale forging or melting applications.

The principle of induction heating is illustrated in Figure 1-1. When a copper coil is wound around a conducting workpiece and an alternating current is passed through the coil, a magnetic field is established that also causes a current to flow in the load. Linear induction coils enable the coils to correspond to the particular shape of the workpiece. The passage of current through the electrical resistance of the workpiece causes the workpiece to heat up. Distribution of the induced current through the load cross section is not uniform, and, consequently, heating is not uniform. The current decreases exponentially in magnitude from the surface to the center of the workpiece. The depth to which the current flows depends on the load resistivity, its permeability and the frequency of the alternating current. For steel, the two key factors are its magnetic properties and the a.c. power source frequency.

To obtain high heating efficiency, the diameter or cross section of the workpiece should be at least three times the heating penetration (the depth to which 87% of the induced heat is developed). Too high a rate increases the distance heat must flow and requires either a large heat time to permit heat to soak to the center, or a greater temperature differential between the surface and center of the workpiece. Table 1-3 shows a.c. current frequencies normally used to heat steel billets and bars to forging temperature as a function of cross section size.

EXISTING APPLICATIONS

Forging and Melting Applications

Induction heaters for forging and melting applications are generally large-sized coil or linear induction machines. For heating steel from ambient temperature to 2350°F in rolling mill preheating operations, induction

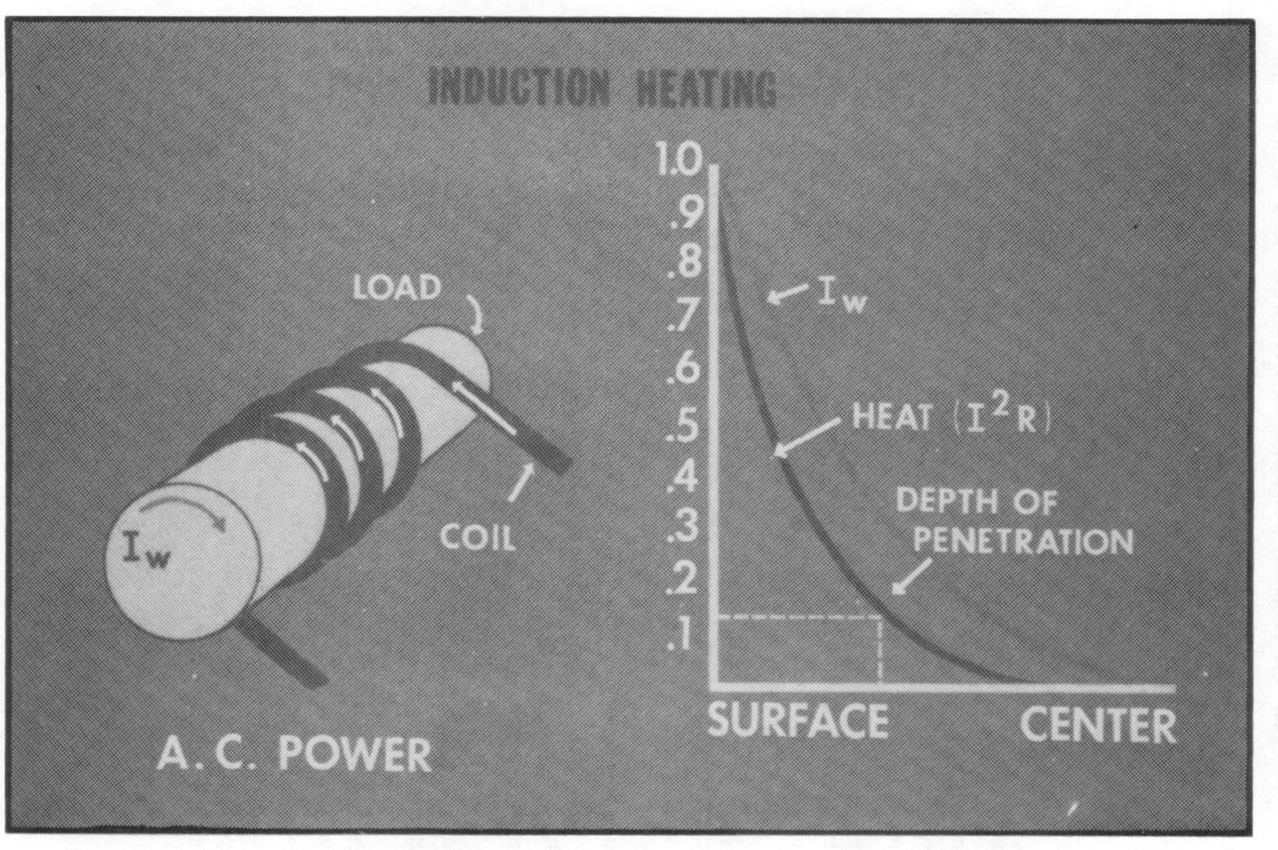

Figure 1-1. Principle of induction heating.[1]

Table 1-3. Frequency Selection Chart–Heating Steel for Forging[1]

Cross Section (in.)	(70-1300°F)	(1300-2200°F)
Over 6	60 Hz	60 Hz
4-6	60 Hz	180 Hz
2-4	60 Hz	1 kHz
1-2	180 Hz	3 kHz
½-1	1 kHz	10 kHz

heaters have been built up to a capacity of 600 tons/hr, requiring a power supply capable of delivering 200,000 kWh.[2] Such large-capacity machines require considerable automation equipment for automatic handling of the production line and power monitoring to adjust the load power factor. In addition, sophisticated switching devices are required to switch the large power loads on and off without unbalancing or damaging the system.

In the rolling mill preheating application, the heater consists of four coils, each connected in parallel. An autotransformer can raise or lower the voltage of each individual coil $\pm$10% in steps of 2%. This permits regulation of heat input to different sections of the slab to obtain uniform temperature. A circuit diagram for a 20-MW heater of this type is shown in Figure 1-2.

In Westinghouse induction heaters, regulation of heat input is achieved by varying the voltage by means of: (1) saturable reactors in series with the input powerline; (2) three silicon control rectifiers and associated firing controls; or (3) a dropper tube (vacuum diode) in the high-voltage d.c. line to the oscillator. A schematic diagram of these controls is given in Figure 1-3.

The manual operation of such a rolling mill plant is inconceivable; no human being could operate such a vast and complex set of equipment without making serious mistakes. This specific installation includes five sets of control systems including slab handling control (digital); heater control (digital); static power switches (silicon control rectifiers); slab temperature control (analog); and process computer (digital). The controls during normal heating perform the following functions:

1. Slab Handling Control
    - Operates the gantry cranes;
    - Depiles slabs;
    - Charges heaters;
    - Deposits heated slabs on the mill, approaches table and sends them to the mill.

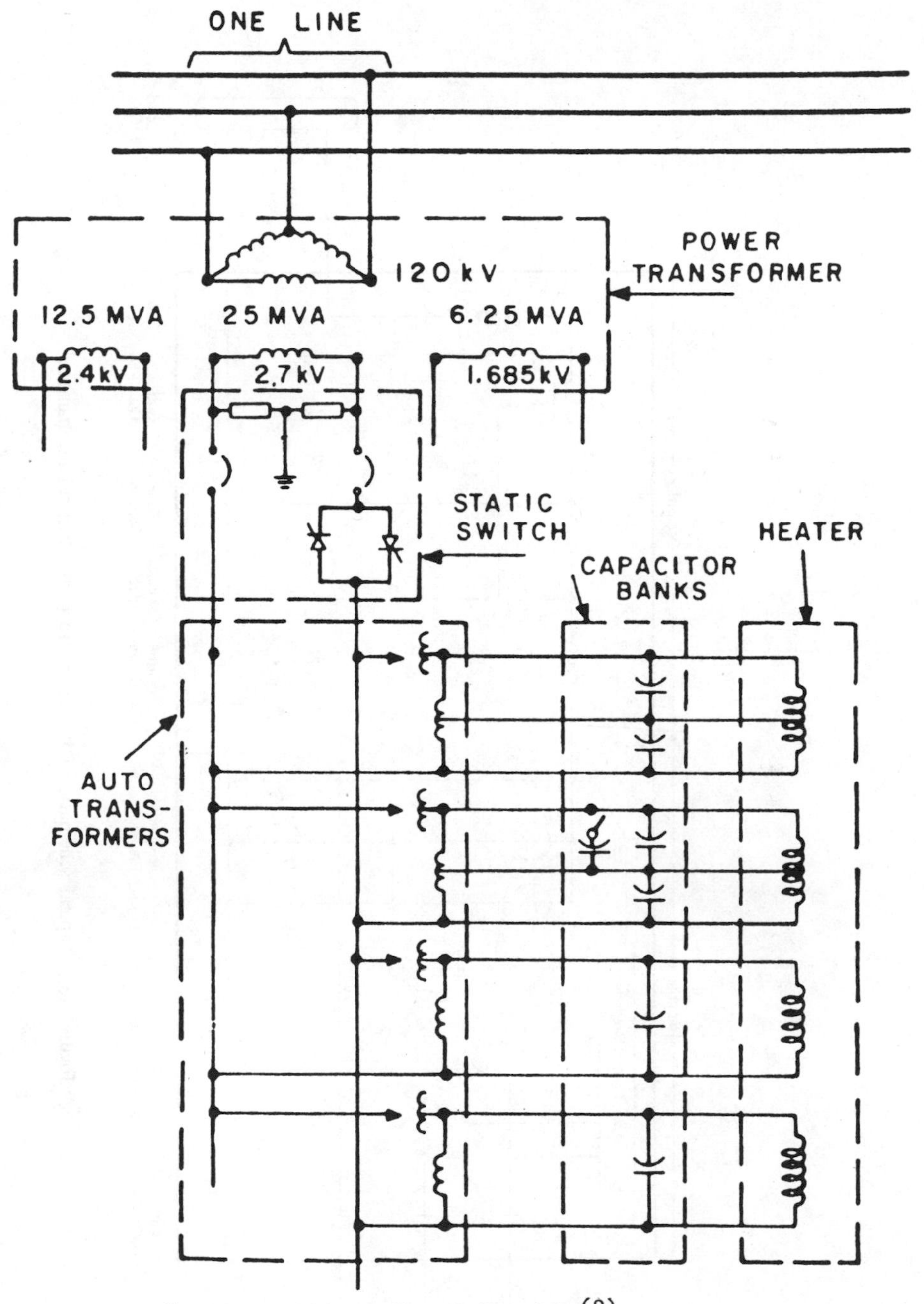

Source: Bijwaard and Sorokin(2)

Figure 1-2. Electrical circuit for one of six heater lines.[2]

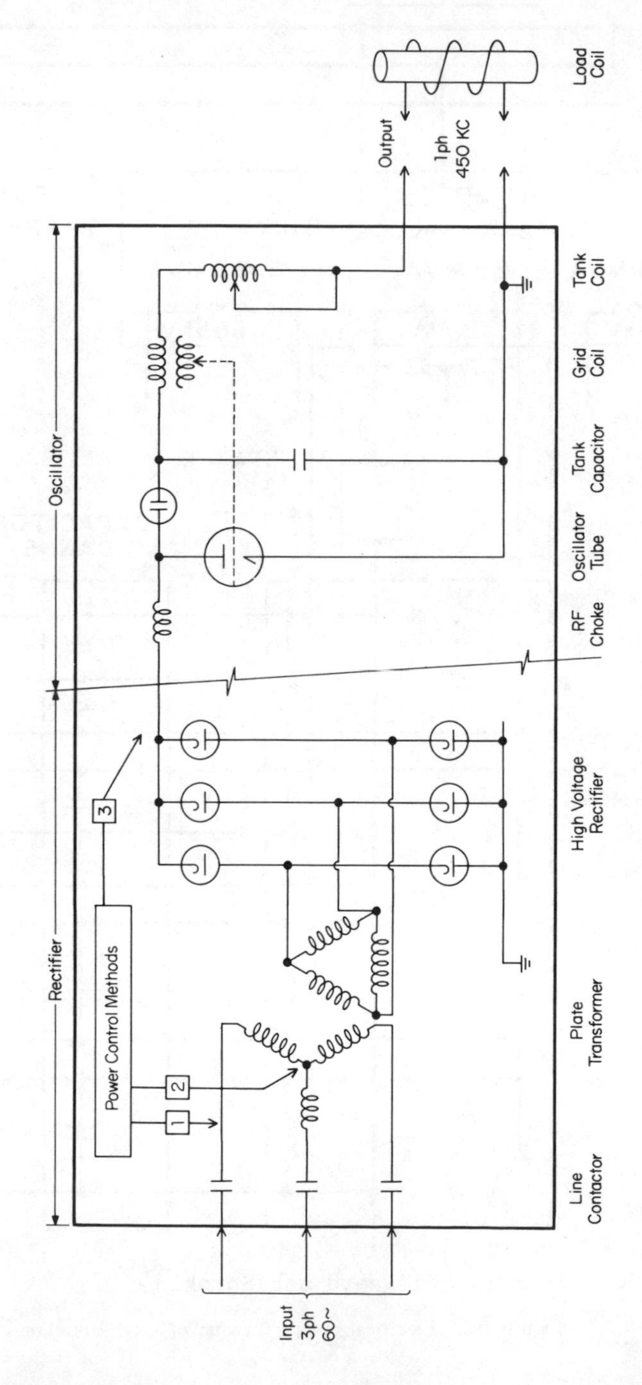

Figure 1-3. Typical schematic of Westinghouse power control mechanism.[1]

2. Heater Control
   - Changes tap settings and sets capacitor switches for each slab width;
   - Advances and retracts thermocouples;
   - Signals static switches to turn heaters on and off after checking the permissive circuits.
3. Static Switches
   - Switches currents on and off under lagging or leading conditions;
   - Detects and clears line-to-line faults;
   - Detects and clears line-to-ground faults;
4. Slab Temperature Control
   - By means of two proximity-type thermocouples per heater, slab surface temperatures are provided to the computer during the heating cycle. Low- and high-temperature adjustable limit switches are connected to recorder servos.
5. Process Computer
   - Demand limit control;
   - Phase balance control;
   - Slab tracking and coil identification;
   - Logging.

In addition to these systems, there are motor control centers, an annunciator system, a closed-circuit TV monitoring system, and an extensive protective relay system. Figure 1-4 shows the automatic handling and control scheme for the entire slab reheat plant.

For melting applications, the heating rate is generally altered by regulating the power supplied to the coils as in the rolling mill applications. However, for one type of induction furnace, Pillar Corporation has provided for automatic monitoring of the power supply load. Virtually all other Pillar induction heaters use automatic monitoring equipment to regulate the heating rate through the power supplied to the coils. Pillar also utilizes more than one frequency to heat below and above the curie temperature (180 Hz, 640 Hz) for heating red brass and bronze to 2150°F. The switch from one frequency to the other is accomplished using solid state, serviconductor equipment.

Recently, Pillar supplied induction melting equipment to West Germany to hold liquid steel before casting. This equipment utilizes disposable crucibles that are filled with the exact charge to be cast. The entire contents are cast at one time avoiding any potential slab problems that might arise by casting only part of a crucible at one time. If it is desired to cast only small quantities of steel, generally a large power capacity is provided to assure a short melting or heating time for the charge of the crucible.[3]

Lindberg manufactures several lines of induction furnaces to melt and hold metal before casting. By using a two-chamber system, close temperature uniformity is maintained and charged metal has little effect on ladeled

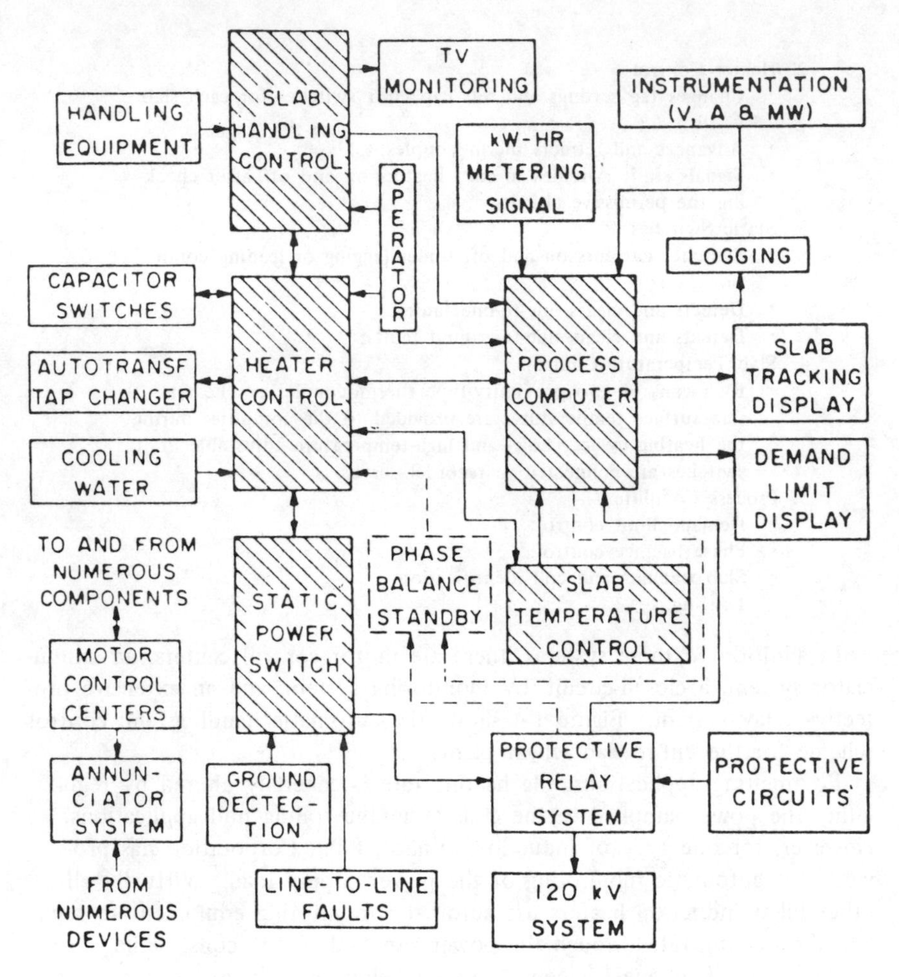

Figure 1-4. Block diagram of automatic handling and control scheme for slab reheat plant.[1]

metal. This results in fewer oxides, and less sludge production. Ladling can be continued even while the channels are being rodded. This design also saves time on shutdowns to replace crucibles or to clean sidewalls, as required by other furnace designs.[4]

Most large induction heaters are designed and constructed based on the specific application and part geometry required. Manufacturers of large-scale induction furnaces maintain design and applications staffs to develop induction heating equipment optimized for the application desired. These manufacturers include Ajax Magnethermic Corporation Westinghouse Electric Corporation; Induction Process Equipment; Lindberg; and Pillar.

Technical data for a few standard induction furnaces for forging of steel are presented in Table 1-4. The examples listed for Pillar and Induction Process Equipment Corporation represent the limits of capacity of the standard equipment detailed in the literature published by the abovementioned companies. Information on larger and smaller units is available upon specific request of these companies. Induction furnace specifications of the other manufacturers of large-scale equipment are also available upon request.

Power Requirements for Forging and Melting Applications

Capacity and power requirements of Pillar induction furnaces for melting applications are given in Table 1-5. The furnace sizes listed are merely representative of the wide range of sizes and capacities available. Manufacturers will supply technical data and costs of equipment for specific applications.

Examples of currently operating large-scale forging equipment include:

- A pusher-type in-line slug heater installed for a major auto manufacturer, which produces 8000 lb/hr of circular 4-in radius by 2- to 4-in-long steel slugs. The five coils of the heater are supplied by 1500 kW of line frequency and 1500 kW of 180 Hz power.
- An installation to heat stainless and alloy steel billets for forging turbine blade preforms. The billets varied from 2.5 in to 6 in. in diameter. A very precise and uniform temperature of 215°F was required. Frequencies of 60 Hz and 1 KHz were utilized with a combined power level of 750 kW.
- A long bar heater for hot shearing and forging railroad bearing races. It heats at 2-4 in. diameter by 20-foot bars at a rate of 14,000 lb/hr. It is powered with 1250 kW of line frequency and 1250 kW of 1 kHz power.
- A 10-ton furnace for melting with a capacity of 5200 pounds of aluminum. It uses 200 kW of input power to produce a melt rate of 1000 lb/hr at 1250°F. It operates at 240 KVA at a frequency of 60 Hz, nominal 230 V or 460 V.[1,5]

Recently, Westinghouse introduced a new type of bar heater—the transverse walking beam heater. In this type of induction heater, bars are placed on walking beam rails. They are then literally walked through the heating line with the bar length transverse or 90°F to the direction of motion.

The walking beam concept provides a natural means of starting production with an empty bar heating line and provides continual production even with interrupted bar availability. When production stoppage is desired, bar feeding ceases to the infeed; in-process bars continue to heat and are fed directly to a forge press. If infeed bar interruption occurs due to

Table 1-4. Technical Data for Forging Induction Furnaces

Motor Generator Manufacturer and Furnace Model	Input from Mains		RF Power (kW)	Frequency (Hz)	Water (gal/min)	Machine Dimensions		
	KVA	Voltage				Wide	Deep	High
Pillar	180	460	150	180-540	7	42	48	90
	2145	460	1800	180-540	54	180	48	93
Induction Process Equipment	-	440, 460, 480	400	1000,3000	-	36	48	96
	-	440, 460 480	1600	1000,3000	-	36	192	96

Table 1-5. Melting Rates, Pillar Mark 14 and Mark V[5]

Pillar Mark 14 Power Supply, 100% Solid State– 180 or 540 Hz Melting Rates			
Metal & Pour Temp.	**Freq.**	**Nominal kW**	**Typ. #hr.**
Red Brass and Bronze to 2150°F		250	1630
		350	2280
	180	500	3255
	540	700	4560
		1000	6510
		250	1630
	180	350	2280
		500	3255
		700	4560
Cast Iron to 2650°F		250	1025
		350	1425
	540	500	2050
		700	2865
		1000	4095
Cast Iron from 2650°F		250	1025
		350	1435
	180	500	2050
		700	2965
		1000	4095
Steel to 3000°F		250	870
	540	350	1210
		500	1730
		700	2425
		1000	3465
Aluminum Capacity (lb)			
Aluminum to 1400°F		250	1025
		350	1435
	540	500	2050
		700	2865
			4095

Pillar Mark V Power Supply, 100% Solid State 1 & 3 Melting Rates			
Metal & Pour Temp.	**Freq.**	**Nominal kW**	**Typ. #hr.**
Red Brass to 2150°F		45	290
	3kHz	90	570
		135	860
		180	1140
		125	790
	1kHz	175	1110
		250	1590
		375	2385
Iron to 2650°F		45	140
		90	310
	3kHz	135	470
		180	630
		125	435
	1kHz	175	610
		250	875
Steel to 3000°F		45	145
		90	305
	3kHz	135	460
		180	610
		125	420
		175	590
		250	850
	1kHz	375	1260
		500	1690
		750	2520
		1000	3380
Aluminum Capacity (lb)			
Aluminum to 1400°F		45	140
		90	310
	3kHz	135	470
		180	630
		125	435
	1kHz	175	610
		250	875

Note: For other metals, multiply melt time for red brass by following figures: copper 1.1, yellow brass .85, silver 170, gold .40. All melt rates are based on hot-lining, properly charged furnace, and Pillar-designed or approved power transmission system. Cold charge takes approximately 10% longer. Vacuum melting takes approximately 100% longer.

availability or flaws during normal operation, an end-to-end bar forging machine must be shut down since a line stoppage without power removal results in melted bars. In the walking bar heater the bars may be walked out of the heater and fed directly to the press.

In the walking bar heater, the rails that carry the bars separate them so they are not in physical contact. This insulates the bars to prevent arcing and insures that parts do not weld together preventing handling problems and reducing scrap.[1]

General Cost Guidelines for Forging and Melting Applications

Since production requirements are known, the equipment costs can be estimated. These vary depending on the handling and auxiliary equipment required. However, typical costs for induction heating equipment for forging applications are given in Table 1-6 as a function of frequency and power requirements. For forging applications, the frequency of the power source is dependent on the diameter or cross section of the workpiece. Typical frequency requirements for steel workpieces are given in Table 1-3.

Table 1-6. Equipment Efficiency and Cost[6]

Class of Equipment	Estimating Efficiency (%)	$/kW
60-cycle	60-70	60-70
180-cycle	50-60	90-110
960-cycle	45-50	105-120
3,000-cycle	45-50	120-145
10,000-cycle	45-50	145-165
450 kcycle (RF)	40	200-225

NOTE:

1. Efficiency is overall thermal energy in work required divided by power from the incoming line.
2. Heating is from 70°F to 2150°F except in the case of 450 kilocycles equipment where maximum is 1200°F.
3. Costs include allowance for work handling apparatus.

To heat 6 tons/hr of high-volume 3.5-in bars would require 2000 kW of power capacity comprised of 1000 kW at line frequency and 1000 kW at 1 KHz. The total cost for this package (Table 1-4) would be $250,000. However, the savings in energy cost that can be realized by introducing induction equipment more than compensates.

Energy costs have typically comprised one-third to one-half the overall operating costs of heating for forging applications. When heating steel to forging temperature, average overall heating efficiency is approximately 65% in induction furnaces. This means 6 pounds of steel may be heated per kilowatt hour of electricity consumed or one ton of product would consume 333 kWh of electricity. Where high-frequency power supplies are used on smaller-sized billets and bars, these values are reduced by approximately 10-15% or to 5-5.5 lb/kWh. Estimates for the energy required to melt ferrous metals are given in Table 1-7.

Table 1-7. Energy Required to Melt Ferrous Metals[7]

	Cupola	Electric Induction	Electric Arc
Heat Equipment Efficiency			
To preheat and melt to 2300°F	60%	60%	75%
To superheat from 2300°F to 2700°F	7	60	25
Millions of Btu to Preheat, Melt and Superheat 1 Ton of Cast Iron			
Theoretical	1.10	1.10	1.10
Actual			
Preheat, Melt and	1.60	1.59	1.28
Superheat	2.04	0.24	0.57
Total	3.64	1.83	1.85

Although the energy input to an electric furnace is approximately half that of the cupola, if the energy consumed in generating the power is counted, the electric induction furnace above consumes, in effect, 5.49 million Btu, 1.5 times as much as the 3.64 million Btu consumed by the cupola.

A comparison of electricity costs for induction furnaces with fuel and gas costs for fired furnaces is given in Figure 1-5. The comparison is based on induction furnace efficiency of 65% and fossil fuel furnace efficiency of 20%. This efficiency indicates 3.5 million Btu are required per ton of heated product. This will vary depending on whether a more efficient closed-type rotary furnace is used or if the less-efficient open-type slot furnace is employed. The curve of Figure 1-5 indicates electricity costs of 1¢/kWh are equal to natural gas costs at 90¢/million Btu

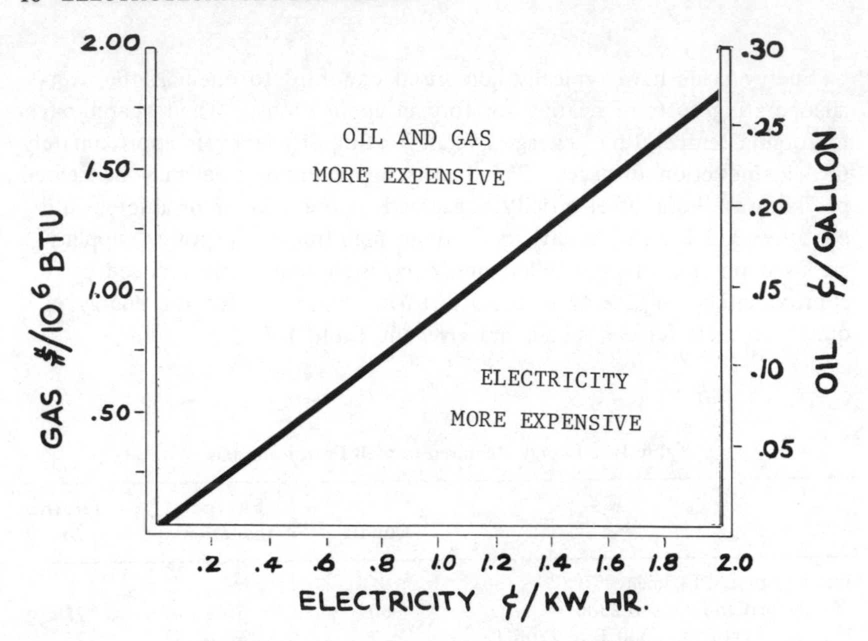

Figure 1-5. Heating for forging. Process heating energy cost comparison.[1]

or 13 ¢/gal of heavy petroleum or refining residual fuel oil. By picking any point on the curve, one can readily determine how energy costs compare to this given set of conversion conditions. Where fossil fuel costs are above the curve for a chosen cost of electricity in ¢/kWh, the fossil energy costs would be higher than using electric induction, conversely where they are below, fossil fuel costs would be lower.

The following is an example of total costs and cost savings for a specific application as prepared by Westinghouse Electric Corporation.[6]

Example

Heat 5-in RCS bar at rate of 20 tons/hr from 70°-2150°F. Heat required in bar:

$$\text{Btu/hr} = \text{lb/hr x specific heat x temperature rise}$$

$$= 20 \times 200 \times 1.16 \times 2080 = 13.3 \times 10^6$$

$$\text{kW required} = \frac{13.3 \times 10^6 \text{ Btu/hr}}{3413 \text{ Btu/kWh}}$$

Step 1. Select frequency to use from Table 1-6. For below-curie heating use 60 cycles. For above-curie heating use 180 cycles.

Step 2. Determine efficiencies from Table 1-6. For 60 cycles use 65%; for 180 cycles use 55%.

Step 3. Determine kW of 60-cycle power and kW of 180-cycle power. (Rule: On dual-frequency applications use 50% of each frequency.)

Step 4. Establish kW of each frequency.

$$\text{60-cycle kW} = \frac{1950}{.65} = 3000$$

$$\text{180-cycle kW} = \frac{1950}{.55} = 3555$$

Step 5. Determine capital equipment costs from Table 1-6. Cycle equipment costs \$60-70/kW and 180-cycle equipment costs \$90-110/kW.

60-cycle equipment cost = 3000 kW x \$ 65/kW = \$195,000

180-cycle equipment cost = 3555 kW x \$100/kW = 355,000

TOTAL EQUIPMENT COST = \$550,000

Step 6. Determine length of induction heating line from Figure 1-6. It takes 7.0 min to heat a 5-in RCS bar to 2150°F.

Bar weight per foot = 5 x 5 x 12 in x 0.293 lb/in^3 = 84 lb

Ft/min at 40,000 lb/hr = 40,000 lb/hr = 475 ft/hr = 7.92 ft/min

Line length = 7.92 ft/min x 7 min = 55.5 ft

56 ft in one line or 29 ft for each of two lines or 19 ft for each of three lines.

Step 7. Establish operating costs using the actual examples shown in Figure 1-5 where electricity cost is 0.8 ¢/kWh.

Operating cost = 6555 kW x \$0.008/kWh

= \$52.44/hr = \$2.64/ton

Note: From Figure 1-5, the corresponding gas cost is \$0.95/million Btu.

Step 8. Determine scale savings using induction.

Note: Scale loss in a furnace is 3% as compared to 1% with induction.

Furnace scale loss = 3% of 20 tons = 0.6 ton/hr

Induction scale loss = 1% of 20 tons = 0.2 ton/hr

Value of hot rolled carbon bar of merchant quality = \$80.00/ton

Furnace scale loss = 0.6 ton/hr x \$80.00 = \$48.00/hr

Induction scale loss = 0.2 ton/hr x \$80.00/ton = \$16.00/hr

Savings Using Induction = \$32.00/hr

Step 9. Determine realized cost reduction in ¢/kWh of electricity afforded by savings in scale, using induction.

$$\text{Scale savings using induction} = \frac{\$32.00/\text{hr}}{20\ \text{ton/hr}} = \$1.60$$

Actual operating cost using induction equals \$2.64/ton - \$1.60/ton = \$1.40/ton

Small-Scale Applications

Induction heating may be applied to a variety of small-scale applications including soldering, annealing, tempering, brazing, shrink-filling, bonding, crystal growing and sputtering. Induction heating is used to braze the heat to a generator housing to produce a high-strength, high-integrity joint for use in high-reliability aircraft and missles. Rapid and localized induction heating permits soldering of parts containing heat-sensitive elements such as instrument cans, transistor seals and electronic components. It may also be used to bond knife blades to plastic handles. The load coil is positioned around the plastic handle and knife blade. Only the metal tang is heated by induction, which in turn heats the plastic sufficiently to flow about the tang. A secure bond is formed without affecting the surface of the plastic handle or the previously tempered knife blade.

Shrink-fitting is another small-scale application of induction heating. For example, a cam shaft, collar and offset collar are heated to expand the inside diameter sufficiently so the shaft may be readily inserted. Glass-to-glass and glass-to-metal seals are also accomplished by induction heating by placing a preoxidized glass-coated Kovar ring between the pieces. The induction-heated ring softens the glass in the joint area and causes plastic flow to produce a seal.

Small-scale uses generally require specialized equipment designs for optimal process performance. Such equipment requires large initial investments in process equipment but the savings in energy, time and floor space can be considerable. Manufacturers of induction furnaces for small-scale uses may be contacted individually regarding specific applications for design and cost estimates.

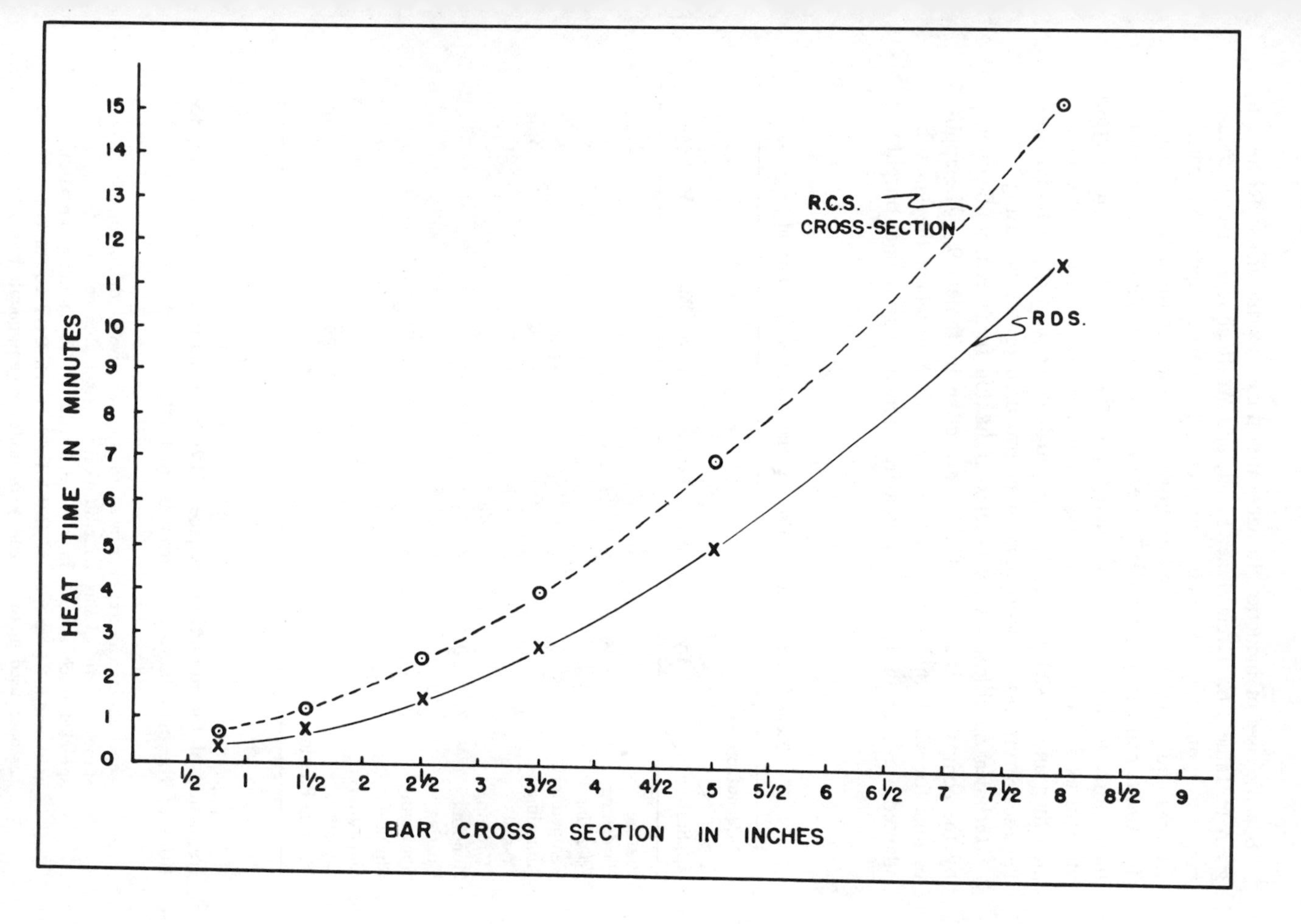

Figure 1-6. Heat time vs. bar cross section, through heating from room temperature to rolling temperature.

Specific uses of induction furnaces mentioned by the manufacturers are given in Table 1-6. Cycle-Dyne, Lepel, and Westinghouse stress a wide range of applications for their equipment.[6-8] Leco Corporation manufactures equipment to be used exclusively in the determination of carbon and sulfur content of materials.[9] Centorr Associates manufactures equipment primarily intended for vacuum applications, especially vacuum pressing and extruding. Their equipment may also be used for tempering, annealing and crystal growing.[10] Although Pillar primarily manufactures large-scale furnaces, their equipment is also used for shrink-fitting.[5]

Technical specifications for coreless induction furnaces manufactured by the companies listed in Table 1-8 are given in Table 1-9. The applications for which these furnaces may be used or to which they may be adapted are given in Table 1-10 as a function of the nominal output

Table 1-8. Estimates of Uses for Manufacturer's Equipment

Application \ Manufacturer	Cycle-Dyne	Lepel	Leco	Centorr	Pillar	Westinghouse
Brazing	X	X				X
Hardening	X	X				X
Soldering	X					X
Tempering				X		X
Annealing				X		X
Crystal Growing	X			X		
Bonding	X	X				
Coating		X				
Sputtering		X				
Shrink-Fitting		X			X	X
Quantitative Analysis			X			
Vacuum Pressing				X		

frequency of the induction furnace. Typical examples of specific applications of the manufacturer's furnaces include:

- Lepel equipment has been used for the selective hardening of gear sprocket teeth. Using a medium-carbon steel, the teeth of the sprocket reach 1600°F in 5 sec using the 10 kW inductor generator at a nominal frequency of 450 kHz.[11] Cycle-Dyne equipment has been used in the assembly of multiple components in a single

Table 1-9. Generator Specifications for Small-Scale Applications[5,8-11]

Generator Model	Power Input–KVA	3-Phase Voltage	R. F. Power Output (kW)	Frequency		Water Req. (gal/ton)	Cabinet Requirements (in.)		
				kHz	mHz		Wide	Deep	High
Lepel									
T10-T80	25.4-192	230, 460, 550	10-80	150-450	2.5-8	8-30	36-52	54-80	76
T1000-T200	240-480	460	100-200	150-450	2.5-5	35-60	52	80	76
T250-T500	590-1200	460	250-500	150-450	-	60-120	52	80	74
Cycle-Dyne									
AA series	1.9	120	1	450	27	5-50 psi	16	15	24
B series	45-120	200-250				3-6			
		400-500	20-60	450	3	gal/min	37	60	78
C series	11-34	200-250				3-6			
		400-500	5-15	450	3	gal/min	27	28	54
Leco									
521 or 523 series	1.0	115	-	-	14	-	16	30	24
621 or 623 series	1.0	220	-	-	14	-	16	30	24
763 series	1.0	115/220	-	-	14	-	16	30	24
Centorr									
600-6	-	-	30	4.2		variable flow–50 psi	-	-	-
Pillar									
1	4.2	230,460	3	maximum 50	-	0.75 gal/min	24	18	56
2	8	230,460	5	50	-	0.75	19	30	42
3	16	230,460	10	50	-	1.5	19	30	42
4	40	230,460	25	50	-	3.5	34	24	75
5	80	230,460	50	50	-	7.5	34	48	75

Table 1-10. Potential Uses for Induction Furnaces as a Function of Nominal Output Frequency[6]

Frequency[a]	Application
80-200 kHz	Deep heating for hardening and forging plated parts.
180-400 kHz	Expitaxial growth; crystal growing; zone processing.
250-450 kHz	General purpose heating; surface hardening and joining operations.
2.5-5 MHz	Plasma processes; crystal growing; zone processing; heating thin parts.
15-30 MHz 30-50 MHz	Research and special applications.

[a]Specific frequency will be determined by the generator, load coil and load being heated.

operation. A 7.5 kW generator has been used to produce two simultaneous sanitary brazes in 15 seconds in the assembly of a deep fat fryer.[8]

- Leco equipment is designed solely to carry out quantitative analyses of carbon and sulfur content of materials. Complete carbon combustion is accomplished in 40 sec; complete sulfur combustion in 50 sec through the application of high temperature (1650°F CO) and high-pressure oxygen (to 20 psi).[9]

Data regarding specific small-scale applications for equipment manufactured by Centorr, Pillar and Westinghouse are available upon request to the manufacturer. (All manufacturers expressed a willingness and desire to discuss design and cost specifications with potential induction furnace users.)

POTENTIAL APPLICATIONS

Induction heaters may be used for heating or melting applications over a wide range of technologies. Solid state induction heaters generally may be used to replace motor generator heating equipment with a resultant increase in heating efficiency. Since induction heaters usually require large initial investments in equipment, some manufacturers have been reluctant to introduce induction heaters. The anticipated energy savings, especially in view of recent energy situations, make induction heaters increasingly attractive.

ASSESSMENT OF THE FUTURE

Market Factors

Over the last 10 years, major technological developments have centered on static power supplies and larger mass heating applications. Conventional rotating motor generators operating at frequencies up to 10 kHz have been largely replaced by solid state power supplies. It is expected that solid state device improvements, increased reliability and energy savings potential will advance sufficiently to convert most high-frequency heating to solid state by 1985.[12] Emphasis will be on large mass heating such as slab, ingots and blooms, and on in-line processing applications and special furnace designs to reduce overall energy consumption. It will include developments in digital, programable and computer-controlled applications to optimize heating. Induction melting will expand into larger ratings of 100 tons or more and several hundred-ton holding units to be used in conjunction with advanced-design automatic-pressure paving machines.

Consumption of induction heating equipment in the United States has been tabulated by the U.S. Department of Commerce in terms of value of total units installed. Table 1-11 shows its estimates for industry shipments of various categories of heating equipment and the portion that is induction equipment. It may be noted that induction equipment has significantly increased its share of the heating equipment market from 1958 to 1974. Figure 1-7 indicates the cumulative units of induction heaters installed from 1958 to 1974 for three types of induction heaters: (1) molten metal induction furnaces; (2) heat processing and metal treating induction heaters; and (3) induction heaters for hot-forming, forging, rolling, piercing and extruding. Table 1-12 indicates that in 1974 the total cumulative value of installed equipment for all three categories of induction heaters was over $1 billion, and total capacity ranged from 8 x 10^6 kW. Estimates of future consumption for 1980 and 1985 have been given by G.F. Bobart[12] (Table 1-13). Use of induction heating equipment has increased steadily since 1965 and is expected to expand significantly by 1980 and 1985.

Competing Technologies in Electric Process Heating

Several technologies are emerging as alternatives to coreless induction heating. Infrared heating, to date, has centered primarily on low-temperature ovens for heating to 200-1200°F, for curing and drying applications. Higher temperature furnaces have been used selectively for heat-treating and even light-forging operations. Technology development in long-life, lower-cost, high-temperature elements is continuing.

Table 1-11. Department of Commerce Data on Industry Shipments of Heating Equipment in 1958, 1960, 1965, 1972 and 1974 ($ thousands)

	1958	1960	1965	1972	1974
Molten metal melting, refining and holding furnaces					
Total	22,909	21,113	58,924	51,184	72,900
Induction	4,446	7,478	24,713	27,184	55,550
Ferrous	NSA	NSA	17,875	25,994	NAS
Nonferrous	NSA	NSA	6,838	12,300	NAS
Hot-forming, forging, rolling, piercing and extruding					
Total	18,804	17,668	33,108	12,702	17,800
Induction	2,407	2,596	3,662	4,938	7,300
Ferrous	1,752	1,609	2,301	NSA	NSA
Nonferrous	655	987	1,361	NSA	NSA
Heat processing and metal treating					
Total	95,877	131,871	159,118	153,031	215,550
Induction	10,436	15,580	15,780	19,661	29,100
Ferrous	7,410	9,682	13,290	17,203	NSA
Nonferrous	3,026	3,898	2,490	2,458	NSA

Microwave process heating has been used primarily in curing or drying small-specialty nonconductive products, and operates at frequencies of 2000-3000 MHz, requiring Klystron and magnetron oscillators with wave-guide transmission systems and requiring high capital equipment costs. Advances in solid state technology have allowed low to medium-power units to be produced at reduced costs. Small, totally solid state, self-contained units for industrial applications will be developed.

For a variety of applications in heat treating metals, including guarding and hardening, laser beams are beginning to be used. It is currently cost-competitive and in some cases significantly cheaper than induction heating, carburizing and nitriding.

In heat treating, the laser beam irradiates the metal surface and is absorbed in an extremely thin surface layer. When the laser beam is moved rapidly across the metal surface, heat is conducted out of the original surface region and into the base material. Thus, quenching occurs by conduction of heat into the interior of the metal.

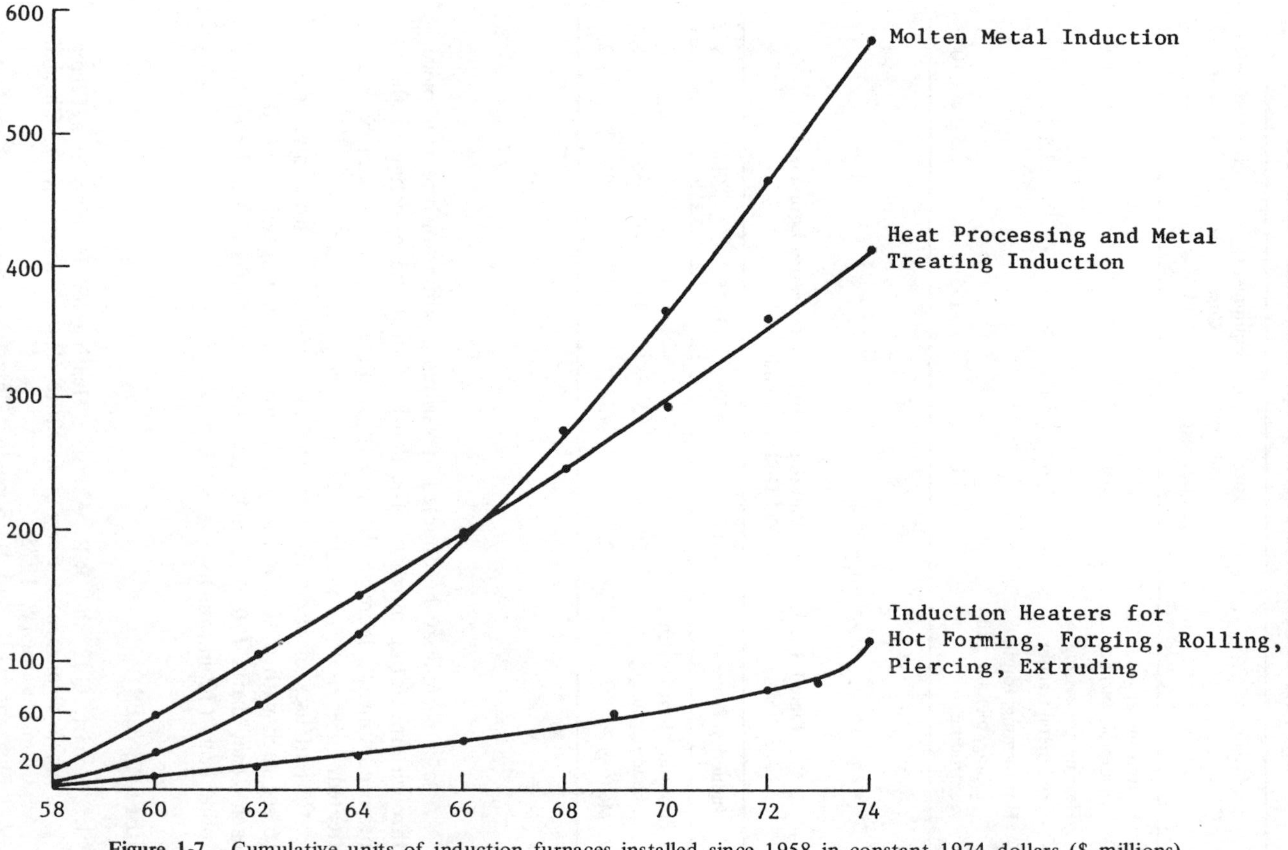

Figure 1-7. Cumulative units of induction furnaces installed since 1958 in constant 1974 dollars ($ millions).

Table 1-12. Cumulative Capacity and Value of Induction Furnaces to 1974

	Total Value ($ millions)	Equipment Cost ($/kW)	Total Installed Capacity (kW)
1. Molten metal induction furnace	579	90-145	$4\text{-}6.4 \times 10^6$
2. Heating, processing and metal tracing	396	70-120	$3.5\text{-}5.5 \times 10^6$
3. Hot-forming, forging, rolling, piercing and extruding	91	90-165	$.55\text{-}1.0 \times 10^6$

Table 1-13. U.S. Consumption of Induction Heating Equipment in 1974 Dollars ($ millions)[7]

Induction Process	1965	1970	1975	1980	1985
1. Forming, heat treating, bedding, curing	20	25	35	35	85
2. Melting	25	45	60	90	125

A laser beam may be directed at various positions on the workpiece through articulating mirrors. The depth of field and directing of the beam combine to increase the variety of piece shapes that can be heat treated.

Current uses of lasers for heat treating include cast iron valve seats (6 sec each with a 10 kW laser); cast iron piston ring grooves (50 sec each); cam shafts (70 parts/hr); 4.140 H steel shafts (30 in./min); and spline gears (30 in./min).

REFERENCES

1. Bobart, G.F. and R.R. Akers. "Heating for Forging in a Changing Environment," *Forging Institute Heating Symposium*, Chicago, Illinois, January, 1975.
2. Bijwaard, G.B. and H. Sorokin. "Electrical Systems for Automatic Handling and Electrical Heating of Slabs" *IEEE Trans. Ind. Appl.* 8:735-743 (1972).

3. Pillar Corporation. Private Communication between Feldman and G. Miller.
4. Lindberg-Tempress, Newton Square, Pennsylvania, 10973. Sales literature.
5. Pillar Corporation, 7000 West Walker Avenue, Milwaukee, Wisconsin. Sales literature.
6. Westinghouse Electric Corporation, Industrial Process Division, Rt. 32, Sykesville, Maryland, 21784. Sales literature.
7. Lounie, H.W., Jr. "Comparing Melting Energy Costs," *Foundry* (December 1967).
8. Cycle-Dyne, Inc. 134-20, Jamaica Avenue, Jamaica, New York, 11418. Sales literature.
9. Leco Corporation, 3000 Lakeview Avenue, St. Joseph, Missouri, 49085. Sales literature.
10. Centorr Associates, Inc., Rt. 28, Suncook, New Hampshire, 03275. Sales literature.
11. Lepel High-Frequency Labs, 59-21 Queens-Midtown Expressway, Maspeth, New York, 11378. Sales literature.
12. Bobart, G.F. "Future Technologies in Electric Process Heating," Industrial Applications Society, Annual Report (1975).
13. Kern, R.F. "Heat Treating Responds to New Varied Pressure," *Iron Age* (May 13, 1974).

UNCITED REFERENCES

"Cutting the Cost of Induction Heating," *Manuf. Eng. Managemt.* 71:32 (October 1973).

Dallas, D.B. "Energy-Saving System for Induction Annealing," *Manuf. Eng. Managemt.* 75:20-22 (July 1975).

Dewan S.B. and G. Havas. "A Solid State Supply for Induction Heating and Melting," *IEEE Trans. Ind. Applic.* IGA-5:449-454 (September/October 1970).

Fallone, A. "Can you Cut Costs with Channel Induction Melting?" *Foundry* 102:74-78 (December 1974).

Habel, A. "NC Induction Hardening Machine Heat and Quarches Gears," *Am. Machinist* 75:20-22 (July 1975).

Havas, G. and R.A. Sominer. "A High-Frequency Power Supply for Induction Heating and Melting," *IEEE Trans. Ind. Elec. Control Instrument.* IECI-17:321-326 (June 1970).

"Induction Heat Hardens Rails, Extends Life," *Elec. World* 178:109 November 1972).

"Induction Heating Machine Automates Billet Handling," *Iron Age* 211:30 (March 1973).

Lozinskii, M.G. *Industrial Applications of Induction Heating* (New York: Pergamon Press, 1969).

Pelley, B.R. "Latest Developments in Static High-Frequency Power Sources for Induction Heating," *IEEE Trans. Ind. Elec. Control Instrument.* IECI-17:297-312 (June 1970).

Peschel, W.F. "Load Power Matching of High-Frequency Power Supplies for Induction Heating," *IEEE Trans. Ind. Applic.* IA-10(3) (May/June 1974).

Ross, N.V. "A System for Induction Heating of Large Steel Slabs," *IEEE Trans. Ind. Applic.* IGA-6:449-454 (September/October 1970).

Segsworth, R.S. and S.B. Dewan. "Power Supply Systems for Induction Furnaces," *IEEE Conf. Records,* Fifth Annual Meeting, IGA, 279-283 (1970).

Segsworth, R.S. and S.B. Dewan. "Thyristor Power Units for Induction Heating and Melting," *IEEE Conf. Records,* IGA, 616-620 (October 1967).

Vaughan, J.T. and F.W. Williamson. "Design of Induction Heating Coils for Cylindrical Magnetic Loads," *AIEE Trans.* 65:887-892 (1946).

2

RADIATION CURING*

INTRODUCTION

Radiation curing of polymers involves the use of given wavelengths of light or electron beams to speed the curing mechanism, the dissociation of free radicals, from the minutes required in conventional curing to mere micoseconds. Radiation curing is used for coatings on wood stocks, motor vehicles, metal and plastic containers, and for inks used for packaging, printing and specialty items.

Advantages of radiation curing for the two largest markets in 1976–coatings and inks–include:

- rapid drying speeds (seconds or less);
- reduction or elimination of organic solvents, thus eliminating air pollution and incineration problems;
- significant reduction or elimination of fossil energy-heated drying ovens and incinerators;
- coating of heat-sensitive materials (plastics);
- increase in production rates;
- more efficient use of polymeric coating materials because of less penetration of flowing material into substrates; and
- savings in space of application equipment.

There are several disadvantages to radiation curing. Coating equipment upstream of the curing equipment, for example, may not be compatible or may negate the time advantage of radiation curing. Radiation does not travel around bends, thus requiring complex configurations of curing equipment for unusual three-dimensional shapes. Also, the chemistry has not been perfected, thus some formulations are severely limited in application. Some present coatings have mediocre chemical and weathering

---

*By: Marcel Barbier, John Vlahakis, Robert Ouellette, Robert Pikul, Rip Rice

resistance, flexibility, abrasion resistance and poor film adhesion to impervious substrates. The technology itself poses certain problems, including use of toxic ingredients, fears for radiation effects on workers and radiation damage to certain substrates. In addition, there is certain universal psychological inertia to accepting any new industrial process.

Radiation curing includes three technologies: ultraviolet (UV), infrared (IR), and electron beam (EB) curing. In UV curing the material being processed, the substrate, is coated with a radiation-curable formulation, then passed through a drying oven containing UV lamps. The medium-pressure mercury lamps dominate this market, although low- and high-pressure mercury lamps also can be used. The most popular unit today is the 200 W/in., medium-pressure, mercury-argon bulb with a claimed 2,000-hr operating life and 63% radiation output-power input efficiency. The cost is about $250 per lamp. Even with multiple UV lamps being required in each dryer, plus safety features on the UV curing oven to keep extraneous radiation from escaping, total unit cost for UV curing is only several thousand dollars per production line.

Recent introduction of infrared curing has further lowered lamp costs. IR bulbs cost less than $100 each, and IR radiation that escapes is much less of a hazard to workers than UV. In addition, standard solvent-based and/or water-based coatings can be cured by IR.

Electron-beam accelerators (EB), on the other hand, are very high in capital cost–on the order of $200,000 to more than $1,000,000 per production line. Nevertheless, EB accelerators, providing several thousand times the radiation energy provided by UV lamps, can do some things ultraviolet and infrared cannot.

Existing applications of radiation curing have resulted in a sizeable market for the technology. Total radiation-cured coatings in the U.S. were 9 million pounds of formulations in 1975, growing from 5 million pounds in 1973, and expected to grow to 32.7 million pounds a year by 1986. The total came from three major market applications: wood coatings; containers, closures and metal decorating; and automotive or motor vehicles.

UV-cured wood coatings in the U.S. in 1974 consumed 6.803 metric tons, about one tenth in clear top coats, and 90% in filler coats.

UV-cured coatings for the containers, closures and metal decorating markets consumed 0.70 million pounds of formulated coatings.

For motor vehicles, the U.S. consumed 100,000 gal of formulation in 1975, all by the Ford Motor Co., and all cured by EB accelerators. In Japan, 59.66 metric tons of coatings were cured by EB accelerators in 1974, of which 38 metric tons were used in the automotive and appliance markets.

Total U.S. consumption of UV-cured inks was 1.6 million pounds in 1975, divided between packaging inks (567 metric tons) and all other printing (240.36 metric tons). Canada used 65 metric tons for UV printing in both applications, for a North American total of 872 metric tons. There are no known applications of EB-cured inks in North America at this time. There is no visible new competitive market for radiation-cured inks, its biggest drawback at present being its high price.

In Japan, 40 metric tons of UV-curing inks were consumed in 1975, double the 1974 use.

The total U.S. market for all sensitized printing plates in 1975 was $1 billion. UV-cured printing plates are now making inroads into this market, but the exact figures are confidential to the three major companies involved. An estimate of the current position of UV-cured materials in this market is 2% or a market value for UV curing materials alone of $20,000,000 in the U.S.

Japan had total sensitized printing plate sales in 1975 of $1.5 million. Using a 20% capture estimate, the market for UV curing materials in this application would have been $300,000.

In the U.S., printed circuit boards used $40 million in total inks in 1975. Assuming that 10% of the market has been captured by UV-curing inks at this time, the market value for UV-curing inks in this application was $4 million in the U.S.

A summary of the existing markets for radiation curing is given in Table 2-1.

Table 2-1. Summary of Market Data for Actual Applications of Radiation Curing

Actual Submarket	Maximum Market Size	Capture	Actual Market Size	Remarks
Coatings			9 million lb	
Wood coatings			6.8 metric tons	Corresponds to 1.05 million gal
Containers, closures and metal decorating			0.7 million lb	
Motor vehicles			100,000 gal	
Inks (UV-cured)			1.6 million lb	$5.3 million market at the average sales price of $3/lb
Packaging			567 metric tons	
Other printing			240.4 metric tons	
Sensitized printing plates	$1 billion	2%	$20,000,000	
Printed circuit boards	$40 million	10%	$4,000,000	

Major potential market areas include expansions on or variations of existing applications. In addition, UV-curing coatings for vinyl floor tiles are in their infancy, with only one production line known in Japan, and only pilot units known in the U.S. Pressure-sensitive adhesives manufacturers are just beginning to study radiation-curing polymers for this application, and no production lines are as yet known to be operating. A summary of potential applications of radiation curing is given in Table 2-2.

Table 2-2. Summary of Market Data for Potential Applications of Radiation Curing

Potential Submarket	Potential Market Size	Remarks
Coil Coatings	5.4 million gallons	Further chemical research is required to overcome high viscosity and poor adhesion to metal surfaces.
Electronic Circuit Boards	$40 million	Energy requirements can be reduced up to 90% over conventional heat-curing systems.
Vinyl Floor Tiles	9 million gallons	Penetration of this market is not expected within the next decade.
Pressure-Sensitive Adhesives	109 metric tons in the U.S.	Economic benefits arise by using radiation-cured over conventional polymers.

The energy savings obtained by radiation curing are based on the difference between the costs associated with the length of time the classical materials have to be heated at a given temperature less the costs of installing and utilizing the radiation-curing techniques. Further economic savings are obtained by the much smaller-sized equipment required for radiation curing, plus the greatly increased speed at which coated materials can be processed, thus lowering the capital and operating cost per unit of production. A comparison of energy consumption and costs between radiation curing and conventional techniques is given in Table 2-3.

In the U.S. in 1975, 98% of all radiation curing was done by ultraviolet. There were some 150 sheet offset and web offset UV drying units in operation in the printing industry alone at the end of 1975, with 12 more lines in Canada. The industry estimated that an additional 8-10

Table 2-3. Comparison of Energy Consumption and Costs, Radiation vs. Conventional Curing

Application	Radiation Curing	Conventional Technique	Savings	Remarks
Coatings				
Metal cans	6.6 Btu/can (UV curing)	23.0 Btu/can (thermal process)	16.4 Btu/can	
Curing of polyethylene cable insulation	2.5¢/lb (by electron beam)	6.3¢/lb (using peroxide)	3.8¢/lb	Includes total costs
Curing of PVC extruded onto copper wire	$0.50/lb	$2.00/lb	$1.50/lb	Includes total cost of producing wire
Inks				
Printing Inks	$2.91/lb (UV curing)	$1.71/lb (thermal)	246% less energy than conventional methods	
Pressure-Sensitive Adhesive Tape	$13,000	$130,000 (thermal oven)	$117,000 in capital costs for a 300 ft/ft/min unit	EB curing capital cost would be $200,000

units were being installed each month in replacing lithographic printing alone. This represented an additional $1.44 million equipment market for 1976, which was expected to continue at this rate into 1977.

In wood coatings, some 70 UV dryer units are in commercial operation in the U.S. Japan is known to have 25 UV wood coating lines as well.

Taking the known UV lines in operation today in the U.S. and Canada as 300 (162 in all printing, 70 in wood coating, 68 in other applications and on pilot lines), at an average price of $12,000/UV oven, this equates to $3.6 million worth of UV curing ovens operating today.

Assuming an average of 4 UV bulbs per UV oven with an operating life of one year, and at a unit price of $250 each, the 300 UV ovens are using 1200 UV bulbs, and the replacement market for UV bulbs for UV curing of inks and coatings at the end of 1975 was $300,000. Addition of 120 UV ovens/year will require 480 UV bulbs/yr, and creates an additional market of $120,000/yr for UV bulbs.

There are no known wood coating lines in North America using EB curing. W. R. Grace in Duncan, South Carolina has an estimated six accelerators for producing heat-shrinkable packaging film; Western Electric has six accelerators for producing telephone wire; Johnson & Johnson has an estimated two, Deering Milliken an estimated two, Surface Activation Corporation and its sole customer has two, and Ford Motor Company has nine accelerators. These 27 units are for major consumer products (except for the SAC textile products). An estimated 23 additional units are scattered in specialty applications not covered in this study, or in research laboratories or pilot plant development lines, making an estimated total of 50 EB units in use today in North America.

At an average cost of $200,000 each, these 50 EB units represent a capital market of $10 million. Future sales of such high-capital items are expected to be slow over the next several years, and should lag the economic recovery. No more than an estimated 12-24 units will be sold by the four U.S. manufacturers during 1976, representing a sales value of $2.4-$4.8 million.

Although IR curing represents a serious threat to UV curing, for economic reasons, not many IR production lines have as yet been installed. Four are known in Japan (for wood coatings) and two lines are in operation in the U.S. for the same use.

GENERAL PRINCIPLES

Technology

Polymers that cure by free-radical mechanisms are the basic raw materials for radiation curing. Classical curing of these polymers has been accomplished by heat or by absorption of sunlight. Both processes dissociate the contained free-radical source into active free radicals, which initiate polymerization of the curable polymer. The speed at which free radicals are formed and their concentration, determine the properties of the cured polymeric material.[1-3]

Both classical approaches to free-radical source dissociation require rather lengthy periods of time; in addition, the thermal heating is energy inefficient. Use of ultraviolet, electron beam or infrared energy, however, can greatly speed the formation of free radicals from the source.

Radiation curing utilizes the principle of selecting a given wavelength or sequence of wavelengths of light or electron beams, which will speed the dissociation of the free-radical source from minutes to microseconds. The material being processed, the substrate, is coated with the radiation-curable formulation by knife edge, rollers, screens, etc., then passed through a drying oven that contains UV or IR lamps, or accelerators.

Polymer systems most widely studied and developed to date for radiation curing include: acrylates, polyesters (cross-linked with styrene), thiols (cross-linked with unsaturated compounds), epoxies, urethanes, vinyls and various mixtures of these polymers with monomers.[4] Typical free-radical sources (photoinitiators) are incorporated into these polymers, along with pigments and photosensitizers, which act to increase the rate of dissociation of the free-radical source when the formulation is exposed to the proper radiation.

Because the technology currently provides a flowable liquid formulation that can be converted quickly into a nonflowing, adherent solid, major applications of radiation-cured polymers developed to date are as coatings and inks. Because of the speed of radiation curing, production times for applying these coatings and inks can be greatly reduced, thus greatly reducing unit cost of production per unit of capital investment. Curing times and characteristics of radiation-cured polymer formulations can be changed by changing the balance of photosensitizer, the pigment, incorporation of extenders, viscosity improvers, or even the use of small amounts of solvents.[5]

Early radiation-cured formulation photoinitiators were also sensitive to atmospheric oxygen. Consequently, surfaces could be depleted of some of their constituents by atmospheric oxygen, giving rise to incomplete surface cures. To avoid this situation, equipment was modified to provide a blanket of inert gas (usually nitrogen). This technique adds additional capital and operating cost to the process. Recently, modifications in polymer formulations, notably by W. R. Grace & Co., provided formulations that are no longer sensitive to atmospheric oxygen, and obviate the need for inert gas. Use of inert gas with UV curing, however, has the advantage of eliminating the undesirable generation of ozone, and increases the amount of radiation energy available for production. This, in turn, can increase the speed of production.

Curing Equipment

Early in 1976 in the United States, more than 98% of all commercial radiation-curing equipment was based on ultraviolet radiation, with the balance being electron beam accelerators.

EB accelerators will make very little progress between 1975 and 1980 –they are simply too expensive to penetrate UV-cured coating and ink markets, employing today's state-of-the-art technology. However, with improvements in the chemistry of radiation-cured formulations (faster curing times, improvements in properties, etc.) the EB curing concept is correct, and equipment available today is satisfactory. When these formulations appear, it will probably be five years before EB accelerator sales will grow.

UV Lamps

Curing equipment most widely used has been ultraviolet lamps providing wavelengths of 300-400 nanometers (nm). It is estimated that 98% of all commercial photochemical radiation equipment in operation in the U.S. uses ultraviolet lamps as their radiation source. Normal life expectancy is 2,000 hr, or one year of one shift a day operation, according to the suppliers, but 1,500 hr according to users. Use of lower-wavelength UV bulbs in a first curing of the surface layers, followed by a second curing using the 300-400 nm UV bulbs for the body of the polymer, is a technique for avoiding the need for inert gas blanketing.

The 200 W/in., medium-pressure, mercury-argon lamp with its 2,000-hr working life and 63% efficiency (radiation output-power input) is the most popular unit today. Lengths are 1-7 ft. Because longer working lives and greater selectivity of wavelength cuts are desired, newer types of UV lamps are being introduced.

Newer Radiation Equipment

Electrodeless UV lamps, pulsed xenon and pulsed xenon-mercury lamps are being marketed for radiation curing. The electrodeless UV lamp (Fusion Systems Corporation, 11810 Parklawn Drive, Rockville, Maryland) is claimed to have a 50% greater operating life than UV lamps (3,000 vs 2,000 hr), to operate at lower temperatures and to operate at 29% lower line operating costs. In addition, Fusion Systems claims their bulb will cure UV coatings twice as fast and that one of their bulbs will replace 1.5-3.5 conventional 200 W/in. mercury vapor lamps.[6] Fusion Systems also claims that a dryer with one 20-in. Fusion Systems radio frequency bulb would cost $14,600 compared to a conventional UV dryer with three lamps at $14,000.

Pulsed xenon lamps are intermediate power sources of UV radiation and have the potential for being the lowest cost lamps. In single lamp lots, conventional xenon lamps cost $125-$150 each, but in lots over 100, they can be purchased for under $100 each. The probable reason that pulsed xenon lamps are not used more widely is they simply are unknown to the UV coatings technologist at present. Only Autocoat, Inc., Middleton, Massachusetts, is known to use pulsed xenon lamps to UV cure inks on plastic food containers.

Pulsed xenon lamp manufacturers seem to be ignorant of the technology of radiation inks and coatings. One potential problem with their use is the time interval of the pulse. Perhaps the coating speeds will be too fast to obtain uniform cure, even though the total power output during the pulsing period is much higher than for UV lamps.

Since pulsed xenon lamps have only now begun to be used to dry UV inks and coatings, it is simply too early to tell how much penetration they will make in these applications.

Safety

UV radiation below 220 nm and EB radiation equipment will produce ozone rapidly from atmospheric oxygen. As the radiation wavelength increases, less and less ozone will be generated. Current regulations by the U.S. Occupational Safety & Health Agency (OSHA) specify the time-weighted average concentration of ozone in the plant atmosphere over an 8-hr working day to be 0.1 ppm (by volume). Recently proposed regulations would set the level of 0.05 ppm (measured at any time) as an "action level." Once the ambient ozone concentration reaches this action level, the employer would be required to institute a lengthy (and costly) series of analytical measurements for ozone in various locations in his plant, plus institute a series of periodic medical examinations for his employees.

Because ambient ozone levels in many large U.S. cities are many times above OSHA's maximum allowable level of 0.1 ppm, it is believed that the action level of 0.05 ppm is meaningless. Health authorities may agree with the logic of this interpretation, but will counter that known health effects data indicate that exposure of human beings to levels of ozone above 0.05 ppm for lengthy periods of time does produce detrimental effects.

The user of UV and EB radiation equipment must, nevertheless, assure that the equipment he purchases will shield his workers from UV radiation (which can be accomplished by proper choice of materials and engineering design) and not produce excessive quantities of ozone.

The use of UV bulbs that produce sufficiently high wavelengths as not to produce ozone could be considered; however, radiation at these longer wavelengths is weaker, and therefore more UV bulbs must be installed per line length to accomplish UV cures in the same length of time. More bulbs will add to the operational cost.

Excess ozone can be destroyed quickly by collecting the off-gases from UV dryers and passing them through an aqueous system containing a reducing agent, such as potassium iodide, sodium thiosulfate, or even activated carbon. Passage of dryer off-gases directly through activated carbon is not recommended, since heats of oxidation of carbon are exothermic and high. Should sufficient ozone be present, the carbon can become so hot as to start burning, and this could result in deflagration or even explosion, depending upon the quantities of ozone and carbon involved, rate of gas flow, degree of packing of the carbon, and a number of other parameters.

Shorter wavelength UV lamps can be used, but with a nitrogen blanket in the dryer. Ozone will not be produced with high-energey UV radiation passes through nitrogen. It has been found that this nitrogen blanket dramatically increases the drying rates of wet UV coatings and inks. Radiation energy formerly going to produce ozone now is used to cure the ink or coating. Additionally, once ozone is created, it will itself absorb UV radiation at a higher wavelength, thus sapping additional energy away from the curing of the coating or ink.

The benefits of eliminating ozone formation and obtaining much higher rates of cure (with less photoinitiator consumption) must be traded off against the cost of nitrogen gas, which currently is about 32¢/lb for an installation using one million cubic feet a month.

Another method of controlling ozone is merely to vent it to the atmosphere, by means of the air fans drawing air into the dryer to cool the UV lamps. This is routine practice today, but as installations become larger, air pollution regulations become stricter, and more materials effects data quantifying the effects of atmospheric ozone upon plants and animals become known, catalytic destruction before discharge probably will be required. Already the Los Angeles County Air Pollution Control Board is "suggesting" zero discharge of ozone from ozone contacting chambers in water and wastewater treatment pilot plants operating in that area of the U.S. because ambient ozone monitors placed strategically throughout the county might be interfered with by local discharges of ozone.

EXISTING APPLICATIONS

Coatings

Flat Wood Stock (Particleboard)

Although Europe consumed 30,000 metric tons of UV coatings for flat wood stock in 1974, the giant North American flat wood stock industry consumed only 6,803 metric tons of formulated, UV-cured wood coatings. The comparatively low U.S. consumption is most likely due to the fact that the European UV wood-coating technology is simply too slow for the scale of U.S. production facilities. Most European UV-cured wood coatings contain a small amount of incompatible wax, which exudes out of the wet coating to form an oxygen-impermeable barrier on the surface, preventing contact of air with the still wet coating. This wax takes time to exude out, and causes European curing times to be at least 30 seconds long. The U.S. wood stock industry, however, needs two-second photochemical cures to maintain productivity.

Conventional wet coatings are applied to particleboard panels and thermally dried at speeds of 80-150 ft/min. A nitrogen blanket will add costs to the process as will the use of EB curing, and the U.S. wood processing industry does not want to incur these extra costs on their production lines at this time. Infrared curing, however, is being studied closely by this industry because of lower costs for bulbs and because the coatings are conventional, and lower in cost.

The 6,803 metric tons of formulated UV-cured coatings consumed in the U.S. for this application in 1974 converts to 1.05 million gallons. This included 100,000 gallons (10%) of clear top coats and 950,000 gallons (90%) of filler coats.

The clear top coats weigh about 8 lb/gal and consumed about 900,000 lb of unsaturated prepolymers, styrene monomers and initiators. Filler coats weigh about 15 lb/gal and consumed about 8,100,000 lb of prepolymers and monomers. In addition, inorganic filler material consumption was 2,721 metric tons. Both coatings consumed between 45 and 59 metric tons of benzoin and benzoin ether photoinitiators.

Between 1967 and 1972, American particleboard production grew at the annual average rate of 23.4%/yr; during this time, coating consumption on particleboard grew at only 7.85%/yr according to the U.S. Department of Commerce. Not all particleboard is coated (floor underlay), and most of what is, is coated on one side only.

During 1974 there were an estimated 70 commercial UV-curing wood lines in the U.S. Recovery of the U.S. particleboard industry from the recent recession was projected to occur by 1977, about a year behind the rest of the U.S. economy. A 30% jump in shipments was projected for 1977. Thereafter, the annual growth rates would be no higher than 6.62%/yr (1977-78), no lower than 4.53%/yr (1984-85). The average annual growth rate between 1977 and 1985 would be only 5.45%.

Assuming that radiation-curable coatings will be developed that will contain no wax and will cure much faster than is now possible, and that UV-curable coatings are not made obsolete by an unforeseen new coating technology, combined wood coatings are projected to grow annually 6.42% from 1976 to 1986; conventional and infrared coatings are projected to grow at 6.91% annually over the same period; and UV-cured coatings will grow 0.69%/yr over the same period. These coatings should grow from 9.53 million gallons in 1974 to 17.29 million gallons in 1986, while UV-cured coatings will be essentially the same in 1986 as they were in 1974 (1.05 million gallons), although peaking during 1981 at 2.22 million gallons.

In this application, IR coatings represent a real threat to UV-cured coatings. EB coatings are not considered to be a major product for this market because the coatings are thin enough to be cured by the less costly UV and infrared radiation.[7]

For most wood coating lines, UV lamps are fast enough. EB accelerators and their accompanying radiation-shielding equipment cost 5-20 times more than the corresponding UV installation. Thus, the only potential for EB in the flat composition wood market is to cure at very high line speeds (30-40 m/min and up). At present, coatings are not available to cure at these speeds, thus the problem for EB in this market rests squarely upon the chemist to develop faster curing coating compositions. At the same time, new infrared thermal drying equipment is entering the market, making the future for EB curing in this application even less encouraging.

Containers, Closures and Metal Decorating

The market for containers, closures and metal decorating includes all finishes used to coat metal cans, crowns, closures and collapsible tubes. It also includes other specialty metal decorative finishes applied by lithography and silk-screen printing on various metal substrates, such as decorative nameplates.

Two technical innovations have affected major and minor can manufacturers in recent years all over the world. These are the growing use of "black plate steel" (which replaces tin plate on the basis of economics and must rely on organic coatings for chemical resistance) and the two-piece seamless can (replacing the three-piece can). In 1975, 25% of all cans manufactured in the U.S. were two-piece, and it is predicted that 50% of all metal cans will be of two-piece construction before 1980.

Three-piece cans generally are coated and printed before construction, that is, on flat stock–ideal for radiation curing. Two-piece cans have what will become their interior surface coated with a heat-cured epoxy. The exterior surface is base coated, printed and top coated after construction. Uniform radiation curing of these exterior steps can be accomplished by rotating the can surface to receive the radiation. Thus, targets for radiation curing are both the two- and three-piece metal cans.

In the U.S., however, there is an environmental movement that can makers fear may cripple the metal can industry. Two states already have passed legislation banning the sale of all disposable beer and soft-drink containers (metal or glass). The U.S. Environmental Protection Agency advises that it will be asking Congress to pass legislation requiring a 5-cent deposit on all glass and metal containers sold in federal government facilities. Should this environmental movement continue to gain momentun, the returnable glass bottle will likely gain in sales and reduce the markets won to date by metal cans.

The inventor of the two-piece can, Coors Brewery (the fourth largest brewery in the U.S.) has ceased fighting the movement and instead has developed a machine for supermarkets and liquor stores that would crush returned cans into shippable bales and automatically return the deposit to the consumer. The energy required to recycle aluminum is much less than to produce the same quantity of aluminum from virgin bauxite ore. Coors has also developed and is preparing to test market a new all-plastic beer bottle.

Thus, although the can industry is one of the best potentials for radiation-cured coatings, the future of the metal can industry currently is in some doubt, at least in the U.S. During 1975, metal can shares of the beer and soft drink markets had dropped to 60 and 30%, respectively, according to the Can Institute, from their former, more-commanding shares just a few years earlier.

The use of natural gas to fuel ovens that dry industrial coatings will be curtailed and eventually eliminated in the United States because of the natural gas shortage. Since radiation-cured coatings and inks require reduced or no natural gas consumption, the technology is applicable to many industries where thermally heated ovens now are used. Commercial space heating, which includes drying of industrial coatings, was the sixth largest user of energy in the U.S. during 1968, according to a 1972 report from the Office of Science and Technology. Japan faces the same situation in supplies of natural gas.

The Can Manufacturers Institute of the United States recently reported that the metal can and closure industry currently is using between 20 and 25 billion cubic feet (bcf) of natural gas, or equivalent fuels, annually. Of this, 5 bcf are used in afterburners to incinerate organic solvents that otherwise would be vented to the atmosphere. If afterburners were to be used at all plants (to comply with the requirements of the Environmental Protection Agency's Clean Air Act), the industry would consume an additional 6 bcf (or a total of 31 bcf) of natural gas.[8]

In 1974, American manufacturers of metal containers, closures and crowns consumed an estimated 113,378 metric tons of various organic solvents in the coating and printing of metal containers. Photochemically reactive coating and ink formulations have the potential to reduce this amount of solvent to zero, thus eliminating the air pollution problems and, at the same time, avoiding the need to incinerate, which is energy (natural gas) consuming.

Radiation-cured coatings will not be used for interior can coatings, mainly because the ingredients of the coatings are too toxic, and no manufacturer wants to petition the Food and Drug Administration for approval in this application. However, exterior-based coats, inks and top-coat varnishes are all excellent markets for radiation-cured coatings.

With radiation curing, however, one liter of radiation-cured coating will replace 3.2 liters of conventional solvent-based coating, thus shrinking the total potential volume demand by 69%. With about 0.3 k/l of polymer binder, the exterior finishes market of 1975 indicates a potential market for 17,007 metric tons of radiation-cured prepolymers and monomers. Assuming an annual coatings growth in the U.S. of 4% by 1986, a potential market of 26,167 metric tons will exist.

Because of the 69% reduction in volume of coatings sold, however, the vendor of coatings will have to raise his price per liter for radiation-cured coating by a factor of approximately three, just to maintain his dollar flow constant.

All other categories of metal decorative coatings reached 9 million gallons of conventional coatings in 1975. At the same time, radiation-cured coatings in this category reached 2.81 million gallons. This volume is expected to grow to 5.34 million gallons by 1986, but assumes a 100% replacement of conventional, solvent-based coatings.

There are still some major technical deficiencies with UV-cured coatings for the metal substrate, can closure, crown, etc., markets. These are:

- insufficient adhesion to metals;
- poor flow-out and leveling characteristics;
- marginal color stability when thermally baked;
- marginal film flexibility and film toughness;
- low pigment-loading capacity, resulting in marginal hiding properties;
- high coefficient of friction of cured film;
- toxic, volatile acrylic monomers, thus making FDA approval for contact with food improbable;
- high cost–100-150% more than conventional coatings; and
- radiation health hazards if UV dryers are misused.

Most of these deficiencies can be alleviated only by changing the chemistry of the radiation-curable polymer system; thus the future depends upon the chemist. Since the two-piece can-making system is replacing the three-piece process on the basis of economics, it would be ideal if all coatings and inks for the two-piece system could be radiation curable. But since the internal coating still must be heat cured (because of toxicity of radiation-cured compositions), the best technique that can be visualized until chemical breakthroughs are attained involves thermally baking the fully coated can at the very last step.

Nevertheless, an estimated 30 commercial UV lines were applying inks and clear lacquers in the U.S. in the general can, closure, crown and metal decorative market at the end of 1975. The severe recession and tight capital markets slowed growth all through 1975, and during the last quarter of 1974.

Energy Savings. Coors' engineers report that 246% more energy is required to thermally dry conventional inks than to cure new inks with UV light. Operating their can manufacturing lines at 600 cans/min, on a 7 hr/day, 250-day work year, the UV dryer would save over 1 billion Btus a year. Such a production line could print 63 million beer cans annually.[9]

If all 37.2 billion beer and soft drink cans manufactured in the U.S. in 1972 had been printed and cured with UV dryers, a potential 59 billion Btus would have been saved in drying inks alone. If all external coatings on beverage cans were radiation cured, it is estimated that more than 100 billion Btus could have been conserved. An estimate of energy savings from Coors Brewery based on their experience with UV curing is given in Table 2-4.

Table 2-4. Energy Comparison for Thermal Drying/UV Curing

In-Plant Ratios	Energy Savings
Natural Gas Btu/UV Light Btu	86% fewer Btus
Oven Blowers/UV Blowers	93% fewer Btus
Oven Conveyors/UV Conveyers	75% fewer Btus
Gas In-Plant Electric Utility Original Source Ratios:	
Natural Gas Btu/UV Light Btu	68% fewer Btus
Oven Blowers/UV Blowers	93% fewer Btus
Oven Conveyers/UV Conveyers	74% fewer Btus

Market Projections. Assuming that natural gas shortages and price increases will be the dominating forces of change in the United States, that EPA pressures will mount substantially, that major technical breakthroughs will occur, but late in the decade and not be exploited until after 1980; that the metal can market will continue to lose market share to large glass and plastic bottles in soft drinks; and that the entire industry will only grow an average of 4%/yr, UV-cured coatings for all metal decorative coating applications, including cans, closures and crowns are expected to grow in the U.S. from 0.70 million pounds in 1975 (an insignificant fraction) to 37.39 million pounds a year in 1986, capturing 50% of the market.

To accomplish this 50% penetration by 1986, the market for UV-cured coatings is projected to grow at an average annual rate of 57.3%, from 1976 to 1980, then at an average annual rate of 22.9%.

Motor Vehicles

American automobile manufacturers annually consume 46.69 million gallons of conventional waterborne and solvent-based coatings. The Ford Motor Company is the only automotive company in the U.S. to use electron beam curing for many of its parts, and this operation consumed 100,000 gallons (0.2% of the total) in the fourth year after commercialization of this process by Ford.[10] Since 1971, use of all EB-cured coatings in the U.S. has grown only at 4.7%/yr, far less than the growth rates for other new coatings, such as electrodeposition, waterborne primers, and powder coatings.

In the Ford operation, heat-sensitive plastic substrates are spray coated. The requirement for spray coating, in turn, necessitates the use of some organic solvent to reduce the viscosity of the coating sufficiently to allow spraying.

The plastic parts being coated are truly three-dimensional in nature, and are not flat plates later shaped and formed. Ford scientists report that they have no problems curing three-dimensional objects with EB, but doubt that they would be able to attain the same results from the weaker strength UV light radiation.

Ford's operation includes six electron beam accelerators on their recently installed Electrocure II line, and three are still employed on the Electrocure I line. Having three accelerators on one side of the item and three on the other allows the coated part to be run through the radiation chamber without being turned. Internal sales of EB-cured coating formulations within Ford (the coatings were developed by Ford) were estimated at $65 million in 1974. General Motors, American Motors and Chrysler appeared to make few developments in radiation-cured coatings during 1975.

Ford has licensed Suedex N.V. in Europe and Kansai Paint in Japan to produce and sell its electron beam, radiation coating technology. In mid-1975, Energy Sciences, Inc. was licensed by Ford to offer electron beam accelerators on a global basis, to cure thin-film, radiation coatings that fall under Ford's Electrocure process patents. No other commercial UV or EB use of radiation-cured coatings was known in the North American automotive industry in 1975.

Inks

All Printing Inks

UV-cured inks and overprint varnishes are used in two general market areas–packaging consumption and printing trade consumption. Actual consumptions for 1975 and those projected through 2000 are given in Table 2-5. Values are in fixed 1975 dollars.

Table 2-5. Use of UV-Cured Inks and Overprint Varnishes in the U.S.

Year	Sales Price ($/lb)	Packaging (metric tons)	Printing Trade (metric tons)	Total Value (million $)
1975	3.00	567	240.36	5.33
1980	2.17	6,136	2,630	41.95
1985	2.05	10,494	4,499	68.08
1990	2.05	15,274	6,549	99.31
1995	1.96	20,118	8,621	124.21
2000	1.96	24,907	10,685	153.92

Sales of UV-cured inks in Canada during 1975 were 65 metric tons, making a total of 872 metric tons sold in the U.S. and Canada that year.

The average weighted price of a 344-lb lot of conventional printing ink in the U.S. in 1975 was $1.71/lb versus $2.91/lb for UV ink.

During 1975 there were 100 sheet-fed offset UV dryer lines in the U.S. and 8 in Canada, and 50 web offset UV ink dryers in the U.S. and 4 in Canada, for a total of 162 operational UV ink dryers. As 1975 ended, an estimated 8-10 new lines were being installed each month, and the UV-dryer population was expected to grow by 70% in 1976. Sales were expected to double in 1977 over those of 1976, and thereafter grow at a slower rate.

The single most powerful force of change benefiting UV inks on heat-set web offset printing presses is the natural gas shortage in North America, which will probably cause the phasing out of thermal ovens over the long term. Conversely, the current high cost of UV-cured inks and their toxicity are the two strongest deterrents to faster acceptance.

About 50% of all installed UV dryers for UV-cured inks in the U.S. are owned by 25 companies and used for metal nameplates, metal cans, plastic tubes, containers, product literature, label stock, calendars, playing cards, brochures, crowns, closures, folding cartons, Mylar labels, plastic containers, books and other publications, paper webs and newspaper inserts.

Folding carton and publication printing markets are more widely dispersed among an additional estimated 70-80 other companies using UV dryers on sheet-fed and web offset presses.

Packaging Market Inks

Packaging inks of all types make up about 40% of all inks consumed in the U.S. ($280 million in 1975). Three major packaging submarkets are identified as prime candidates for UV ink penetration. These are: (1) metal and plastic containers; (2) folding paper cartons; and (3) labels.

The largest single potential for UV inks in containers is the soft drink bottles and is expected to grow dramatically over the next 10 years at the expense of large glass bottles (and some metal cans). Coca Cola is working hard on processes to print their new all-plastic bottle with UV inks.

Folding paperboard cartons, used to package food, beverages, soaps and detergents, tobacco, cosmetics and medicines represent the second, prime, short-term packaging submarket for UV-cured inks and overprint varnishes. Printed labels are the third, prime, short-term submarket for UV-cured inks.

Only on metal containers and web offset paperboard cartons are heat-set lithographic inks dried thermally. Because of the natural gas situation in North America, market penetration of UV-inks is virtually assured. It is estimated that by 1980, 70% of all UV-cured inks will be used in these packaging markets, which now consume 40% of all inks produced in the United States. The remaining 30% of UV-cured inks will be sold to the publication and commercial printing markets. Thus, in 1980 it is projected that 6122 metric tons of UV-cured inks will be sold to the packaging markets, while 2630 tons will be sold to the publication and commercial printing markets.

It is the current consensus of the industry that it is still too early to predict the rates of penetration of UV inks into the separate packaging submarkets of metal cans, crowns and closures, folding paperboard cartons, corrugated boxes, plastic containers and labels. However, the metal decorating and folding paperboard carton markets are unanimously considered to be the largest potential, and should represent 65-80% of the 1980 packaging market. To reach this level, UV-cured packaging inks must average 61% annual growth, which is considered probable under normal economic conditions. Some ink suppliers now feel that they "are about one year behind this projection, but are still optimistic about being able to reach the 1980 target."[11]

Publication, Commercial and Specialty Printing Inks

Publication inks are used to print newspapers, magazines and books, and consume 20% of the 60% of nonpackaging inks in the U.S. Forty of the 60% is used in the commercial field for printing most other reading matter (commercial flyers, direct mail ads, brochures, catalogues, business forms, instruction manuals, menus, message items, etc.).

Lithographic inks have not penetrated these markets to the extent they have penetrated the packaging market, thus UV-cured inks (which replace conventional lithographic inks because of energy savings) have less of a potential base from which to increase their penetration.

The printing industry possesses only 67% of the web offset presses that the packaging industry does, but they have more sheet-fed offset presses. The principal penetration of UV-cured inks will be to replace conventional lithographic inks in both sheet-fed and web offset printing. This industry is also a smaller consumer of natural gas than the packaging industry, so there is less pressure on it to conserve energy. The industry is largely controlled by small entrepreneurial printers who have neither the time nor the capital to invest heavily in UV-ink technology. When the new technology is proven and ink prices drop (1980 and beyond), however, the small printer will probably adopt the technology. Projections for UV-cured inks in 1980 are given in Table 2-6. Average annual growth for UV inks in all printing trades is estimated in Table 2-7.

Table 2-6. Projected Markets for UV Inks in the U.S.

Year	Ink Application	Metric Tons	Ink Value (millions)
1980	Newspaper supplements	399	$ 1.89
1980	Commercial publications	1982	9.44
1980	Specialty printing	263	1.25
	1980 Total		$12.58
1985	Newspaper supplements	676	3.06
1985	Commercial publications	3147	14.28
1985	Specialty printing	676	3.06
	1985 Total		$20.40

The projected $12.5-million UV-ink market in 1980 will probably support only 5-6 UV-ink suppliers, and Sun Chemical and Inmont are the two recognized leaders today.

Table 2-7. Average Projected Annual Growth of UV Inks in Printing (U.S.)

Period	Growth Rate (%/yr)
1975-1980	61.5
1975-1985	34
1980-1985	11.2
1980-1990	9.5
1975-2000	16.4

EB-cured inks will have no future for two major reasons. First, EB-cured inks have no advantages in properties over conventional or UV-cured inks. Second, inks are not generally highly pigmented, thus there is no need for the very much greater energy available from EB accelerators.

EB Application

Western Electric manufactures telephone wire by radiation curing polyvinyl chloride extruded onto copper wire, allowing them to produce wire at $0.05/lb, as opposed to $2.00/lb by the earlier technique of wrapping the wire in an insulating fabric, then slowly extruding PVC over it. Six EB accelerators have a maximum annual capacity to manufacture 5.2272 billion ft/yr (12,000 ft/min x 22 hr/day x 330 days/yr) of wire.

Johnson & Johnson has long used EB sterilization of prewrapped bandages, and Deering Milliken used an EB accelerator to produce a proprietary permanent press fabric. In addition, W. R. Grace & Co., in Duncan, South Carolina, has long been using EB curing in the production of their Cryovac, heat-shrinkable, packaging film (polyethylene-based).

These companies are among the top 80 in the U.S. and are leaders in their fields. The EB-cured products manufactured by them all have a high value-added surcharge, and are aimed at vast retail markets. Virtually every American is a customer or potential consumer of these four products. Such ubiquitous consumer products, with their high value-added surcharges, can balance the high capital costs for EB accelerators, which cost between $200,000 and more than $1 million per line. There are few similar, ubiquitous, high-margin consumer products that are potential candidates to be coated or printed with EB-cured coatings or inks.

In lines where EB curing of thick, heavily pigmented coatings already has been economically justified, EB curing of inks can be performed, and there are some important specialty coating applications for EB curing. These include a nonyellowing, clear, "wet-look" topcoat with outstanding chemical and abrasion resistance for vinyl-asbestos tile. In addition, a spectrum of pigmented coatings with the outstanding properties of today's conventional coil coatings could justify 10-11 EB accelerators in the coated aluminum and steel coil coating market.

An EB accelerator probably is the only radiation source that could cure coil coatings heavily pigmented with titanium dioxide, a notorious absorber of UV light. Coated fabrics could justify a few EB accelerator lines, if the EB-cured coatings were equal in properties to today's polyurethane coatings. The textile industry consumes a great amount of natural gas in many thermal ovens, and any cost-effective method of reducing the amount of gas needed will be of interest.

Silicone release papers are being studied with EB curing to increase the rates of cure during manufacture. This industry uses large quantities of natural gas to force-dry the solvent-based adhesives used to manufacture tapes and labels, and vents millions of gallons of solvents into the atmosphere. However, hot-melt adhesives already are being used to reduce the amount of natural gas used, and eliminate solvent discharges.

A problem in this technology is that hot-melt equipment can coat silicone release paper at much faster speeds than the silicone paper can be manufactured upstream from the hot-melt extruder-coater. A promising concept is to photochemically cure both the silicone release paper and photochemically reactive liquid adhesives. In this case, both the hot-melt adhesive (and the need for any heat) plus the organic solvents are eliminated.

A small company in Westbury, New York–Surface Activation Corporation (SAC)–has developed a process whereby acrylate monomer is photochemically grafted onto pure polyester fabric. The fabric is passed through a chamber containing argon at low pressure and acrylate monomer vapor. Exposure to EB radiation breaks the argon into ions and electrons, which drift to the fabric or web. Radiant energy breaks C-C bonds in the polyester and allows the acrylate monomer to attach itself at these points, then to polymerize. This treatment provides a more comfortable feel to the permanent press fabric, and the process is claimed to be cheaper than chemically treating a cotton-polyester blend. To date SAC has sold one machine and a license to Spring Mills (North Carolina), which produces and markets pure polyester gabardine sports garments made by the process under the Refresca label.

In the United States, there are only four manufacturers of EB accelerators: High Voltage Engineering; Radiation Dynamics; Energy Sciences; and Systems, Science & Software.[7] Energy Sciences has a license from Ford to market their EB-curing technology for automotive purposes. Systems, Science & Software has an agreement with PPG for EB curing of their coating systems.

A comparison of the capital and operating costs of three of High Voltage Engineering's (HVE) three EB accelerators and conventional thermally heated ovens is given in Table 2-8. Note that capital costs for HVE's smaller models are about the same as conventional ovens, but that utilities costs are much lower.

Table 2-8. Electron Beam vs. Conventional Curing[a]

	High Voltage Engineering System			
	GSI-1	GSI-2	ICT-7	Typical Oven
Equipment Type	300kV-20mA	300kV-50mA	300kV-100mA	150-ft heating section
Installed Cost	$100,000	$125,000	$175,000	30-ft cooling section
Length (floor space)	10	10	15	180
Start-Up Time (min)	1	1	1	90
	Operating Costs (4000 hr/yr)			
Amortization (10-yr straight line)	$10,000	$12,500	$17,500	$15,000
Floor Space ($2/ft^2)	200	200	300	3,600
Utilities (3¢/kWh)	1,100	2,800	5,600	56,000
Maintenance	3,000	4,000	6,000	1,000
TOTAL COST	$14,300	$19,500	$29,400	$75,600

[a]Source: High Voltage Engineering, Burlington, Mass.

Infrared Curing

Infrared radiation is an important new threat to UV curing, and its use has been pioneered in Italy.[12] In fact, Italian wood spokesmen believe that UV-cured wood coatings have reached their penetration zenith, and that new wood coating lines will be cured with warm air and infrared lamps.

IR-cured coatings are strictly thermally cured. There are no photochemical reactions involved, only thermal evaporation of organic solvents from conventional solvent-based coatings and new, high-solids coatings, or water evaporation from conventional latices and new waterborne coatings. Thermal energy reaches the wet substrate directly from the IR radiation, but also from metal reflectors. The IR filament itself produces operating temperatures of about 4000°F.

The by-product heated air is used to dry wet, coating films thicker than 2.5 mils. Hot air from the lamps is drawn to the front end of the IR line to a "flash-off" chamber. Here most solvents and/or water evaporate smoothly, prior to exposure of the coating directly to the IR radiation. In this manner, thick coatings can be dried without blistering.

IR radiation threatens UV radiation because of lower costs and because IR-curable coatings can be used in conjunction with proven, conventional coating technology.

IR bulbs available at $28-100 each compare with UV bulbs at $20-1000 each. Coatings for IR curing are conventional, selling for about half the price of UV coatings. IR electrical energy requirements are claimed to be about half those for UV, and are less than natural gas energy costs.

Conventional solvent-based and latex coatings, with their lower costs and wider spectrum of desirable properties can be IR cured. Use of conventional coatings eliminates the viscosity and toxicity problems associated with UV-cured coatings. Adhesion problems to metal are eliminated. In addition, IR radiation is oblivious to whether the film is clear, filled or pigmented. No shielding is required with IR dryers as with UV dryers.

The major limitation of IR curing is space for curing coatings thicker than 2.5 mils. To avoid blistering in these thicker coatings, the flash-off chamber is used, which must be at least 20 ft long (6 meters). For coatings of less than 2.5 mils thickness, only 1.0-1.5 sec IR radiation exposure are required to cure, in a 4-foot radiation chamber.

In the U.S., Research, Inc., Minneapolis, Minnesota, seems to be the only company active in producing IR equipment for coatings.

Other Applications

Coil Coatings

Coatings for electrical coils are usually clear varnishes. Coil coaters are huge consumers of natural gas, using it to thermally dry solvent-based coatings at line speeds of 75-120 m/min. In 1973, coil coaters consumed 5.38 million gallons of various organic solvents (40.32 million pounds). This was either vented to the atmosphere or incinerated at punitive costs.

On the other hand, the chemistry of radiation-curable coatings is not yet versatile enough to satisfy the coil coaters. Such coatings are made with a great variety of properties, and radiation-curable formulations are not as yet proven. High inherent viscosities and poorer adhesion to metal also mitigate against radiation curing of these coatings. There is no UV or EB curing line operating commercially in the U.S. at this time on coil coatings. Instead, research activity is being focussed on waterborne coatings, high solids (exempt solvents), liquid coatings and two-part, reactive, liquid coatings. The former two have begun penetrating the coil coating market.

At least two infrared curing lines were recently installed in the U.S. for commercial coil coating operations. One is a 120-foot (36 meters) line, curing waterborne top coats, running at 83 kW and costing only $13,000.

Given today's state-of-the-art, it is unlikely that radiation-cured coatings will ever be used on coil coating lines, except as specialty UV-cured inks and coatings. With chemical breakthroughs, however, this market could be penetrated within 5 years.

Sensitized Printing Plates

In 1969 the total market for all types of sensitized printing plates, including UV-cured plates, was $675 million. In 1975 the market was estimated to be $1 billion in commercial printing and publications alone. W. R. Grace & Co., DuPont and Polychrome are the three largest suppliers of sensitized printing plates in the U.S., but the market cannot be estimated at present for those plates processed by UV curing.

Electronic Circuit Boards

Dynacure Corporation, Santa Fe Springs, California, and W. R. Grace & Co., Columbia, Maryland, are the major U.S. suppliers competing for the $40 million market for silk-screened inks used in the manufacture of printed circuit boards. Dynacure entered the market first (1974) with a complete system of UV-dryer and UV-printing ink.

Conductive substrates are first silk-screen printed upon in the exact configuration of the ultimate circuitry. The radiation-curable ink (or photoresist) is now applied dried, and unprinted, bare portions of the conductive substrate are removed by chemical etching. The temporary and chemically resistant ink is washed away in an aqueous bath, exposing the circuitry, which is thermally dried and ready for assembly or shipment.

Solder-resistant photoresists, UV-cured decorative finishes and circuit coding colored inks are also sold by Dynacure and Grace. Dynacure claims that its system requires less than 30% of the plant space of comparable infrared or gas-fired ovens, and reduces energy requirements by as much as 90% over conventional heat-curing systems.

POTENTIAL APPLICATIONS

Pressure-Sensitive Adhesives

Pressure-sensitive adhesives is the single largest potential market for radiation-cured polymers. Photochemically cured adhesives could potentially replace the 217,687 metric tons of polymers used worldwide in 1974 to manufacture pressure-sensitive tapes and labels (54,000 tons in Europe, 109,000 in North America, and 54,000 in Japan).

Pressure-sensitive (PS) tapes and labels could cost less to manufacture by radiation curing. At the volume of polymers used, selling prices of well under $1.00/lb can be visualized for formulated EB-curable adhesives.

Energy Sciences has estimated the economic benefits to be attained by using EB accelerators and UV curing compared with thermal ovens (the conventional method) to cure photochemical pressure-sensitive adhesives. Data in Table 2-9 are based on the conventional oven using an equivalent of 1.055 million J/ft^3 at a gas cost of $1.00/1000 ft^3, power costs of 2¢/kWh, UV-cure requirements of 1 J/cm^2 and electron beam cure requirements of 0.25 J/cm^2.

Table 2-9. Economics, Photochemical vs. Thermal Production Lines Pressure-Sensitive Adhesive Tape and Labels[a] [300 ft/min (90 m/min)][13]

	Electron Beam	UV Cure	(Solvent-Based Adhesive)
Line speed			
(ft/min)	300	300	300
(m/min)	90	90	90
Beam			
(kW)	6.3	18.6	NA
Power consumption			
(kW)	40.2	510	5,100
Operating costs			
($)			
Power	.81	10.20	18.00
Maintenance	3.50	4.50	2.70
Nitrogen gas	2.75	NA	NA
Water	.20	NA	NA
Total Costs	7.26	14.70	20.70
Capital Costs[a]	$200,000	$13,000	$130,000

[a]Estimate of range average from R. Rice.

Energy Sciences further claims that a thermal oven would use gas at a rate of 3,000,000 watts, a UV unit would use electrical energy at the rate of 500,000 watts, and the Electrocurtain EB unit would use electrical energy at the rate of 30,000 watts. Energy Sciences claims that 70% of all electrical energy goes into the liquid adhesive and that 70% of this energy is used to cross link the adhesives–none is used to heat the substrate.[13]

Pressure-sensitive wallpaper and decorative papers represent other potentials for EB curing, as do coated paper films for various packaging applications.

Vinyl Floor Tiles

EB accelerator manufacturers were asked to cure vinyl top coats for the production of floor tiles, and have complied. However, none expected to receive orders for accelerators for this application.

Union Carbide's Linde Photocure Systems Division, Indianapolis, Indiana, reports that a 3-mil, clear, urethane acrylate coating was just too slow in curing for commercial production (50 ft/min). However, when a nitrogen blanket was used, a commercially acceptable line speed of 300 ft/min could be attained.

Although it is too early to predict the market for radiation-cured coatings for this market, if all 1.445 billion square feet of vinyl-asbestos tile produced in the U.S. in 1970 had been coated with a 3-mil topcoat (UV-cured), the market would have been 9.007 million gallons of radiation-cured coating. EB generators can have their place in this market because relatively thick coatings are used. It is projected that no significant penetration of this potential market will be achieved before 1985.

ASSESSMENT OF THE FUTURE

Economic Considerations

Motivations

The major motivation for switching from conventional materials to radiation-cured formulations is the cost savings for natural gas which is used for thermally heating and curing conventional coatings and inks and for incineration after drying organic solvents. Additional advantages of radiation curing include smaller-sized overall production lines, increased production rates, lower total production costs, and the ability to coat or print on heat-sensitive substrates (UV and EB curing). Many other additional process advantages accrue to radiation curing, but these depend on the specific application involved.

Energy and Cost Savings

Various estimates have been made of the energy that can be saved by employing radiation curing materials in place of conventional, thermally cured coatings and inks. In coatings, Coors Brewery has estimated a potential 59 billion Btu in energy savings for the beer and soft drink can market.

The Can Manufacturers Institute estimates that if the metal and closures industry were required to incinerate all organic solvents from conventional coatings and if all then switched to UV or EB curing, the use of 11 bcf/yr of natural gas could be eliminated.

Cities Service has estimated total energy consumption for conventional manufacture of metal cans at 23.0 Btu per can, whereas the same can manufactured by UV curing requires 6.6 Btu per can.[14]

Radiation Dynamics estimates that in the EB curing of polyethylene cable insulation, total costs (including amortized capital) for the conventional extrusion plus steam process are 6.3¢/lb vs 2.5¢/lb by EB curing. Of the 3.8¢/lb differential, 0.8¢/lb is in the higher cost for conventional insulation, and 3.0¢/lb is the cost for peroxide (for the conventional formulation, but not needed in the EB cross-linked material).[15]

Thermogenics, of New York, reported that a 38-in., 6-UV bulb dryer uses 50 kWh of electricity. At a mean cost of $0.03/kWh, the power cost of this UV dryer is $1.50/hr.[16]

TEC Systems, Inc., a supplier of UV curing materials for the web offset printing industry, estimates that for a 38 in.-wide web offset press capable of printing speeds at 1000 ft/min, total operating costs of a conventional, heat-set, high-velocity system are $50,327, vs. $33,532 for the UV curing system. These costs include installation and amortized capital costs.[5]

DeBell & Richardson estimates that for radiation curing of adhesives, to cure a 1-mil pigmented coating 48 in. wide at the rate of 300 ft/min would require plant operating costs of $20.70/hr for thermal oven curing, $14.70/hr for UV curing and $7.26/hr for electron beam curing. These are plant operational costs only, and do not include capital costs. Included in these plant operational costs, however, are the following costs: thermal oven processing, requiring 5100 kWh at a power cost of $18.00/hr; or UV curing requiring 510 kWh at a power cost of $10.20/hr; or EB curing requiring 40.2 kWh at a power cost of $0.81/hr.[17]

In 1974, the O'Brien Corp., a metal can maker, estimated the cost to operate a UV dryer at $2.00/hr ($1.40 for electricity and $0.60 for lamp amortization), vs. a gas oven at $2.75/hr ($2.00 for the gas and $0.75 for the electricity). Considering that the original cost of a UV dryer would be about one-fifth that of a gas dryer and that the cost of a gas-fired incinerator is about $130,000, with an operational cost of about $5.00/hr, the choice of UV curing in a new installation becomes apparent.[18]

Reliance Universal, Inc., reported that the first furniture manufacturer to use UV coatings in the U.S. for coating particleboard replaced his single-headed sander (which presands the board before filling) with a multiheaded sander. This cut his UV filler usage from 55 to 30 gal/day.

The Future of Radiation Curing

Ultraviolet radiation-curing technology was introduced commercially in the United States in the late 1960s. Acceptance in terms of better product at lower costs had been gained prior to the natural gas shortage in the areas of lithographic printing inks, two-piece metal can coating and printing and fabrication of polymeric printing plates.

Electron beam radiation is commercial practice to produce heat-shrinkable polyethylene packaging film, to sterilize prepackaged bandages, for production of telephone wire, insulation of cables, and coating of automobile parts. However, each practice is by a single company, and EB is not an industry-wide practice.

Major obstacles to both UV and EB curing are the rate of cure (faster speeds are desired) and improvement in properties of cured coatings, particularly with respect to metal adhesion. Formulations are desired that are nontoxic, allowing their use for food packaging.

Assuming that faster curing formulations are developed, which will allow faster production rates, the ability to coat the fomulation prior to curing will become paramount. If the coating is too viscous to flow out and level at high production speeds, the advantages of increased rates of cure will have been lost.

The most promising applications for radiation-cured coatings are: (1) pressure-sensitive adhesives; (2) coatings for vinyl-asbestos floor tiles; and (3) printed circuit boards.

Coil coatings represent an attractive market, but require major formulating breakthroughs, since a spectrum of properties is desired from many coil coating compositions.

Competition for UV and EB curing is arising from infrared curing. IR-curable coatings are conventional, and only the upstream part of installed coating systems needs to be modified. By contrast, UV and EB lines all require complete production line replacement. In addition, competition for radiation-curable formulations is arising from waterborne, high-solids and powder coatings. These are lower in cost and can be used in conventional production equipment. Globally, all radiation-cured ink and overprint varnish markets are expected to grow from their small 1975 base at a rate of 20% through 1986.

Radiation-cured inks will grow faster and larger in North America than in Europe because the former has many more web offset presses to convert; also, the U.S. uses more natural gas in these operations, and the gas shortage will force conversion to processes that will not require it.

By 1986, it is projected that radiation-cured coatings and lacquers will replace no more than 4% of today's solvent-based coatings, since the technology is inappropriate for 64% of industrial coatings. New wood coating lines will be dominated with infrared drying lamps.

RADIATION CURING IN JAPAN

In Japan, radiation curing is being used for several different applications, notably for wood and motor vehicle coatings and for inks.

Wood Coatings

There are about 25 commercial UV-dryer lines in operation in Japanese wood plants. Most are small factories that buy raw wood products and finish them prior to cutting and assembly. Of 165.84 metric tons of radiation-cured wood coatings consumed in Japan in 1974, 153.84 were cured by UV and 12.00 by EB. Wood coatings cured by UV represented 51.28% of all wood coatings in Japan during 1974.

Despite the 1975 recession, Japan's use of UV-cured wood coatings was projected to grow by 23% during 1975. It is further predicted that UV-cured wood coatings will grow in Japan at an annual rate of 25% between 1975 and 1980, and at 20% between 1980 and 1985.

There are only five companies in the world using electron beam accelerators to cure photoreactive wood coatings, and one of these is in Japan.[5] Three are in Europe (Varsseveld, Suedex, N.V., Netherlands; Bruynzell, W. Germany; and Parisot, France), and one is in Brazil (Bergamo); there are none in North America. In Japan, Eitai Sangyo Co. cures wet films at the rate of 80 m/min (maximum speed). Film thicknesses range from 8.66-59.05 mils, although thinner films also can be cured by EB processing. Only some 5 metric tons a year of EB-cured coatings are now applied by Eitai.

Interest in IR coatings also is being explored in Japan by four large users of conventional wood coatings and UV wood coatings. These are: Eitai Sangyo, Dantani Sangyo, Ookura Kogyo and Kishimoto Toso.

Motor Vehicles

In 1971, Ford Motor and Kansai Paint established a joint venture—Nippon Electrocure, Ltd.—to exploit Ford's Electrocure process. In 1973, Kansai installed an EB pilot plant consisting of an electron beam accelerator (300 kV × 100 mA) and a conveyer capable of handling parts up to 100 mm in height and 500 mm in width at line speeds of 5-30 m/min.

Nippon Oils & Fats Co. also has an EB pilot plant line (500 kV × 100 mA) with running speeds of 1-6 m/min, which can handle parts 1600 mm high and 600 mm wide. Kansai and Nippon Oils & Fats are reported to dominate the small sales volume of EB cured coatings to the Japanese motor vehicle industry.

Before 1972, the only EB radiation-cured coatings sold in Japan were polyester/styrene wood coatings. Since 1970, this market has stayed constant at 36 ton/yr. In 1972, 24 metric tons of EB-cured polymers were sold, principally to the motor-vehicle and electrical-appliance industries. Consumption of these coatings was equally divided between epoxy and polyester acrylates, and sales remained constant between 1972 and 1973. It is estimated that 26 metric tons of polyester acrylates and 12 tons of epoxy acrylates were consumed in Japan in 1974 and again in 1975 by the motor-vehicle and electrical-appliance industries.

Suzuki Car Manufacturing Co. has had a semicommercial EB curing line in operation since May, 1973, consisting of an EB accelerator (300 kV x 10 mA) and two overhead conveyer lines capable of line speeds of 3-4 m/min. EB coatings are used for a variety of its motorcycle parts, including polypropylene and asbestos flame covers and fenders, headlight housings, gasoline tanks and some steel parts. Suzuki is reported to believe that EB curing will eventually result in a 15% reduction in total costs compared with conventional bake-type coatings and lacquers. Honda is reportedly using EB coatings in a limited way to coat plastic covers for its motorcycles.

Using 1974 as the base year, when 59.66 metric tons of formulated EB-cured coatings were consumed in Japan, it is believed that consumption in 1975 remained constant, but that the market will grow to 235.50 tons in 1979, 525.95 tons in 1980, and to 847.80 tons in 1985.

The entire Japanese motor vehicle production, by all companies, is roughly twice that of Ford Motor Co. alone. Ford has consumed as much as 363 metric tons of EB-cured coatings on its Electrocure lines during any one year.

Inks

Japanese consumption of UV-cured inks in 1974 was 20 metric tons, but this doubled in 1975 to 40 metric tons (at a value of $80,813). This consumption represents 6.7% of all EB- and UV-cured formulations in all coatings and inks in Japan, and 8.4% of all formulations for coatings and inks cured solely by UV.

Projections are that these Japanese UV ink markets will increase to 420 metric tons in 1980 and to 1280 metric tons by 1985. Strict anti-air pollution regulations in Japan, coupled with the same critical natural gas shortages as in the U.S. will spur the Japanese to eliminate thermal drying wherever possible.

Dai Nippon Ink & Chemical is the most active major ink company, having negotiated a license from Sun Chemical in 1975, and is reported to have a line of inks dominated by urethane-acrylates. Mitsubishi Rayon, Nippon Oils & Fats, Dai Nippon Toryo and Dai Nichi Seika Kogyo are also active in UV-cured ink development.

Toyo Ink Manufacturing, Tokyo Ink, Sakata Shokai and Morohoshi Ink have active research programs on UV-cured ink formulations. Morohoshi is reportedly supplying UV inks to the first large sheet-fed offset press in Japan (owned by Dai Nippon Printing). This system cost $50,000 and can UV dry up to 8,000 sheets/hr (2.2/sec).

Vinyl Tiles

A Japanese manufacturer of vinyl-asbestos floor tiles is reported to have recently installed a UV line to dry semicommercial quantities of urethane-acrylate, high-gloss, "wet-look," tile top coats. Coatings are cured by 6 mercury lamps under a nitrogen gas blanket to facilitate rapid cure. The clear coatings are applied at a thickness of 100 μ (4 mils) at a line speed of 30 m/min (99 ft/min).

Sensitized Printing Plates

In Japan, nine companies photoprocess relief printing plates. They are: Asahi Chemical (Kaesi), Kansai Paint, Nippon Paint, Teijin, Tokyo Oka, Toray, DuPont, Grace and BASF. Some 80% of photopolymers and plates are manufactured domestically (although the technology was originally introduced by DuPont, Grace and BASF). Asahi is reported to dominate 64-68% of the Japanese market of 5 billion yen ($16.515 million), which represents only 4-5% of the total printing plate market.

By 1980, Japanese spokesmen believe that all relief printing plates, including systems and auxiliary material, will be worth 15 billion yen

($49.545 million), and that 50% of this total will be photosensitive relief plates. It is estimated that photopolymers used to make printing plates were worth $1-5.5 million in 1975 in Japan.

REFERENCES

1. First International Conference on Radiation Curing, Society of Manufacturing Engineers, Dearborn, Michigan, April 10, 1974.
2. Second International Conference on Radiation Curing, Society of Manufacturing Engineers, Dearborn, Michigan, May 7, 1975.
3. Symposium on Polymer Materials for Electronics, American Chemical Society, Organic Coatings & Plastics Chemistry Division, Washington, D.C., August 28, 1975.
4. "New Radiation Cured Printing and Coating Techniques," UCB, SA Division Spécialités Chimiques, Rue D'Anderlecht 33, Anderlechstratt, B-1620 Drogenbos, Brussels Belgium.
5. Banker, W. W. "UV Curing Equipment for the Web Offset Printing Industry," FC75-316, Second International Conference on Radiation Curing, Dearborn, Michigan, May 7, 1975.
6. Fusion Systems Corporation, Rockville, Maryland.
7. Cleland, M. Radiation Dynamics, Inc., Westbury, New York. Private communication.
8. Can Manufacturers Institute, Washington, D.C.
9. Coors Brewing Company, Golden, Colorado.
10. Ford Motor Company, Dearborn, Michigan. Private communication
11. Fitzgerald, M., W. R. Grace Company, Columbia, Maryland. Private communication.
12. Research, Inc., Minneapolis, Minnesota.
13. Energy Sciences, Inc.
14. Pansing, H. E. "Cylindrical Cans," FC74-535, First International Conference on Radiation Curing, Society of Manufacturing Engineers, Dearborn, Michigan, April 10, 1974.
15. Morgenstern, K. M. "Radiation Time Has Arrived for Plastics and Rubber," FC74-540, First International Conference on Radiation Curing, Society of Manufacturing Engineers, Dearborn, Michigan, April 10, 1974.
16. Pray, R. W., "Simplifying Radiation Curing for Web Presses," FC74-513, First International Conference on Radiation Curing, Society of Manufacturing Engineers, Dearborn, Michigan, April 10, 1974.
17. Childs, E. S. "Radiation Curing of Adhesives," FC75-340, Second International Conference on Radiation Curing, Dearborn, Michigan, May 7, 1975.
18. Whitmire, W. E., "Metal Decorating–Ultraviolet-Curable White Basecoats for the Can Industry," FC74-536, First International Conference on Radiation Curing, Society of Manufacturing Engineers, Dearborn, Michigan, April 10, 1974.

UNCITED REFERENCES

Ganyo, R. "Infrared: Hot Contender for Curing Coils and Cans," *Mod. Metals* (October 1975).

Rybny, C. B., *et al.* "UV Radiation Cured Coatings," *J. Paint Technol.* 46:60-69 (1974).

Stolze, E. "Cost Reduction with Electron Beam Curing," BASF-LACKCHEMIE (56) (November 1974).

Vanderboff, J. W. "UV-Cured Printing Inks; A Review," *Am Chem Soc. Div. Org. Coatings Plastics Chem.* **35** (1):124-129 (1975).

Vincent, K. D. "Utilization of Photopolymer Inks to Print on Thin Plastic Fiber by Offset Lithography," First International Conference on Radiation Curing, Dearborn, Michigan, April 10, 1974.

3
CONTINUOUS CASTING*

INTRODUCTION

Continuous casting is a steel-manufacturing technique in which the three conventional processing steps of casting ingots–reheating (soaking) ingots prior to rolling to slabs, blooms, or billets–and the rolling step itself are replaced by a continuous process which achieves the same end. In continuous casting, molten steel is poured from the ladle to continuously produce solidified, metallurgically acceptable, slab, bloom or billet cross sections. These cross sections can then be cut to length and sent to the conventional rolling mill for the production of finished shapes.

The continuous casting technology was first developed just before World War II and was first accepted and applied worldwide in the 1950s. The total worldwide capacity of the installations in 1975 was about 150 million tons of continuously cast steel, about 20% of the total world steelmaking capacity. This installed capacity is expected to grow at an annual rate of about 10.5% and exceed 200 million tons by 1978.

In the United States, 1975 production of continuously cast steel was 9.6 million tons, *i.e.*, 9.1% of the 106-million-ton annual U.S. steel production, a level below the world average. The expectations, however, are that the growth rate in the U.S. will be much higher than the world average so that installed capacity should reach about 80 million tons and be about 50% of annual steel production by 1985.

In Japan, the 1975 production of continuously cast steel was in excess of 30 million tons per year. The annual crude steel production in 1975 was 102.2 million tons by conventional and continuous casting, a rapid growth rate. The percentage of continuous casting capacity was 20.7 in 1973, and 25.2 in 1974. In terms of installed continuous casting tonnage, Japan leads the United States, Canada and the European Economic

---

*By: Marcel Barbier, Charles Bliss

Community countries. In Table 3-1, the continuous casting output and the share of the crude steel output are given for Japan and the U.S. from 1969 through 1975.

Table 3-1. Continuous Casting Output and Share of Crude Steel Output[1]

	Continuous Casting Output (x 1000 tons)	Share of Crude Steel Output (%)
U.S.		
1969	3,651	2.9
1970	4,521	3.8
1971	5,272	4.8
1972	6,973	5.8
1973	9,270	6.8
1974	10,722	8.1
1975	9,653	9.1
Japan		
1969	3,287	4.0
1970	5,226	5.0
1971	9,918	11.2
1972	16,473	17.0
1973	23,714	20.7
1974	29,400	25.1
1975	30,663	30 (estimate)

The rapid growth experienced by continuous casting since its initial acceptance as a steel-making practice is a result of the marked reduction in manufacturing costs arising from its use. This cost saving is accompanied by a significant reduction in energy consumption which has importance apart from its economic implications.

Cost savings for continuous casting arise from the reduction in the amount of recycled scrap from continuous casting as compared to the recycled scrap generated in the conventional ingot-to-semifinished shape practice, and in the elimination of the soaking furnaces for reheating ingots. A reduction in recycled scrap permits an increased productivity from the equipment in the processing steps from raw materials to semifinished blooms, blooms, slabs and billets. Increased productivity is reflected as reductions in metallurgical coke consumption in the blast furnace route to steel-making, or in electrical energy consumption in the electric furnace route. A comparison of the energy consumption for continuous and conventional casting is given in Table 3-2.

Table 3-2. Comparison of Energy Consumption for Continuous Casting and Conventional (Ingot) Casting

	Continuous Casting		Conventional (Ingot) Casting and Breakdown		Continuous Casting	Conventional (Ingot) Casting and Breakdown
Ratio of Hot Metal to Scrap in BOF Charge	70/30	90/10	70/30	90/10	–	–
Net Energy Consumption per ton of Production (10^6 Btus)	8.4	10.8	11.3	14.1	6.7	9.6
Savings Over Conventional (% of Conventional Casting Consumption)	25.7	23.4	–	–	30.2	–

Typical costs in the United States for the production of slabs by conventional ingot casting, soaking and breakdown on a scale of two million tons per year production is estimated at $204.50. A comparable cost utilizing continuous casting is $185.60. Continuous casting thus offers production cost savings of $18.90 per ton of slab, or about 9% of the cost by the conventional ingot route. Similar cost savings can be estimated for billet production, but in addition, this advantage is particularly enhanced by a reduction in the requirement for capital. In a new plant, the capital needed for the continuous casting installation is about $9 million as compared with a capital need of $25-35 million for ingot installation.

A vacuum degassing step preceding the continuous casting step avoids the formation of gas bubbles and possible surface imperfections in the continuous cast shapes. The conventional metallic additions to the molten steel for degassing is avoided. Control of temperature and rate of cooling through water sprays permits continuously cast shapes to have acceptable metallurgical properties, to the extent that these shapes could possibly be sent to the finishing rolls without reheating.

Continuous casting is a particularly attractive adjunct to electric furnace steel-making in the small-scale mini-mill concept. The ability to substitute for the scrap charged normally to the electric furnace by prereduced iron pellets enhances this combination and even enables it to increase its cost-effectiveness in larger scales of production. With present direct reduction technology, the concept needs to be based on assured supplies of attractively priced natural gas, although a viable coal process can emerge in the near future.

GENERAL PRINCIPLES

In steel-making, solidification of the liquid steel has conventionally been accomplished by casting the molten steel into molds to produce ingots. After cooling, ingots can be stored until needed, transferred to other plants, or sent to shaping or breakdown mills where they are reheated and rolled into basic, semifinished shapes suited for further rolling to the finished products (Figure 3-1).[2] These basic shapes are billets for bars, rods and tubular products, blooms for structural profiles, and forging blanks, and slabs for plates and sheets. Ingot casting is a batch-type process. The conventional production of raw steel in the form of ingots introduces technical complexities because of the need for reheating and primary rolling into billets, blooms and slabs. Ingot heating and rolling processes are major consumers of energy, time and, therefore, money.

Continuous casting, however, as its name implies, continuously produces the semifinished shapes, and therefore bypasses the primary rolling and hot rolling mills. Its basic steps are:

- delivery of liquid metal to a holding tundish;
- uniform flow of liquid metal from the tundish to the casting strand;
- solidification of a metal skin in a water-cooled mold;
- continuous withdrawal of the solidifying strand casting from the mold;
- heat extraction to cool the strand casting by controlled spray cooling; and
- cutting and removal of the solidified sections.[1]

Continuous casting eliminates the need for ingot casting facilities and soaking pits in front of the primary mill, as well as the breakdown mill itself.

Continuous casting was developed in Europe, predominantly by Junghans in Germany. Although limited efforts were also undertaken in the United States, Russia, Austria, England, and France, progress with continuous casting was slow. The problem of controlling the solidification mechanism of a freezing continuous strand in a mechanical operation was very complex and had previously not been thoroughly investigated.

The earliest work in continuous casting was undertaken on the special-quality, fully killed steels, since these suffered some of the higher cropping losses, or lowest yields, upon being cast into ingots. Because of their interest in this type of steel, Babcock & Wilcox and Allegheny Ludlum were

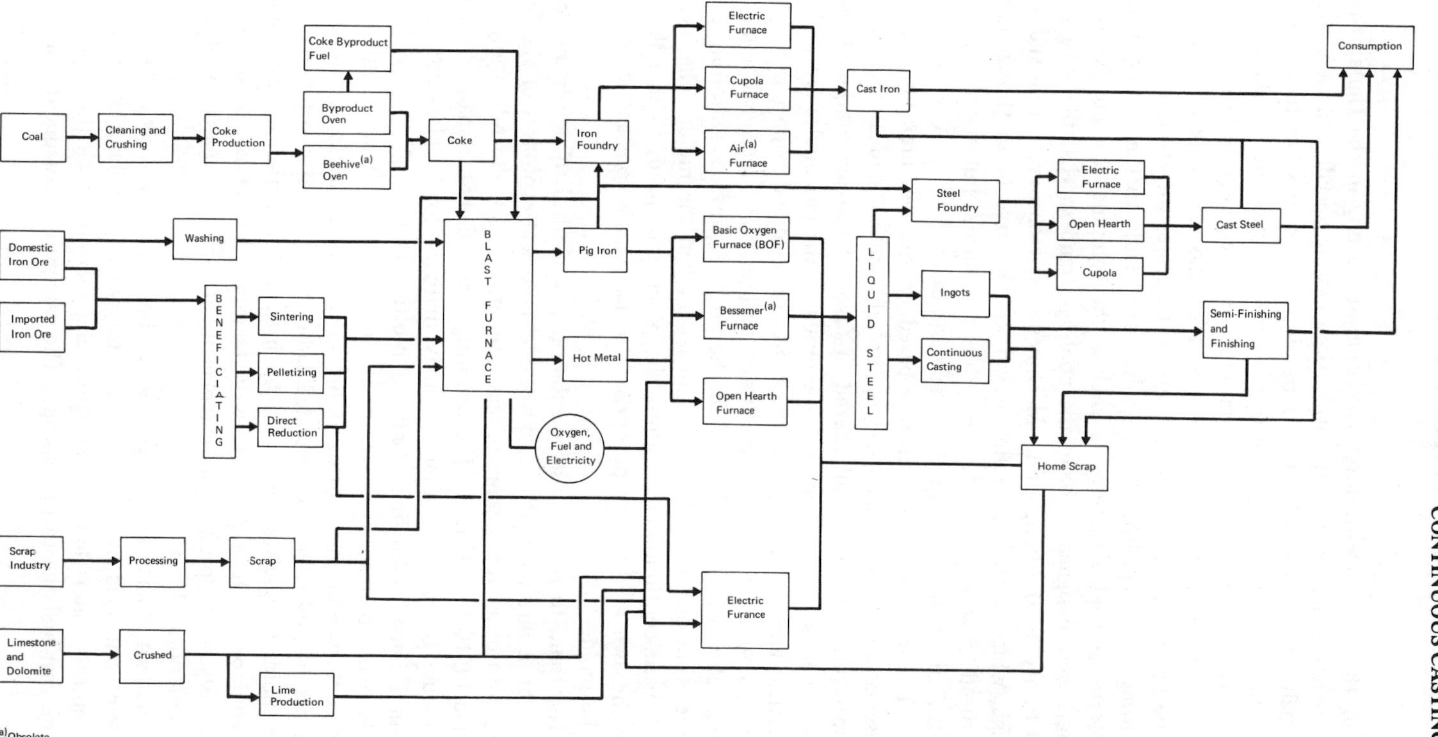

Figure 3-1. Flow diagram of steel-making process.[2]

among the early investigators of continuous casting in the United States. The earliest machines had relatively slow casting speeds and were limited to small rates of throughput. As development efforts continued, however, improvements were made in operating practice and machine design.

The first continuous casting machines were entirely vertical. No bend occurred in the casting strand so that the solid section had to be rotated to the horizontal after being cut off. Therefore, the structures were tall, requiring both high lifts of the molten steel ladle to the casting machine platform, and tall structures to enclose the equipment. Later, continuous casters were designed to bend the cooling strand so that the steel shape was turned to the horizontal. Machines now incorporate a curved mold design, which starts the cooling strand in the curve to the horizontal. Overhead space requirements were thus minimized, but complexity in the control of solidification of the strand increased.

The whole process control is based on the measurement, control and conservation of the temperature of molten steel.[3] Manual control of the temperature tends to be minimized in favor of instrumentation, such as spectrometers and pyrometers.[4] A series of thermocouples are placed in a vertical line on the upper part of the water-cooled copper mold, with their hot junctions close to the inside surface. The registered temperatures change abruptly at the level of the liquid steel surface, indicating both the temperature of the molten steel and its level in the mold. The level signal is used continuously to adjust the pinch roll speed, so that the casting is withdrawn at the rate appropriate to maintain a constant level of steel in the mold.

The trend to on-line instrumentation and control of continuous casting has been a direct outgrowth of the technological evolution of the basic oxygen-furnace steel-making process and its critical, short-duration refining period of 10-15 minutes. The resulting high-speed steel-making cyclic rate of about 50 minutes is almost exactly matched to the allowable time of about 1 hour for holding a ladle of molten steel.

To avoid premature cooling of the molten steel during continuous casting, conservation and control of the steel temperature in the mold requires the use of preheated and insulated ladles and preheating of the tundish. The molds themselves are cooled by internal circulation of water and may be equipped with several nozzles to facilitate casting several strands simultaneously, or to provide rapid replacement of a nozzle that is plugged or has deteriorated.

The cast strands are cooled during their withdrawal by a strand cooling system controlled by a digital computer, which calculates and automatically controls water flows to the spray chamber to match the type of steel to be cast and the casting speed. The computer subsequently monitors the temperature of the cast strand at various points and modifies the flow of water to maintain the appropriate metal temperature.

A typical continuous casting installation is the two single-strand slab machines at the Ravenscraig plant of the British Steel Corporation, Motherwell, Scotland (designed by CONCAST of West Germany). The plant layout is shown in Figure 3-2. Continuous reheating furnaces (not shown) may be added to improve the control of steel temperature before the rolling mills. A schematic diagram of a modern integrated steel mill, in which both ingot casting and continuous casting are employed, is given in Figure 3-3. The diagram is based on the Fukuyama (Japan) works of the Nippon Steel Corporation. The continuous casting machine, in parallel with ingot casting, is fed by the output of a basic-oxygen steel-making furnace, and the solidified continuous steel strand is cut as required by a flame cutter. The steel from continuous casting can be delivered to four basic lines of mills, depending upon its configuration as bloom or slab:

- *A Billet Mill* includes a weighing mill, a continuous finishing mill, a flying shear, a billet shear and a cooling bed, to produce billets to specifications. No reheating is done to this line.
- *A Shape Mill* includes a continuous reheating furnace (walking beam type), a scale-breaker, a breakdown mill, hot saws and a cooling bed; and produces H beams, sheet piling and general structural shapes.
- *A Plate Mill* includes a continuous or batch-reheating furnace, a scale-breaker, a roughing mill, a reversing finishing mill, a hot leveler, a cooling bed, side and end shears, a flame planer, heat treating equipment, ultrasonic tester and a shot blaster to produce rails and plates.
- *A Continuous Hot Mill* includes a continuous reheating furnace, a scale-breaker, roughing and finishing mills, a run-out table, coilers, skinpass mill to produce hot-rolled sheets and coils, hot-rolled slit-edge strip and hot-rolled pickled sheets and coils.

A part of the steel from the continuous hot mill is diverted, after cooling, to a continuous pickling line (hydrochloric acid) and then to a cold reduction mill. After this rolling step, the steel is electrolytically cleaned, annealed, tempered, tin- or zinc-coated, and sheared or coiled to produce a variety of products. Another part of the production is diverted from the plate mill and trimmed, formed and tested by a variety of operations to produce large diameter welded pipes.

An important auxiliary process step contributing to successful continuous casting is vacuum degassing of the liquid steel from the basic oxygen process prior to its being poured into the continuous casting machine. Vacuum degassing decreases the levels of dissolved hydrogen and oxygen

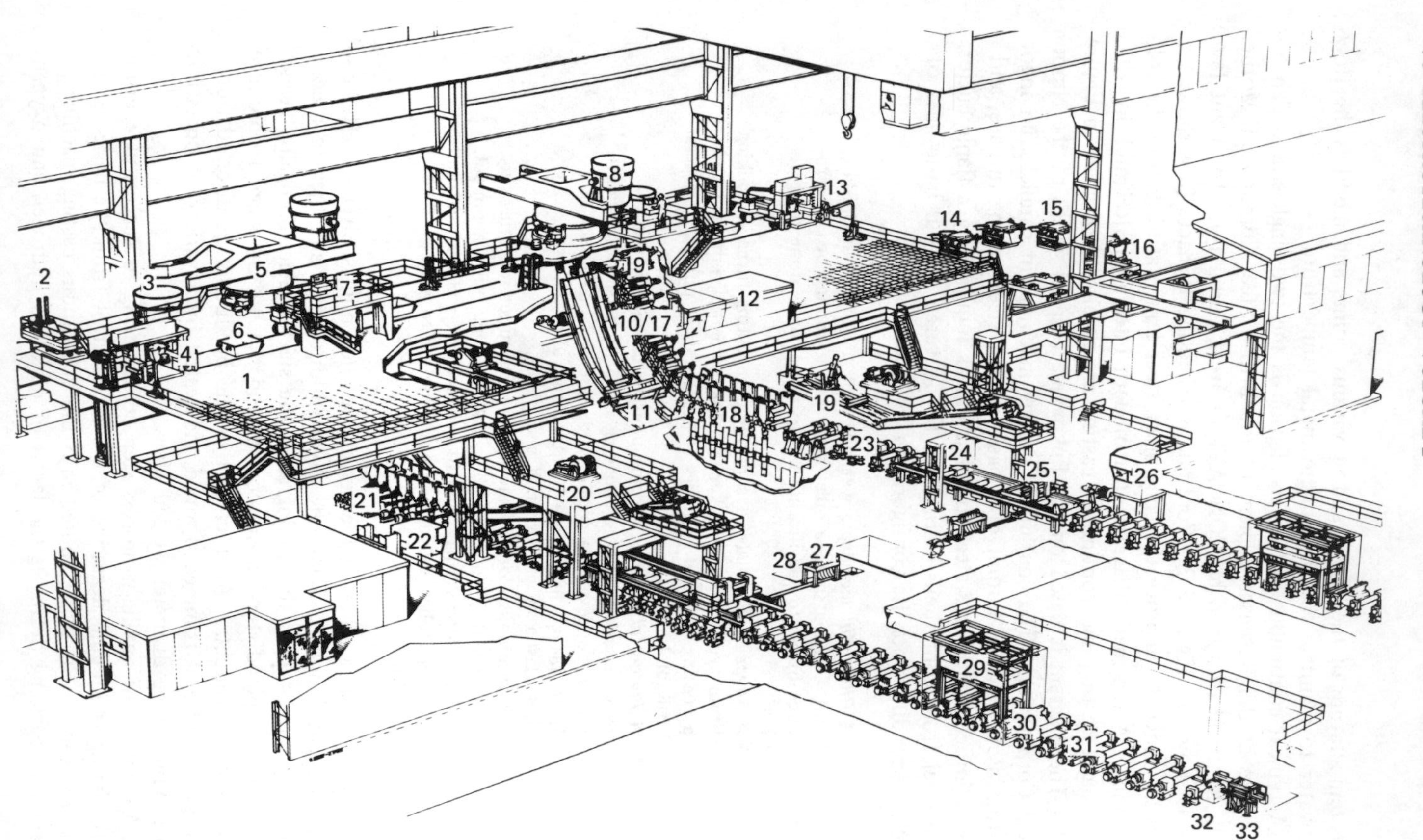

Figure 3-2. Continuous casting plant.[5]

Figure 3-2. Continuous Casting Plant.[5] Ravenscraig Works, British Steel Corporation, Motherwell, Scotland. Two Single-Strand Slab Machines.

Legend:

1. Casting floor
2. Argon flushing station
3. Emergency ladle
4. Tundish car
5. Ladle turntable
6. Slab box
7. Ladle operator's platform
8. Ladle
9. Mould and top zone assembly
10. Segment change car and winch unit
11. Casting floor pulpit
12. Strand guides
13. Tundish preheat station
14. Tundish cooling and gunning stations
15. Tundish relining stations
16. Tundish wrecking stands
17. Spray chamber
18. Withdrawal and straightening unit
19. Dummy bar receiver
20. Hoist winch for dummy bar
21. Drives for withdrawal and straightening unit
22. Withdrawal drive control cabin
23. Pre cut-off roller table
24. Cut-off roller table
25. Oxypropane cutter
26. Cut-off control cabin
27. Crop end handling unit
28. Crop end pit
29. In-line weighing unit
30. Disappearing stop
31. Discharge roller table
32. End stop
33. Slab identification unit

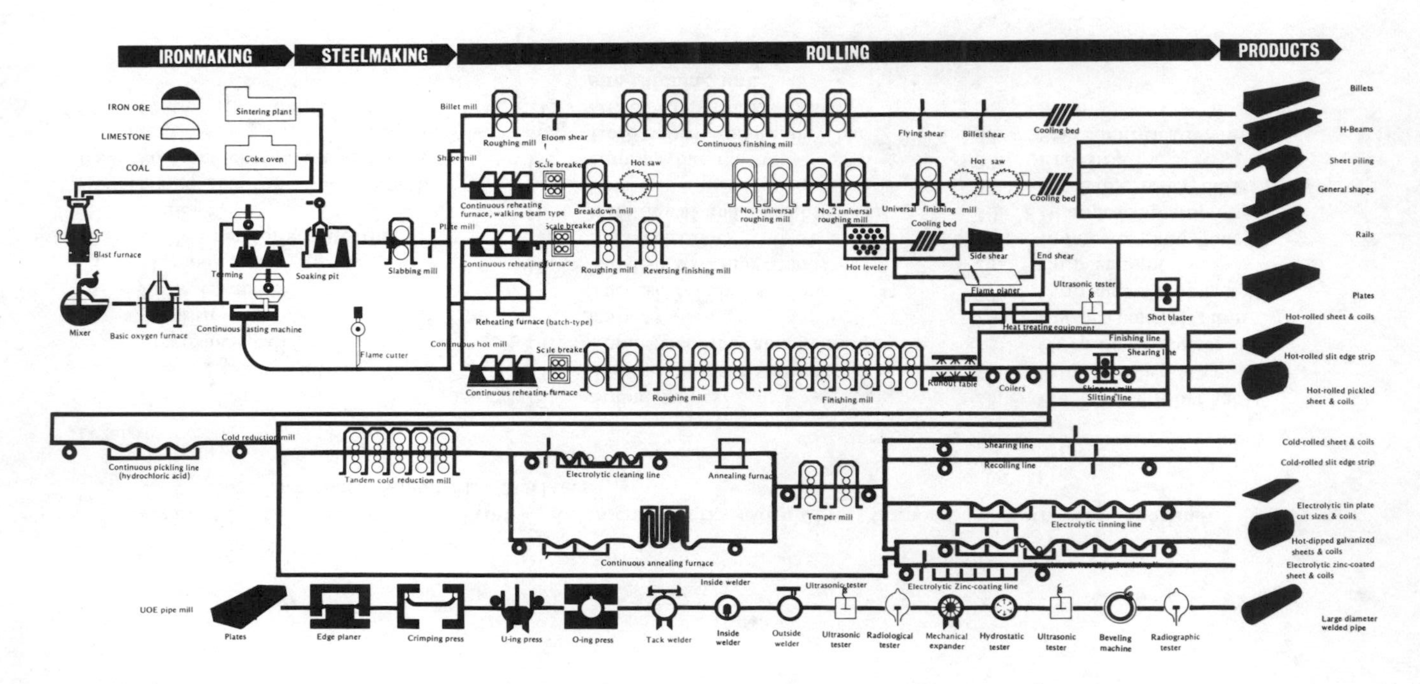

Figure 3-3. Production flow chart of an integrated plant.[6]

in the melt to desired levels. The same decrease of dissolved oxygen might also be achieved by the addition of appropriate ferroalloys, but this solution could result in the formation of metallic oxide particles within the melt and subsequent surface defects in downstream processing of the solid section.

With vacuum degassing, continuous casting is effective in producing a rimming grade of steel, the type used for deep-drawing mill products, especially for automotive bodies and domestic appliances. This grade of steel is desired for these applications because the ingot has good surface characteristics and a clean skin, making it eminently suitable for producing flat-rolled products with a low occurrence of surface defects.

Deep-draw flat product is now being rolled from continuously cast killed steel in the United States and has met all product quality requirements. Continuous slab casting for ultimate reduction to tin-plate is practiced in Germany, Japan and the U.S.S.R., but it is said that the resulting material would not meet U.S. can makers' quality specifications.

There is a great deal more experience with continuous casting of killed steel billets (square) than of rimming steel slabs for flat-rolled products. The yield of killed steel billets readily attains 95-97%; deep-drawing flat steel has slightly lower yields, about 94-96%. Production of slab by continuous casting in high volume, exclusive of the cost of the metal, has actual direct costs no greater and possibly somewhat lower than when done conventionally. The chief virtue in continuous casting is expected to be entirely in the improvement of quality, coupled with the gain in yield. When the cost of raw material (molten steel) is charged to both operations, appropriate allowances made for recycling recovery of yield losses in ingot casting and breakdown, and charges levied for degassing of the melt as part of the continuous casting procedure, the continuous casting operation can show significant cost advantage over the conventional process.

EXISTING APPLICATIONS

Impressive growth in applications in the past 20 years has resulted from recognition that continuous casting reduces energy consumption, eliminates some process steps, increases product yields, and improves the surface condition and metallurgical quality of the steel, thus saving time, energy and steel production costs. Today, continuous casting is an essential and integral part of modern steel plants. Aggregate worldwide capacity of continuous-casting installations, as of 1975, was about 150 million metric tons. This capacity amounts to approximately 20% of world steel production. The International Iron & Steel Institute looks for this capacity to continue at a growth rate of 10.5% annually, and to exceed 200 million metric tons by 1978. In 1972, installed capacity stood at 75 million metric tons.

The worldwide increase in the number of installations and in the number of strands of equipment supplied by one manufacturer, CONCAST AG, is shown in Figure 3-4. Since 1963, CONCAST has served 209 steel companies in 41 countries on all continents, designing 299 machines with 846 strands in operation and under construction, corresponding to approximately 100 million tons per year of installed capacity, about two thirds of the world's current continuous casting capacity. Table 3-3 lists the CONCAST machines by type of installation. Table 3-4 lists CONCAST machines by source of molten steel and type of steel produced.[7]

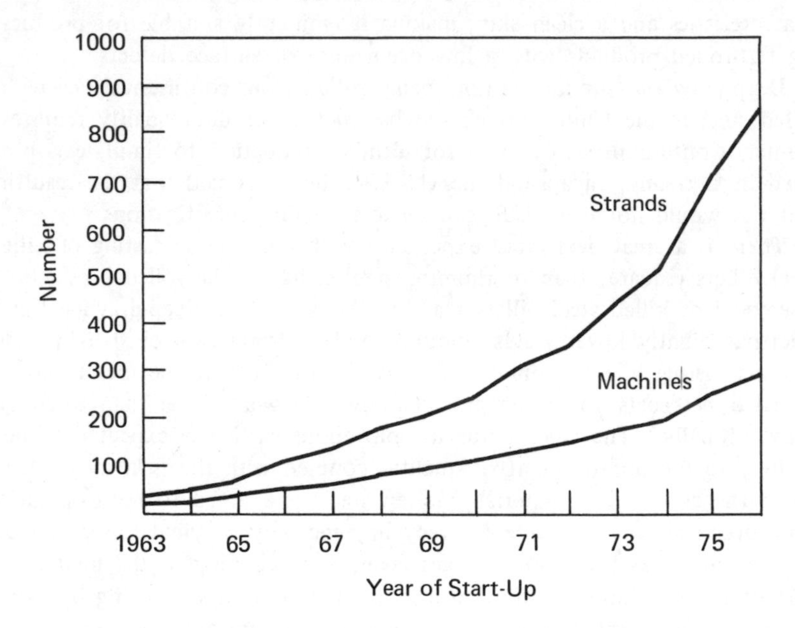

Figure 3-4. Number of CONCAST installations in operation.[5]

Table 3-3. Distribution of Continuous Casting Equipment Built by CONCAST (1963-1975)[5]

Type of Installation Employing Casting Equipment	Number of Machines	Number of Strands
Machines for Slabs	61	88
Machines for Blooms and Shapes	57	203
Machines for Billets:		
1 or 2 strands per installation	71	130
3 or 4 strands per installation	94	328
5 or 8 strands per installation	16	97

Table 3-4. Worldwide Distribution of Machines Designed by CONCAST (1963-1975) by Type of Furnace and Type of Steel[7]

	Source of Steel[a]					Type of Steel[b]			
Type of Equipment	Open Hearth Furnace	Electric Arc Furnace	OBM/LW Converter	Basic Oxygen Furnace	Kaldo Converter	Stainless	Carbon	Alloy	Low Alloy
Slab Machines	2	21	0	37	3	14	49	5	17
Machines for Blooms	2	34	2	15	6	4	47	9	8
1-2 Strand Billet Machines	2	68	0	3	0	1	70	9	7
3-4 Strand Billet Machines	0	82	4	5	0	0	91	9	3
5-8 Strand Billet Machines	2	8	1	4	0	0	15	0	4

[a]Some machines have more than one source of steel.

[b]Some machines handle more than one type of steel.

Table 3-5. Raw Steel Production in the United States (Net Short Tons)[8]

Type of Furnace	Total	Carbon	Alloy (Other than Stainless)	Stainless and Heat Resisting
		1975		
Open Hearth	22,165,002	19,796,106	2,368,896	–
Basic Oxygen Process	71,800,344	64,839,296	6,961,048	–
Electric	22,817,986	15,811,915	5,889,916	1,116,155
Total Raw Steel	116,783,332	100,447,317	15,219,760	1,116,225
Continuous Casting[a]				
AISI-reported	0,640,031	9,730,435	678,561	231,035
Estimated all U.S.		12,000,000		
		1974		
Open Hearth	35,496,834	34,475,202	3,021,632	–
Basic Oxygen Process	81,553,403	74,533,904	7,018,499	–
Electric	28,444,812	19,480,460	6,809,530	2,154,822
Total Raw Steel	145,494,049	126,489,566	16,849,661	2,154,822
Continuous Casting[a]				
AISI-reported	11,818,748	10,891,978	574,419	352,351
Estimated all U.S.		14,000,000		
		1973		
Total Raw Steel	150,798,927	132,746,557	16,163,512	1,888,858
Estimated Continuous Casting[a]		13,000,000		
		1972		
Estimated Continuous Casting		11,000,000		

[a]Included in raw steel figure.

Overall U.S. steel production has grown since World War II at an average rate of 2.5% annually. It is expected that the long-range rate of growth will continue at about the same or a slightly lower rate. Table 3-5 gives the 1974 and 1975 total steel production in the United States by type, and the portion that was produced by continuous casting.[8] In 1975, total U.S. crude steel production was 106 million metric tons. The proportion of U.S. steel continuously cast, roughly 10% or 11 million tons, is below the world average; however, it is experiencing a much faster rate of growth. In 1973, the continuous-casting capacity was about 15 million tons, and installed capacity is expected to be 80 million tons in 1985,[9] thus exceeding 50% of the United States steel billet mill and slab mill production capacities expected by that time. Eventually, continuous casting should account for the major part of total U.S. steel production.

Table 3-6 includes the distribution of the number of units and strands and the installed capacity by the shape produced in the continuous caster. Table 3-7 gives the distribution by types of steel-making furnaces and cast shape.[7,10]

Table 3-6. Present Distribution in the United States of the Continuous Casting Capacity by Shape Produced in the Caster (November, 1975)[7]

Shape	No. of Machines	No. of Strands	Comprehensive Casting Capacity (1000 ton/yr)[a]
Billet	56	124	> 9,034,000
Slab	18	27	>11,869,000
Bloom	3	12	875,000
Unlisted	2	8	
Total	79	171	>21,778,000

[a]Minimum values, as capacities for some installations were not available.

In the United States and much of the rest of the world, continuous casting tends to be the preferred process for carbon-steel merchant bar products. For heavy structurals and plates, the process may be preferred, depending on size, quality and tonnages. For sheet, strip and tin plate, continuous casting is slowly gaining wide acceptance.[11] Thus, the process could account for about one-quarter of U.S. metal steel tonnage produced in the early 1980s.

Table 3-7. Distribution of Continuous Casting Machines in the U.S. by Types of Steel-Making Furnace and Cast Shapes[7]

	Shape			
Type Furnace	Billet	Slab	Bloom	Total
Electric Furnace (including EF/AOD and EF/BOF)	52	9	3	64
Argon Oxygen Decarburation (including EF/AOD)	–	2	–	2
Basic Oxygen Furnace (including EF/BOF)	2	10	–	12
Not listed	2	3	–	5

POTENTIAL APPLICATIONS

The potential for the application of continuous casting is expanding as the maximum dimension of the continuously cast strand increases. A new CONCAST curved mold unit installed at National Steel's Great Lakes Division, Detroit, Michigan in 1976, produces a single strand of steel 104 inches wide by 12 inches thick. After being sheared into 30-foot slabs, the steel is flame cut lengthwise into slabs having the width required by subsequent mills. The delays required for mold changes have thus been eliminated.

New routes in refining technology are being pursued in such areas as ladle metallurgy, nitrogen shrouding of the metal stream from the tundish to the mold, and automatic weight and level control to govern metal accumulation into and through tundish and mold. Equipment under development will improve in-line inspection and thermal monitoring of the cast product and calibration of the rolling mill passes.

Continuous casting is the first step toward a truly continuous steel-making process. The initial process, the continuous casting of ladles back-to-back in regular sequence without interruption, has been demonstrated and is now a frequent practice. Now receiving serious considera-tion is the feeding of the hot continuous cast strand directly to the follow-on rolling mill, which would need to be followed eventually by continuous in-line reheat furnaces. The culmination of continuous steel-making could then be reached by using prereduced ferrous materials for continuous charging of the steel-making furnace preceding the continuous casting operation.

A listing of continuous casting machines under construction, on order, or recently completed in the United States as of early 1976 is given in Table 3-8. According to this source, there will be an additional 12 slab casters and 11 billet and bloom casters. The slab casters will vary between 800,000 and 1,000,000 tons capacity, while the billet and bloom casters will range from 250,000-400,000 tons capacity. Thus, the total tonnage would increase and be in the area of 13-15 million tons. As this source indicates, most of the billet and bloom casters will be limited to mini-mills, which should grow to midi-mills with a capacity in excess of 500,000 tons.[12]

ASSESSMENT OF THE FUTURE

Economic and Energy Analysis

The well-recognized benefits continuous casting offers in the production of steel are the improvement of steel quality and the increased yield of solid steel from the basic steel-making operations. The result is an increase in the shipment of finished steel products at lower cost from a given volume of raw steel production.

In addition to these benefits, continuous casting, in comparison with ingot casting and breakdown, also offers decreased energy demand in steel production. In the past, the energy conservation has only been considered a modest economic benefit; however, current conditions of high energy cost and possible curtailment of energy supplies give this factor substantial importance and make it comparable to other benefits of continuous casting.

Types of Equipment

Continuous casting machines fall into two broad categories: billet (small cross section) or bloom casters (larger cross section), and slab casters. A billet caster will typically cast 2-6 strands simultaneously and most commonly 3-4 strands. The cast strand has a nominally square cross section. The usual size billet strand has a cross-sectional dimension of 4 × 4 in., but the section size can range from as small as 3.25-7 in. on an edge in squares or rectangles. The withdrawal speed of a 4-in. square billet is about 100 in./min, which can result in a strand-casting rate on the order of 10-15 ton/hr per strand. Since the molten steel in the ladle can be kept at casting temperatures for about one hour, the casting machine is designed to have an hourly casting rate to match the furnace heat size. Thus, small electric furnace shops with furnace sizes of 25 tons per heat will have two or three strand-casting machines. One of the largest electric

Table 3-8. Continuous Casting Machines Under Construction, on Order, or Completed in U.S. (1976)[12]

Company	Location	Capacity (net tons)	No. of Machines	Type
National Steel Corp.	Midwest	2,000,000	2	Slab
United States Steel Corp.	Baytown, Texas	1,000,000	2	Slab
Kaiser Steel Corp.	Fontana, California	700,000	1	Slab
Georgetown Steel Corp.	Beaumont, Texas	600,000	2	Billet
CF & I Corp.	Pueblo, Colorado	500,000[a]	1	Billet
Armco Steel Corp.	Kansas City, Missouri	500,000	1	Billet
Ameron-Steel Producing Div.	California	350,000	1	Billet
Lone Star Steel Co.	Daingerfield, Texas	300,000	1	Billet
Allegheny-Ludlum Steel Corp.	Brackenridge, Pennsylvania	250,000	1	Billet
Atlantic Steel Co.	Cartersville, Georgia	250,000	1	Billet
H. K.Porter Co., Inc.	Huntington, West Virginia	250,000	1	Billet
Structural Metals, Inc.	Sequin, Texas	250,000	1	Billet
Tennessee Forging Steel Corp.	Calvert City, Tennessee	250,000	1	Billet
	Hariman, Tennessee	150,000	1	Billet
Florida Steel Corp.	Jacksonville, Florida	200,000	1	Billet
Knoxville Iron Corp.	Knoxville, Tennessee	200,000	1	Billet
North Star Steel Co.	Iowa	150,000	1	Billet
California Steel	Chicago, Illinois	125,000	1	Billet
Nucor Corp.	Darlington, South Carolina	100,000	1	Billet
Ceco Corp.	Birmingham, Alabama	100,000	1	Billet
Roblin Steel Co.	Buffalo, New York	b	1	Billet

[a]Capacity is 500,000 tons; however, it will be operated at closer to 300,000 tons.
[b]Addition of two strands to existing two-strand caster is equivalent to another casting machine.

furnace shops with two furnaces, each rated at 225 tons per heat has a six-strand machine and casts a strand 7 in. square, which is really a bloom rather than a billet.

A slab-casting machine will usually be designed to cast only one or two strands. For carbon steels, the dimensions of a cast slab strand will usually be from 8-12 in. thick and from 3-6 ft wide. The casting speed of the slab strands is on the order of 50 in./min and, depending on slab thickness, casting rates as much as 200-300 ton/hr can be realized. Thus, the casting rate of a slab caster can be readily matched to the heat sizes of a basic oxygen furnace (BOF) or L-D vessels which have capacities most commonly in the range of 180-250 tons per heat in the U.S.

Effects on Processing and Raw Materials

In terms of benefits of continuous casting, the most apparent are the increase in process yield and reduction of raw material requirements. This benefit can then be used as a basis for examining other effects. A comparison of raw material requirements in the production of slabs for continuous casting and conventional ingot casting is given in Table 3-9. The results for billet or bloom production would be very similar. The basis for the comparison is the nominal process yield which should be obtained in general operating experience for each practice. In continuous casting, there is only a single step from molten steel in the ladle to continuous casting production of intermediate mill shapes such as slabs. Nominal yields of 94-96% can be obtained with good practice in continuous slab casting.

With ingot casting, however, several steps are involved. The first is actual teeming of the molten steel from the ladle into the ingot molds to cast the ingots. After the ingots are solidified in the molds, the molds must be stripped off. They are then transferred to the ingot soaking pits and held at temperatures of 1200-1400°C for several hours prior to slab rolling. The soaking pits are fired with coke oven gas, blast furnace gas, natural gas, or oil in various combinations, depending on local plant conditions, and are the major consumers of energy in the slabbing step. At the end of the soaking period, the ingots are immediately rolled in the primary slabbing mill to desired thickness. The rolled slab is cropped to remove ingot defects and sheared to designated dimensions for subsequent rolling operations.

Table 3-9 gives the nominal yields at each of the sequential steps of the slabbing process and their resultant net yield (684.4%). This yield could have a range from 75 to about 90% depending upon the steel and the ingot casting practice. As a result of these slabbing yields, the requirements

Table 3-9. Comparison of Raw Material Requirements in Continuous Casting and Conventional Ingot Casting Practice for Strip Mill Slabs by Integrated Steel Works[9]

Process Step	Continuous Casting			Conventional Ingot Casting		
Equipment	Molten Steel to Slab Casting Machine			Molten Steel to Slab 1) Ingot casting molds & mold stripping 2) Ingot soaking pits 3) Reversing slabbing mill		
Step Yields	Molten Steel to Slab–94%			1) Molten steel to ingot–98% 2) Raw ingot to slab mill–99% 3) Ingot to slab–87%		
New Yield	94%			84.4%		
Recycle Materials	Scrap 3% Scale 1%			Scrap 12% Scale 15%		
Material Losses	2%			2%		
Required Raw Materials (tons/ton slab)						
Slab	1.0			1.0		
Soaked Ingots	NA			1.15		
Raw Ingots	NA			1.16		
Molten Steel	1.06			1.18		
BOF Raw Material Required[a]						
BOF Charge Ratio	**70/30**	**80/20**	**90/10**	**70/30**	**80/20**	**90/10**
Hot metal	0.84	0.96	1.08	0.93	1.06	1.20
Coke–0.58 ton/ton hot metal	0.49	0.56	0.63	0.54	0.62	0.70
Recycle Metal	0.36	0.24	0.12	0.40	0.27	0.13
Slabbing Recycle	0.04	0.04	0.04	0.16	0.16	0.03
Required Makeup Recycle	0.32	0.20	0.08	0.24	0.11	excess

[a]Based on yield, BOF charge to molten steel, of 88%.

for molten steel become 1.06 and 1.18 tons of steel for continuous casting and ingot casting, respectively. These yields mean that production of slabs by continuous casting only required 90% of the raw steel that would be needed for slabs from conventional ingot casting.

These raw steel requirements are reflected in the raw materials needed to produce the steel for the slabs. However, BOF steel-making can be practiced over wide ratios of hot metal and scrap in the charge mix. Therefore, the requisite raw materials per ton of slab are shown in Table 3-9 for BOF charge mix ratios of 70/30, 80/20 and 90/10 hot metal to scrap or recycle iron units.

Effect on Energy Consumption

The comparison of material requirements for production of slabs by continuous casting and conventional ingot casting provides a basis for the corresponding comparison of energy consumption. The principal sources of energy for production of steel slabs are fossil fuels. The consumption of electrical energy in this stage of steel-making is relatively small. Furthermore, the respective electrical energy demand by the two slabbing methods is quite similar so that the difference between them is quite small and to a first approximation, is negligible relative to fossil fuel energy usage.

Energy demands do arise, however, at two process steps—the production of hot metal in the blast furnace, and the heating of the ingots in the soaking pits. The energy for the blast furnace is supplied by coke produced in slot-type by-product coke ovens. In Table 3-10, the energy demand is given for production of steel slabs by continuous casting and conventional ingot casting for BOF steel-making charge mix ratios of 70/30 and 90/10, about the technical limits of the allowable range.

The reduction in steel demand because of the increased yield of continuous casting results in a corresponding energy demand (from coke) of 90% of that for the conventional ingot practice. The energy for firing the soaking pits to heat the cast ingots is usually supplied by a fluid fuel and, frequently, by coke oven gas recovered from the coking operation itself. This energy consumption is commonly in the range of 1.5-2.0 million Btu per short ton of ingot; a consumption rate of 1.65 million Btu per short ton of cast ingot is representative. Since this step is bypassed in continuous casting, the entire block of energy for ingot heating in soaking pits is eliminated. Combined energy demands for the two slabbing practices show that the energy demand for slabs made by continuous casting is about 80% of that of conventional ingot practice. These energy savings amount to 3-4 million Btu per short ton of slab, depending on steel plant operation. At current energy cost of approximately $2 per million Btu, these

Table 3-10. Energy Comparison of Continuous Casting with Conventionally Cast Ingots for Strip Mill Slabs by Integrated Steel-Making[9]

Equipment	1) Slab-Casting Machine		1) Ingot-mold casting 2) Ingot soaking pits 3) Reversing slabbing mill	
Step Yields	1) 94%		1) 98% 2) 99% 3) 87%	
BOF Charge Mix	**70/30**	**90/10**	**70/30**	**90/10**
Raw Material Requirement (tons per ton of slab)				
Ingots	NA	NA	1.15	
Molten Steel	1.06		1.18	
BOF Charge	1.20		1.33	
Hot Metal	0.84	1.08	0.93	1.20
Scrap and Additions	0.36	0.12	0.40	0.13
Fuel Energy Consumption (million Btu per short ton of slab)				
Furnace Coke	12.2	15.6	13.5	17.4
Soaking Pits	NA	NA	1.9	1.9
Total	12.2	15.6	15.4	19.3
Potential Fuel Energy Credits (million Btu per ton of slab)				
Blast Furnace Gas	3.4	4.4	3.7	4.8
BOF Gas	.4	.4	.4	.4
Total	3.8	4.8	4.1	5.2
Net Fuel Consumption per Ton of Slab	8.4	10.8	11.3	14.1

energy savings are equivalent to cost savings of $6 to $8 per short ton of slab for continuous casting.

A similar comparison of the demands for energy for production of billets from electric furnace steel-making for continuous casting and for conventional ingot casting practice is given in Table 3-11. The steel and raw material requirements for production of billets are similar to, but somewhat less than those for slabs because of higher yields of billets by both continuous casting and ingot practice. Again, billets from continuous casting require about 90% of the materials needed for conventional ingot casting practice.

Table 3-11. Energy Comparison of Continuous Casting with Conventionally Cast Ingots for Bar Mill Billets by Electric Furnace Steel-Making

	Continuous Casting	Conventionally Cast Ingots
Equipment	Billet Casting Machine	1) Ingot Mold Casting 2) Ingot Soaking Pits 3) Reversing Blooming Mill and Billet Mill
Step Yields	96%	1) 98% 2) 99% 3) 88%
Raw Material Requirement (ton per ton of billet)		
Ingots per ton of billet	NA	1.14
Molten Steel per ton of billet	1.04	1.17
Furnace Charge per ton of billet (based on yield, furnace charge to molten steel, of 90%)	1.16	1.30
Energy Consumption (million Btu per short ton of billet)		
Steel-making Electrical Energy Consumption[a]	2.0 electric	2.2 electric
Soaking Pits (Thermal Energy)	NA	2.3
Steel Plant Combined	2.0	4.5
Thermal Energy for Power Generation[b]	6.7	7.3
Gross Thermal Energy	6.7	9.6

[a]Based on energy equivalents of 3413 Btu/kWh.
[b]Based on 30% generating efficiency.

For either process, electric furnace steel-making has a nominal electrical energy consumption of 550 kWh per ton of molten steel. Based on the yield of each billet forming process, and the resulting raw steel requirements, the energy requirements become about 2.0 and 2.2 million Btu per short ton of billet for continuous casting and ingot casting, respectively. Again, the cast ingots for billets must also be heated in soaking pits. Generally, billet ingots tend to be smaller than slab ingots and have less-efficient soaking pit furnace practices, so that energy consumption is closer to 2 million Btu per ingot ton. The resulting fuel energy demand becomes about 2.3 million Btu per short ton of billet.

Thus, billets made by conventional ingot casting practice consume a total of 4.5 million Btu of electrical and thermal energy per short ton. In comparison, those made by continuous casting consume about 2.0 million Btu of electrical energy only per short ton, or less than half the total energy of billets from cast ingots. If provision is made for the conversion of thermal energy to electrical energy at a central power station with a 30% conversion efficiency, the gross thermal energy demands become 6.7 and 9.6 million Btu per short ton, respectively. On this basis, billets from continuous casting consume about 70% of the energy costs of approximately $2 million Btu, and the decrease of energy consumption amounts to a cost savings on the order of $5 per ton of billet for continuous casting.

Comparison of Costs

Prospective cost savings can be derived from decreased energy consumption in production of crude steel intermediate shapes by continuous casting rather than the conventional practice of casting and primary mill rolling of ingots. The relative significance of these savings can be seen from consideration of the nominal costs for the basic production of the steel itself. Estimates of these costs are given in Table 3-12 for two types of steel plants. One is an integrated steel works with coke ovens, blast furnace and BOF, producing 2 million short tons of molten or raw steel annually, the nominal minimum economic size. The other is a semi-integrated plant with electric furnaces producing about 500,000 tons annually, the size of an established steel plant that is likely to use conventional ingot casting and breakdown to billets. Smaller-capacity plants would use continuous billet casting, except for very small or unmodernized plants which would cast small-sized ingot to billet dimensions. These estimates show that the integrated steel plant has a nominal cost advantage initially of almost $20 for materials. This decreases to about $10 after operating costs are considered. The largest single cost element for either type of plant is raw materials, which currently amount to about $100/ton. The largest cost difference

Table 3-12. Estimated Total Cost[a] of Raw Steel Production as Raw Molten Steel[9]

	Integrated Plant	Nonintegrated Plant
Capacity Equipment	2,000,000 tons/yr	500,000 tons
Equipment	1) Coke oven	
	2) Blast furnace	
	3) BOF	Electric arc furnace
Raw Materials	1) Coal	
	2) Iron ore pellets	
	3) Hot metal and scrap	Scrap
Raw Materials	1) 1,450,000 tons coal	
	2) 2,550,000 tons pellets	
	3) 1,600,000 tons hot metal and 650,000 tons scrap	550,000 tons of scrap
Unit Costs	1) $48/ton	
	2) $46.50/ton Fe	
	3) $90/ton	$90/ton
Costs in Steel/ton	1) $34.80	
	2) $37.20	
	3) $28.20	$99/ton
Other materials	$22.00	$42/ton
Total	$122.20	$141/ton
Other Direct Costs ($/ton steel)		
Labor and Overhead	$22.00	$14.70
Maintenance	5.00	3.00
Power	3.60	11.00
Consumable Supplies	9.65	3.20
Depreciation	4.00	2.60
Total of Other Direct Costs		$34.50
Total Cost Molten Steel	$166.45	$175.50

[a]All unit costs estimated as of April 1976.

arises from other material costs, which are about $20/ton greater for an electric furnace plant than for an integrated steel works. The resulting cumulative costs of raw steel appear to be about $166/ton for the integrated works and about $175/ton for the electric furnace plant.

The comparative costs of continuous cast slabs with ingot cast slabs are shown in Table 3-13. Those cost elements show that the cumulative slab production costs become about $185/ton for continuous casting and about $205/ton for ingot casting. The cost differential of $20/ton arises almost exclusively from the higher yield of continuous casting, which alone contributes a cost saving of $22/ton for continuous casting. Analysis of the cost of billet production would show similar advantages for continuous

Table 3-13. Cost Comparison of Continuous Casting with Conventionally Cast Ingots for Strip Mill Slabs by Integrated Steel-Making (Blast Furnace and BOF)

	Continuous Casting	Conventionally Cast Ingots
Equipment	Twin Strand Slab-Casting Machine	1) Ingot Mold Casting 2) Ingot Soaking Pits 3) Reversing Slabbing Mill
Capacity	1,000,000 tons	2,000,000 tons
Step Yields	94%	1) 98% 2) 99% 3) 87%
Raw Steel Requirements		
Ingots	NA	2,300,000 tons
Molten Steel	1,060,000 tons	2,350,000 tons
Molten Steel Cost per Ton	$166.45	$166.45/ton
Molten Steel Cost per Ton of Slab	$177.10	$197.20/ton
Capital Costs	$65,000,000	$130,000,000
Unit Capital Costs	$65/ton	$65/ton
Depreciation	$3.25/ton	$3.25/ton
Other Direct Costs ($ per short ton of steel)		
Labor and Overhead	$5.30	$7.90
Maintenance	$.85	$2.25
Power	$.80	$1.40
Fuel	$1.00	$3.30
Depreciation	$3.25	$3.25
Total of Other Direct Costs	$11.20	$18.10
Scrap Credits	$2.70	$10.80
Net Operating Cost	$8.50	$7.30
Slab Steel Cost	$177.10	$197.20
Slab Production Cost	$185.60	$204.50

casting over conventional ingot casting practice. However, the investment requirements for continuous casting machines to produce about 300,000 tons of billets annually would be about $4,500,000 for purchase of the casting machine and about as much again for its complete installation and start-up. In comparison, the investment requirements for a billet mill for conversion of cast ingots could be in the range of $25-$35 million. Because these capital costs have become so high, there has not been a new billet mill for cast ingots installed in this country for more than a decade and none is expected in the foreseeable future.[9]

The Future

Continuous casting is an alternative to the casting of ingots and their subsequent breakdown via a slabbing or blooming mill; it eliminates the need for ingot casting facilities and soaking pits in front of the primary mill, as well as the breakdown mill itself. Continuous casting practice shows a significant savings compared to ingot practice, but the amount of savings in terms of time, energy, operating cost and use of raw materials varies a great deal from mill to mill and depends on the grade of steel produced.

An important energy savings is realized because the energy consumption for reheating of the ingot before rolling in conventional mills is eliminated. Also, electricity consumption in the ingot breakdown rolling mill itself is also eliminated. Energy is conserved at the steel-making stage because of the reduction of the cropping scrap. This results from fewer cut-off, butts and sheared materials. Also, productivity is increased, as these materials do not need to go through the process repeatedly, and a higher proportion of the molten steel goes through to finished product. Other advantages to continuous casting include:

- reduction in scrap loss rates;
- reduction in time;
- reduction in labor force;
- reduction in minimum economic size of installation;
- probable higher productivity (output per unit of capital);
- higher process step product yield;
- increased uniformity of product quality;
- improved metallurgical quality of products; and
- higher yield of end product.

The main disadvantage of continuous casting is the possible increased capital investment to replace conventional equipment; it would need to be justified by the savings in operating costs. In addition, available technology restricts the crude steel size available.

Forecasts[6] made in 1973 of continuous casting steel production to 1980 for the United States and Canada, Japan and the European Economic Community countries are presented in Figure 3-5. Continuous casting has grown sharply in recent years, and will continue to increase sharply for the balance of this decade. The growth of continuous casting will mean a smaller supply of revert or home scrap available for any given melting facility. The production estimate for the United States, which is considerably less than rated capacity, may be conservative. Our best current

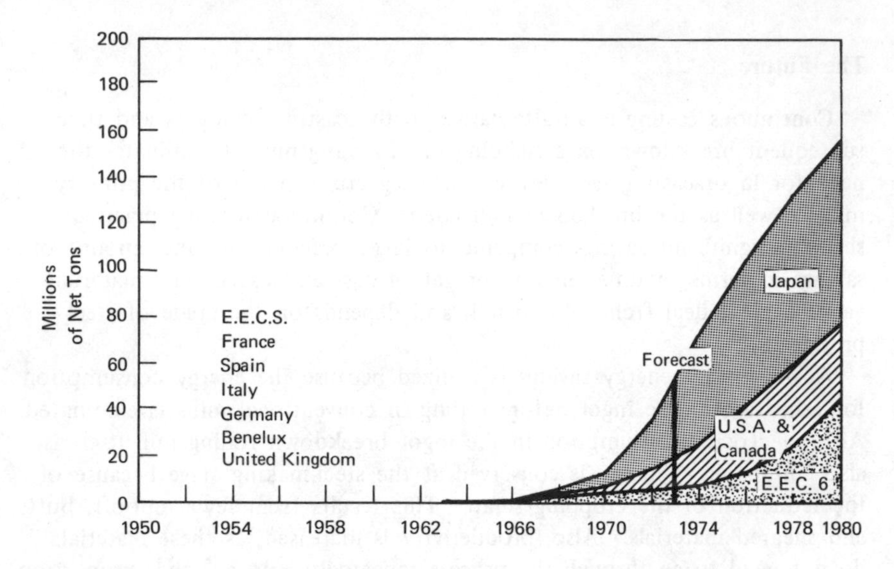

Figure 3-5. Continuous casting production.[10]

estimate is that continuous casting capacity in 1980 should be about 20 million tons for billets and slabs and in 1985 about 30 and 50 million tons, respectively.

CONTINUOUS CASTING IN JAPAN

Advances in the use of continuous casting in Japan have been particularly outstanding during the past 10 years. In 1973, about 90 continuous casters processed 20.7% of Japan's total crude steel production and in 1974, the percentage had grown to 25.2%. In 1975, total Japanese steel production was 102.2 million tons for both conventional and continuous casting. Continuous casting production was 30.66 million tons, *i.e.*, 30% of the total crude steel production. The increase in the number of continuous casting facilities in Japan from 1955 through 1973 can be seen in Figure 3-6. The graph in Figure 3-7 shows the share of continuously cast steel in the total crude steel production from 1970 through 1975 for Japan, with estimates for 1980, for small producers other than the five leading Japanese companies and for Nippon Steel Corporation. Figure 3-7 indicates, for example, that in 1980, 33.8% of Nippon Steel Corporation crude steel production will be by continuous casting; also, 35% of total Japanese crude steel production will be continuous casting, and 50% for the small producers. The five leading companies provide approximately 80% of the total crude steel production. The small companies have a higher percentage of continuous casting than the major ones.

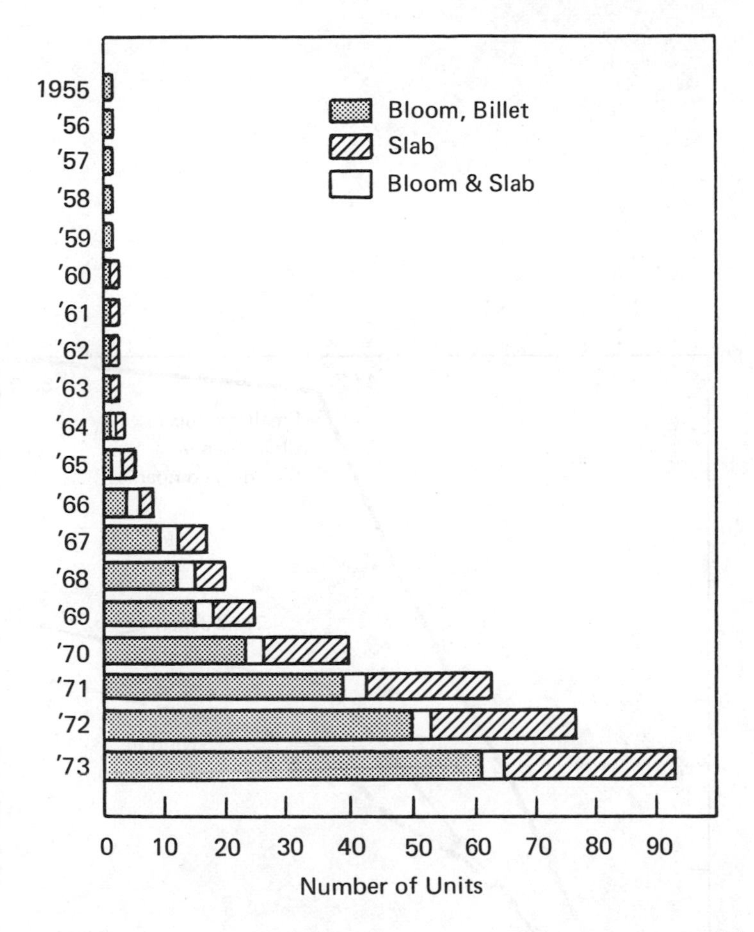

Figure 3-6. Number of continuous casting facilities (1955-1973) in Japan.[13]

Table 3-14 presents data on the distribution of the number of units and strands and the installed capacity in Japan by the shape produced in the continuous caster. Table 3-15 similarly gives the distribution by types of steel-making furnaces and cast shape.

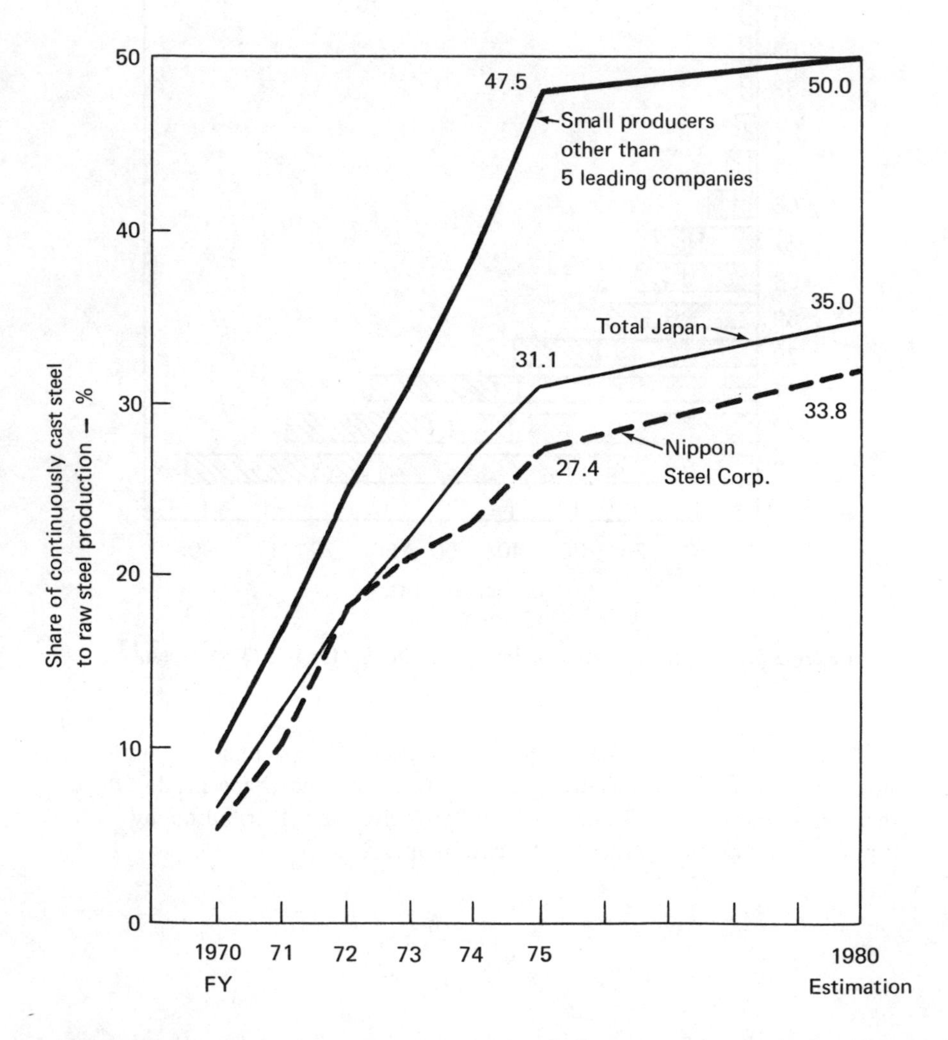

Figure 3-7. Continuous casting steel production in Japan as a percentage of raw steel production.[14]

Table 3-14. Distribution of Continuous Casting Capacity in Japan by Shape Produced in the Caster (1975)[7]

	Units	Strands	Capacity tons/yr[a]
Billet	26	98	> 3,313,000
Slab	30	48	>15,575,000
Bloom	16	58	> 1,807,000
Total	72	204	>20,695,000

[a]Minimum values, as capacities for some installations, were not available.

Table 3-15. Distribution of Continuous Casting Installations in Japan by Types of Steel-Making Furnace and Cast Shapes[7]

	Billet	Slab	Bloom	Total
Electric Furnace (including EF/AOD and EF/BOF)	23	8	12	43
Argon Oxygen Decarburation (including EF/BOF)	0	2	1	3
Basic Oxygen Furnace (including EF/BOF)	3	18	4	25
Open Hearth	0	1	1	2
Uncited	2	–	–	2

The Role of Prereduced Materials In Continuous Casting

The combination of the electric furnace and continuous casting is particularly attractive as a steel-making route because it is independent of metallurgical coal as an energy and reductant source and lends itself to small-scale production with low capital investment requirements (*i.e.*, the mini-mill approach). The disadvantage is that it depends on scrap steel as a source of iron units, and thus depends on a commodity that has a volatile price and sometimes uncertain purity as the fundamental raw material. Prereduced materials, such as can be made from iron-ore pellets, can be substituted for scrap.[15] Thus, there is a potential for prereduced iron materials in the future of continuous casting made by the direct reduction of iron ores.[7] These materials can offer a steady source of iron units of consistent specification at a stable price for electric furnace steel-making. A comparison of scrap price variations with iron-oxide pellet price variation from 1968 to 1974 is shown in Figure 3-8. Prereduced pellets made from oxide pellets should show a similar price stability.

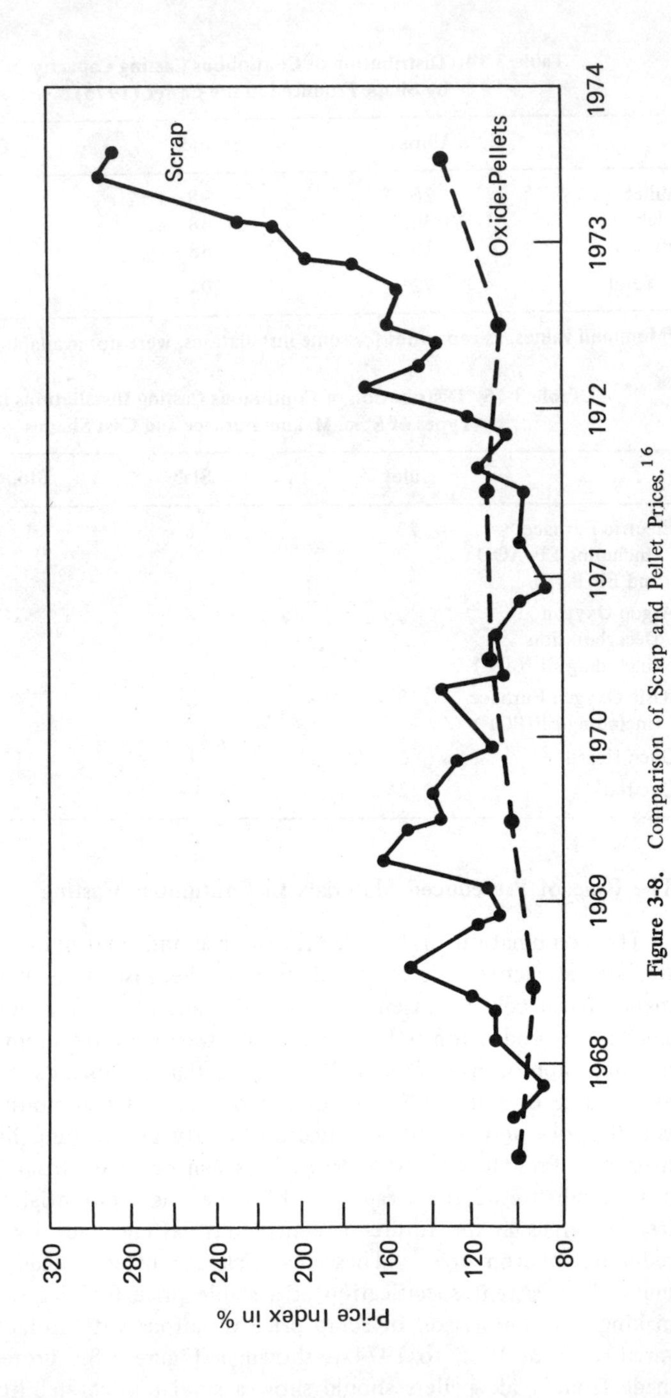

Figure 3-8. Comparison of Scrap and Pellet Prices.[16]

The use of a direct reduction process is at present closely linked with the availability of natural gas, because the most developed direct-reduction processes require this fuel. Solid reduction using coal instead of natural gas has had a difficult start, but the basic problems seem to have been resolved. A period of demonstrated successful operation of the newly built SLRN Plant at the Griffith Mine, Bruce Lake, Ontario, over the next few years, is necessary to enable large-scale, coal-based reduction processes to be demonstrated and find widespread application.

In the SLRN process, a rotary kiln approach is used in which oxide pellets and coal are charged. The process is a combined development of the Steel Company of Canada, Lurgi of Germany, and Republic Steel and National Lead of the U.S. Plants have also been operated in New Zealand, South Korea, South Africa, Brazil and Japan. Total SLRN annual capacity around the world may be around 1 million tons (worldwide capacity of direct reduction plants based on natural gas totals about 25 million tons). A list of direct reduction installations in operation or on order as of February 1976 is given in Table 3-16.

There is also a possible role for nuclear energy as a source of heat in direct reduction.[17] Heat from a high temperature gas-cooled reactor can be used to increase the yield of reducing gases (carbon monoxide and hydrogen) from a unit quantity of natural gas or coal and to produce them at the required 1500-1800°F. Alternative, nuclear-generated electricity could be used to produce a hydrogen reductant by electrolysis of water. The use of pure hydrogen, however, could cause problems because iron ores reduced this way tend to be pyrophoric and to reoxidize easily.

In the MIDREX process, natural gas is reformed to mixtures of CO and H_2 at the desired temperature level, and the gases passed counter-currently in a shaft in which the oxide pellets move downward. In the HyL (Hojalata y Lamina) process, the pellets are reduced in bath reactors. Several reactors are on stream in different stages of reduction so that the reducing gases can be diverted as required to make maximum use of the reducing capability.

The metallized or prereduced products charged to the electric furnace are rather uniformly reduced to a 90-94% range of the iron content present as metallic iron. Approximately 30% of the iron oxide raw material weight is lost in the oxygen removal. On an economic basis, compared to scrap, direct-reduced materials can make small-scale steel plants (mini-mills) cost-effective for selected finished products (rebars and small profiles, for example). Decreases in capital cost can range from 40-50%/ton of product using direct reduction coupled with the electric arc furnace and continuous casting as an alternative to the coke oven/blast furnace/BOF complex. Also mini-mills require fewer highly-trained personnel than do fully integrated steel mills.

Table 3-16. Worldwide Installations of Direct Reduction Plants in Operation and on Order (February 1976)[18]

Location	Process	Start	Rated Capacity (MTPY)
NORTH AMERICA			
1. Georgetown Ferreduction, South Carolina, U.S.A.	Midrex	1972	400,000
2. Oregon Steel, Oregon, U.S.A.	Midrex	1969	300,000
3. Sidbec I, Contrecoeur, Canada	Midrex	1973	350,000
4. Sidbec II, Contrecoeur, Canada	Midrex	1976	630,000
5. Gulf Ferreduction, Texas, U.S.A.	Midrex	1979	1,260,000
6. HECLA, Arizona, U.S.A.	SLRN	1975	65,000
7. STELCO, Bruce Lake, Canada	SLRN	1975	360,000
8. Armco Steel, Texas, U.S.A.	Armco	1973	330,000
SOUTH AMERICA			
1. Acindar, Buenos Aires, Argentina	Midrex	1977	420,000
2. Dalmine Siderca, Campana, Argentina	Midrex	1976	330,000
3. SIDOR I, Mantazas, Venezuela	Midrex	1976	350,000
4. SIDOR II, Mantazas, Venezuela	Midrex	1979	1,200,000
5. USIMINAS, Ipatinga, Brazil	Midrex	1978	800,000
6. USIBA, Aratu, Bahia, Brazil	HyL	1974	220,000
7. FESA I, Monterrey, Mexico	HyL	1957	95,000
8. FESA II, Monterrey, Mexico	HyL	1960	260,000
9. FESA III, Monterrey, Mexico	HyL	1964	420,000
10. TAMSA, Vera Cruz, Mexico	HyL	1967	220,000
11. HyLSAMEX I, Puebla, Mexico	HyL	1969	315,000
12. HyLSAMEX II, Puebla, Mexico	HyL	1976	630,000
13. SIDOR I, Mantazas, Venezuela	HyL	1976	380,000
14. SIDOR II, Mantazas, Venezuela	HyL	1979	1,800,000
15. PIRATINI, Charquedas, Brazil	SLRN	1973	60,000
16. SIDPERU, Chimbote, Peru	SLRN	1976	100,000
17. COSIGUA, Rio de Janiero, Brazil	Purofer	1976	350,000
18. ORINOCO, Puerto Ordaz, Venezuela	HIB	1972	1,050,000
19. FIOR de Venezuela, Mantanzas, Venezuela	FIOR	1976	400,000
WESTERN EUROPE			
1. Hamburger Stahlwerke, Hamburg, West Germany	Midrex	1971	400,000
2. British Steel Corp., Hunterston, Scotland	Midrex	1978	800,000
3. Norddeutsche Ferrowerke GmbH, Emden, West Germany	Midrex	1979	1,200,000
EASTERN EUROPE			
1. Government, Kursk 1, U.S.S.R.	Midrex	1979-81	2,500,000

Table 3-16, continued

Location	Process	Start	Rated Capacity (MTPY)
AFRICA			
1. Highveld, Witbank, South Africa	Highveld/ Lurgi	1968	720,000
2. Dunswart, Benoni, South Africa	Krupp	1973	150,000
3. Tika, Solwezi, Zambia	HyL	1977	250,000
MIDDLE EAST			
1. NISIC, Ahwaz, Iran	Midrex	1976	1,200,000
2. Qatar Steel Company, Ltd., Umm Said, Qatar	Midrex	1977	400,000
3. NISIC, Ahwaz, Iran	HyL	1977	1,000,000
4. Government Khor Alzubair, Iraq	HyL	1977	1,500,000
5. NISIC, Isafan, Iran	Purofer	1976	330,000
ASIA			
1. Krakatau Steel, KotaBaja, Indonesia	HyL	1978	500,000
2. NKK, Fukuyama, Japan	SLRN	1974	350,000
3. Kawasaki, Chiba, Japan	Kawasaki	1974	60,000
4. Kawasaki, Misushima, Japan	Kawasaki	1974	240,000
5. Nippon Steel, Hirohata, Japan	NSC	1975	400,000
OCEANIA			
1. N. A. Steel, Glenbrook, New Zealand	Lurgi/NZS	1975	120,000
		TOTAL	25,215,000

Total Midrex 12,540,000

Total HyL 7,590,000

Midrex Market Penetration as of 2/76. . . 50%

HyL Market Penetration as of 2/76 . . . . .30%

The use of prereduced iron can have a linear dilution effect on the concentration of residual elements in scrap iron and steel such as copper, nickel, chromium, sulfur, phosphorous and tin (Figure 3-9), and thus could offer better control of finished steel composition. Also, higher amounts of continuously fed prereduced iron provide a smoother furnace operation and a reduction of electrode consumption by about 5%.

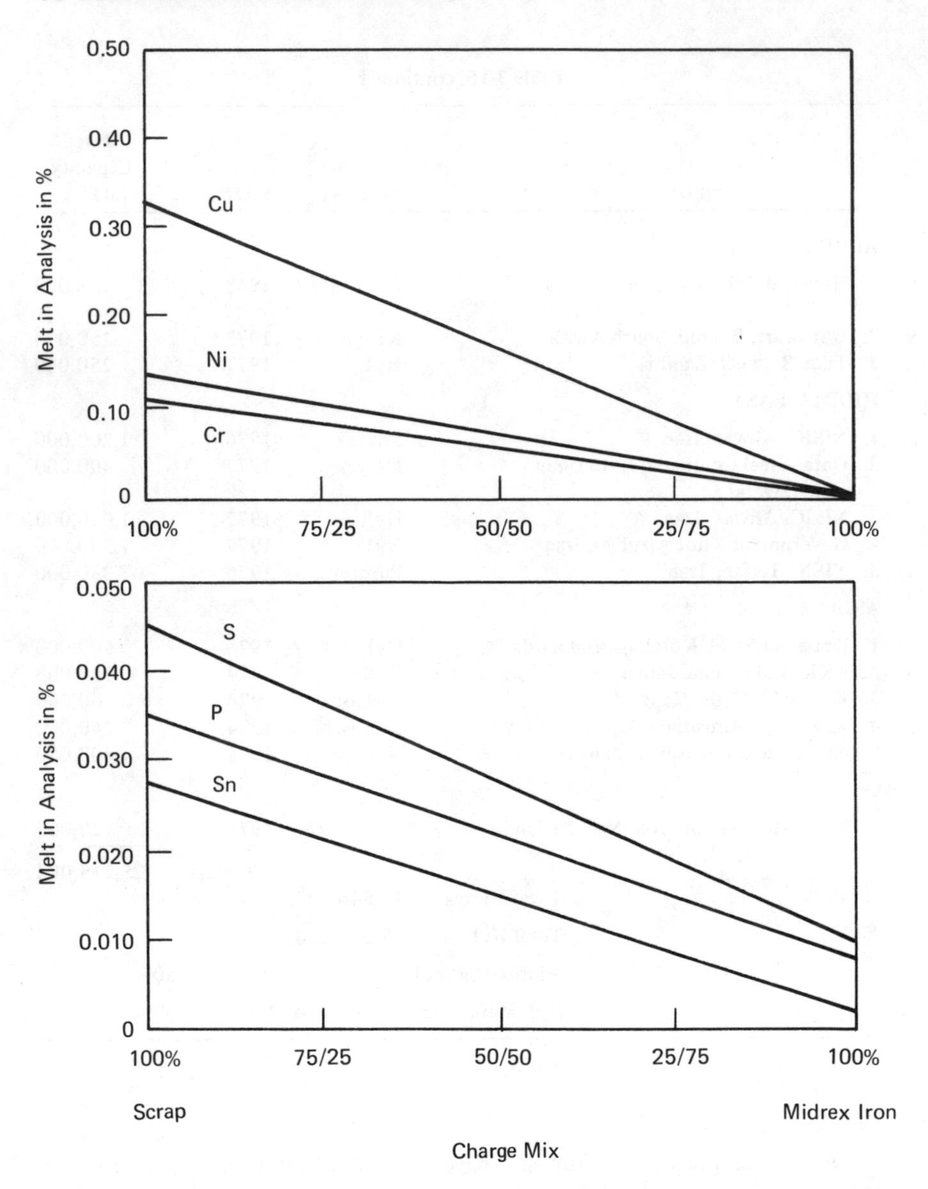

Figure 3-9. Concentration of residual elements vs percentage of prereduced iron.[18]

The ratio of scrap to prereduced iron in the charge can increase productivity 20% for an energy increase of only 5%, (Figure 3-10) for a charge mix ratio of 50% prereduced iron/50% scrap. Yields increased up to 5% when going from scrap to 100% prereduced materials. Energy

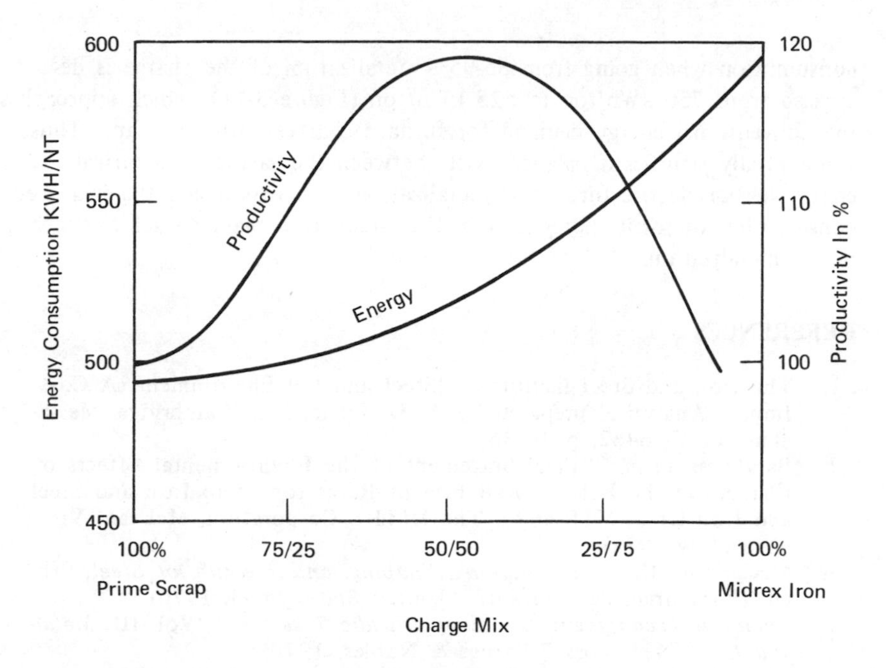

Figure 3-10. Productivity and energy consumption vs percentage of prereduced iron.[18]

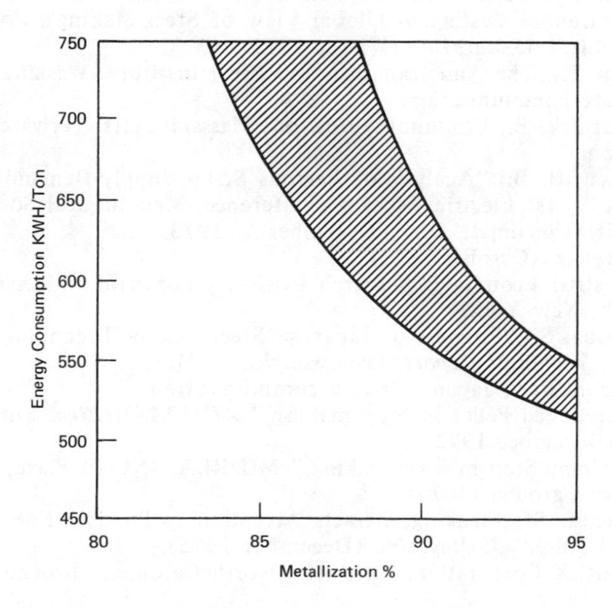

Figure 3-11. Energy consumption vs percentage of prereduced iron.[18]

consumption when going from 85-95% metalization of the charge is decreased from 750 kWh/ton to 525 kWh/ton (Figure 3-11), which approaches the conventional energy demand for furnace charges with all scrap. Thus, a potentially significant balance exists between the saving in electrical energy in the electric furnace as metalization is increased and the increased consumption of fossil energy in the direct-reduction plant to achieve the higher metalization.

REFERENCES

1. The Iron and Steel Institute. "Steel and the Environment, A Cost Impact Analysis," prepared by A. D. Little, Inc., Cambridge, Massachusetts, C-76482, p. III-36.
2. Baratz, B. *et al.* "Final Statement of the Environmental Effects of Changes in the Relative Rail Freight Rates for Scrap Iron and Steel and Iron Ore," MTR-6690, The MITRE Corporation, McLean, Virginia (July 1974).
3. McGannon, H. E., *The Making, Shaping, and Treating of Steel*, 9th ed. (Pittsburgh, Pennsylvania: United States Steel, 1970).
4. *Chemical Technology: An Encyclopedic Treatment*, Vol. III, *Metals and Ores* (New York: Barnes & Nobles, 1970).
5. CONCAST AG., Zurich, Switzerland. Brochure (April 1975).
6. Fukuyama Steel Works, Nippon Kokan. Brochure (November 1972).
7. "Continuous Casting, A Global View of Steel Making's Former Enfant Terrible," *33-Magazine* (Winter 1973).
8. Allen, K., The American Iron and Steel Institute, Washington, D.C. Private communication.
9. Bovarnick, B., Consultant, Newton, Massachusetts. Private communication.
10. Jensen, H. B. "Analysis of Ferrous Scrap Supply-Demand/Balance USA," 31st Electric Furnace Conference, Metallurgical Society of AIME, Cincinnati, Ohio, December 5, 1973.
11. *J. Metals* (October 1973).
12. Industrial Economics Research Institute, Fordham University, New York, New York.
13. Toyoda, S. "Progress of Japanese Steel-making Technology," *Nippon Steel Technical Report*, Overseas No. 5 (March 1974).
14. Nippon Steel, Japan. Private communication.
15. "Prereduced Pellet in Steel-making," *AIME/ASM Elec. Furnace Proc.*, **32** (December 1972).
16. "A Grant Step in Steel-making," MIDREX, INCNB Plaza, Charlotte, North Carolina (1974).
17. "Nuclear Steel-making Attracts Attention as Possible Fuel Saver of the Future," *33-Magazine* (December 1975).
18. MIDREX Corporation, Charlotte, North Carolina. Brochure.

UNCITED REFERENCES

Kono, T. "Trends of Steel Demands in Developing Countries," Research Department, Nippon Steel Corporation, Tokyo (February 1975).
Kono, T. "Changing Corporate Environment and the World Steel Industry," XXII International Meeting of the Institute of Management Science, Nippon Steel Corporation, Tokyo, July, 1975.
Kono, T. "Changing Corporate Environment and Countermeasures of the Steel Industry," Hudson Institute, April 1-4, 1976, H1-2413-BN/2/8.

4

ELECTRIC HEAT TREATMENT OF METALS*

INTRODUCTION

The electric heat treatment equipment manufacturing industry is a healthy growing industry in this country. Continuously increased quality requirements in ferrous and nonferrous metal products have created the need for heat treatment equipment, and sophistication of this equipment and the fuel crisis have made electric heat treatments more desirable and economical. Progress in resistance heater alloys have made it possible to build furnaces operating at higher temperatures and have opened the door to large-sized electric equipment in the steel-making industry. So-called vacuum furnaces, permitting heat treatments at reduced pressures, result in cost reduction of the heat treatment proper and are widely used.

Market data on various types of equipment, including both electric and oil-fired, are summarized in Table 4-1. Electric treatments already have about 50% of the market for basic treatments, and 40% of the market for ovens. Motivation for adopting electric heat treatment is principally economic, depending on manpower and electricity costs vs fossil fuel costs, and ease of operation. Two factors that will encourage the use of electric treatments in the future are the scarcity of natural gas needed for some conventional equipment, and the environmental advantages of the electric treatments. Table 4-2 presents a comparison of the capital costs and energy requirements in electric and fossil-fueled treatments.

GENERAL PRINCIPLES

Electric heat treatments of metals used by industries involve the scientific application and control of heat to change the physical and metallurgical properties of materials. The heating equipment industry serves the

*By: Marcel Barbier

Table 4-1. Summary of Market Data for Electric Heat Treatments[1]

Type of Equipment		Industry Shipments				Percentage of Market		
		Electric Except Induction ($ thousands)	Electric Induction ($ thousands)	Fuel-Fired ($ thousands)	Total ($ thousands)	Electric Except Induction	Electric Induction	Fuel-Fired (%)
Basic heat processing and metal treating	1974	38,800	29,100	73,300	141,200	27.5%	20.6	51.9
	1972	27,800	19,661	50,700	98,161	28.3%	20.3	51.6
Miscellaneous, including vacuum, laboratory bath-type furnaces, equipment without chamber, and for glass, ceramic and cellulose processing	1974	NSA[a]	–	NSA	74,300	NSA	–	NSA
	1972	NSA	–	NSA	54,870	NSA	–	NSA
Industrial process ovens	1974	NSA	–	NSA	53,700	NSA	–	NSA
	1972	18,744	–	28,205	46,949	39.9%	–	60.1

[a]Not separately available.

Table 4-2. Comparison of Capital Cost and Energy Requirements for Electric and Conventional (Gas) Furnaces[2,3]

	Capital Costs		Ranges of Energy Requirements (per short ton of treated product)		
Type of Equipment	Gas-Fired ($/Btu/hr)	Electric ($/kW)	Type of Equipment	Gas-Fired (Btu)	Electric (kWh)
Very large sized, special purpose	0.02-. . . .		(Depending on unit size, material being treated and process)	600,000-4,000,000	175-600
Large production-type furnaces, continuous with some degree of automation	0.04-0.09	400-500			
Bath, box-type, middle sized, unautomated	0.03-0.06	325-375			
Laboratory furnaces, vacuum furnaces, medium-sized ovens	(do not exist)	750-850 1000-2000 200			
Draw furnaces	0.04	200			

whole range of processes from melting metals to altering their inner structure and eventually applying protective coatings. This is a rapidly evolving industry, due to the increased use of unconventional metals and new applications for common ones, which constantly set new requirements on the equipment used.

Most heat treatments are carried out on iron and steel. Major heat treating processes include:

- quenching, a hardening process brought forth by rapid cooling of the metal in water, oil or aqueous solutions of salts, producing a very hard but brittle product;
- annealing, a very slow cooling process in order to produce a softer steel;
- normalizing, which is cooling in air to obtain a finer grained structure. It is used extensively for steels that have been rolled and is also a hardening and toughening process;
- patenting, the passing of wires and rods through a heating tube at austenitizing temperatures and then through a cooling bath, to produce toughness and strength in wire already strain-hardened by drawing;
- tempering, which consists of reheating quenched steel very carefully to a temperature below 730°C to remove brittleness and reestablish some ductility and toughness. It is rather similar to subcritical annealing but more controlled and refined. Variants of this process include martempering (whereby austenitized steel is plunged in a salt bath above the martensitic temperature of 230°C until all of it has reached this temperature, before it is air-cooled with a following temper) and austempering (whereby the steel is quenched in a molten salt bath at about 230°C and for a long period of time so that complete isothermal transformation takes place); and
- age-hardening, which requires temperatures of the order of 60°C for up to 100 hr to speed up precipitation hardening at grain boundaries.

To these basic heat treatments, in which the physical transformation takes place in the bulk of the metal, are added the surface treatments, such as cementation, nitriding and carburizing, which provide surface hardening effects.

Included in the term industrial heating equipment are furnaces, ovens, heaters, heating machines, kilns, induction and dielectric apparatus, industrial combustion equipment and accessories; also included are auxiliary equipment such as quenching apparatus, atmosphere generators, heat exchangers and washers. Excluded are processing equipment for food: kilns for cement, brick, clay, tile, grain, fertilizer and wood; coke ovens, tanks for melting glass, steam and hot water boilers and their accessories, prime power equipment, comfort heating equipment, and oil, chemical and petrochemical processing equipment.

This report focuses on heat processing and metal treating equipment and ovens used in heat treatments. Heat processing and metal treating furnaces and related equipment are the machines that form and alter the microstructure and change metallurgical properties of metals, as other machines form or shape metals to change their outward appearance. Heat treating is essential to give metal parts the characteristics needed to withstand the environment to which they will be exposed under continuous operation. The size and shape of parts, density, physical properties and type of metal or alloy to be processed determine the individual heating and cooling cycles used and the equipment required.

Ovens are used for drying, dehydrating, curing, heat treating and stress relieving. They are generally designed for operating only at temperatures up to about 1200°F. Heat may be derived from steam, electricity, or the combustion of fuel oil or gas. The transfer of heat energy may be accomplished by convection from circulating hot air, by radiation, or by a combination of both.

Induction heating is accomplished when alternating electrical current is passed through a coil creating a magnetic field which induces a current in a metal placed within that field. These induced currents cause a temperature rise in the metal which can be accurately controlled for desired operations.

Dielectric heating is a precise, high-speed heating process for drying, curing or preheating nonmetallic material for a variety of requirements. Any material that is a nonconductor of electricity can be heated dielectrically by using it as the dielectric component of a simple capacitor arrangement. The heating occurs simultaneously throughout the mass and drying takes place essentially from the inside out, permitting processing at extremely rapid rates.

Auxiliary industrial heating equipment comprises a wide variety of standard equipment such as washers, quenching apparatus and calciners which are generally utilized in connection with industrial heating processes.

Combustion equipment includes systems and system components required to convey, measure, control, mix and burn the fuel and air or oxygen required for the production of usable heat in industrial processes. This equipment can be used for combustion of both gas and oil and is adaptable for a variety of heat applications.

These units produce hydrogen, nitrogen, carbon monoxide, lithium gas and other such gaseous substances used to provide the proper atmosphere in the interior of furnaces to prevent such undesirable effects as scaling, pitting, discoloration and decarburization of the metal. For example, if metal parts are heat treated in a normal atmosphere, the gases present in the air can create a scaling condition on the surface of the metal. These undesirable effects are eliminated by substituting a suitable atmosphere in the furnace chambers.

EXISTING APPLICATIONS

Small- to Medium-Sized Heating Furnaces

To depict the present status of the technology, an example of data for a particular line of products from the Leeds and Northrup Company has been analyzed. These data adequately cover most of the heat treating operations for ferrous and nonferrous metals, and also for other items such as glass and rubber. The furnaces are used on moderate-sized products and are themselves in the medium size range. The family of furnaces includes (names used are trade names from Leeds-Northrup):

> VAPOCARB–Forced-circulation heating with controlled temperature, time and atmosphere, primarily for hardening tool steels, but also for carburizing and carbonitriding limited batches of production parts.
>
> HOMOCARB–Forced-circulation heating with controlled temperature, time and atmosphere carbon potential, for a wide variety of heat treating processes (such as case carburizing, hardening, carbon restoration and homogeneous carburizing) at high production rates;
>
> TRICARB–Forced-circulation heating with controlled temperature, atmosphere, carbon potential and quench, for a wide variety of heat treating processes (such as carburizing, homogeneous carburizing, carbonitriding, carbon restoration and hardening) at high production rates;
>
> HOMO NITRIDING–Forced-circulation heating with controlled temperature, time and flow of ammonia gas, for production nitriding to critical specifications; also available in models that automate operation, and program control of temperature and atmosphere for conventional and other process cycles;
>
> HOMO TEMPERING–Forced-circulation heating with controlled temperature, for maximum temperatures to 650°C (for tools), to 425, 760 and 900°C (for production parts), and a special model for dense-load production parts to 760°C, also available in models for precision annealing of optical glass. Horizontal furnace available as a companion furnace to the Tricarb line.
>
> STEAM HOMO–Forced-circulation heating with controlled temperature, time and steam atmosphere, for heat treating wrought ferrous and nonferrous metal parts and powdered iron parts, and for curing parts of molded rubber.

The processes for which these furnaces are applicable are indicated in Table 4-3 together with the materials handled and the temperature range.

For each of these types of furnaces the price, chamber volume, installed electric power and price ($/kW) is given in Table 4-4.

Table 4-3. Processes Using Medium-Sized Furnaces[4]

Process	Vapocarb	Homo-Carb[a]	Tricarb	Homo-Nitriding	1650°F Homo Tempering	1400°F Homo Tempering	1200°F Homo Tempering	800°F Homo Tempering	Horizontal Homo	Steam Homo	Material[b]	Range[a] (°C)
Hardening	X	X	X								1,3	
Normalizing	X	X	X								4,5	1000
Case Carburizing	X	X	X								4,5,8	950
Homogeneous Carburizing		X	X								4,5,8	
Carbon Restoration		X	X								4,5,7,8	
Normalizing		X			X						4-7,9	
Carbonitriding	X	X	X								4,8	
Hardening	X	X	X								1,4-7	
Annealing		X			X						1,4-6,9	
Age Hardening		X	X		X						2	
Annealing		X			X	X					1,4,5	760
Age Hardening		X			X	X					2	
Stress Relief					X	X					4-6,9	
Steam Oxidizing										X	1,4-6,8,9	650
Nitriding				X							1,5,6,19	
Blueing										X	1,4,5,8	
Stress Relief						X	X	X	X		1,4,9-14, 17,18	
Solution Treating						X	X		X	X	16,17	
Annealing					X	X	X	X	X	X	10,11,13, 14,16-18	
Tempering					X	X	X	X	X	X	1,3-8	
Age Hardening						X	X	X	X		12,16,17	
Curing										X	15	150

[a]Maximum operating temperature: 950°C; special models available for operation up to 1038°C.

[b]Material Code Numbers: 1. Tool steel; 2. High-temperature alloy steel; 3. Martensitic chromium stainless steel; 4. Carbon steel; 5. Alloy steel–medium or high carbon; 6. Alloy cast iron; 7. Ferrous investment castings; 8. Powdered iron; 9. Cast iron; 10. Brass; 11. Bronze; 12. Beryllium copper; 13. Copper; 14. Glass; 15. Rubber; 16. Aluminum; 17. Magnesium; 18. Gold; 19. Nitralloy.

Table 4-4. Ranges of Price, Volume and Electrical Energy for Leeds-Northrup Furnaces[4]

Type of Furnace		Price	Volume	kW	$/kW	Average $/kW
Vapocarb	min.	10,500	.2 ft^3	5	2100	1455
	max.	15,000	2.02 ft^3	18.5	811	
Homo	min.	11,700	1.06 ft^3	13	838	555
	max.	36,800	109.08 ft^3	136	271	
Steam Homo	min.	11,800	.90 ft^3	16	738	549
	max.	48,900	109.08 ft^3	136	360	
Homocarb	min.	26,500	1.84 ft^3	33	803	566
	max.	53,700	40.9 ft^3	163	329	
Tricarb	min.	43,000	3.12 ft^3	35	1229	1383
	max.	100,000	6.94 ft^3	65	1538	
Nitrider	min.	23,000	1.43 ft^3	13	1769	1163
	max.	48,500	49.77 ft^3	87	557	
Isovac	min.	32,000	.148 ft^3	15	2133	1503
	max.	131,000	13.13 ft^3	150	873	

Large-Sized and High-Temperature Furnaces

Typical examples of large-scale equipment working at high temperatures were analyzed, based on the experience of super resistance heaters manufacturers with molybdenum disilicide.[5] A summary of the data on the equipment described in the following sections is given in Table 4-5.

Pit Furnace for Heating Stainless Steel Ingots

The first furnace was built at Hagfors Ironworks, Sweden, in 1961. Heating elements with 9-mm heating zone diameter and 18-mm terminal diameter were specially developed and subsequently found to be an essential part for the construction of large industrial furnaces. Accurate temperature control and uniformity of temperature (vital for the subsequent rolling process) make electric furnaces preferable to others for the accurate heating of small charges of special steel qualities.

A second furnace built in 1966 has a much larger power. Steel ingots of three tons each are placed in the furnace at a temperature of 800-900°C, heated for 3 hr at 1250-1275°C and soaked for 2 hr. There is excellent uniformity of temperature both within individual ingots and from one ingot to another.

Billet Heating Furnaces

Billet heating furnaces again feature very good temperature uniformity, and are so compact that they can be adapted to most premises and can be equipped with ceramic muffles or used for heat treatments in protective atmospheres.

Forging Furnaces

With forging furnaces, it is easy to select the accurate temperature, to ensure that the bar material is heated right through and to avoid excessive heating by accurate temperature control. Also, the temperature in front of the furnace is relatively low. Thermal efficiency is cited to be about 30%. Circular hearth furnaces for treatment of heavy billets were also built up to 1050 kW.

Furnaces for Treatments in Protective Atmospheres

The high element consumption reported for the 140-kW unit (1 set/yr) for these furnaces was due to the addition of an excessive amount of town gas to the protective atmosphere, resulting in carbon deposits on the elements. This has since been corrected.

Table 4-5. Summary of Large-Sized and High-Temperature Furnaces Using Super Kanthal Elements[5]

Application	Date of Operation	Electric Power Rating (kW)	Tempera-ture (°C)	Material Treated	Heater Elements Consumption (sets/yr)	Observations
Pit furnace for heating stainless steel ingots	1961 1966	430 1400	1275	Stainless steel 85%, Carbon Steel, 15%	1	Production: 45,000 mt/yr
Billet heating furnace	1957	600		Steel	0.5	
Forging furnaces		36-52	1300	Steel		Energy Consumption: 0.79 kWh/kg steel
Circular hearth furnace for heavy billets	1963	700 1050		Steel	0.3	
Furnaces for treatment in protective atmospheres	1963	140	1200-1280	High-grade steel	1	
Circular hearth furnace with protective atmos-phere	1967	115-700		Heavy duty machine parts	none in first 3 years	

Pot furnaces	1961	156		Lead crystal glass	0.2-0.3	Production 25% larger than in oil-fired furnaces, pot life increased 20%
Glory holes		15	1400	Glass	very low	
Feeder channels		54-88-104/ channel		Glass	0.3	Uniform temperature, accurate regulation
Forebays				Glass for hand forming		
Electrodes for electrode heating (melting and holding of glass)			temperature of glass melt	Glass		
Tunnel furnaces for sintering			1400	Ferrites	Elements' lifetime is 2.5 years	
Tunnel furnaces for firing ceramics	1963 1960	350 200	1250 1420	Sanitary ware, steatite, electro-porcelain	0.38	Production: 250-450 mt/yr
Small furnaces for firing bricks		40	1620	Heavy-duty bricks	0.5-1	
Laboratory furnaces		3.3-11.5	1500-1600			

Pot Furnaces for Melting Lead Crystal Glass

The results obtained with pot furnaces were so good–improved quality of glass and reduction of reject figure–that the first firm to install it in Norway in 1961 decided to replace all of its oil-fired units with electric ones. The automatic control system with different programs for each type of glass simplifies operation. The working conditions are also improved by elimination of flue gases and irritating noise.

Glory Holes

A glory hole is a furnace in which a piece of glass is reheated between the first and second blowing processes. Electric furnaces provide good work-hygiene conditions and have a low element consumption.

Feeder Channels

Electric channels provide at least the same uniformity of glass temperature and more accurate control than gas-heated devices. In the case of boron silicate glass, organic fuels cause boron impoverishment in the glass, which does not occur with electrically heated channels. In the case of fluorine-containing glasses, fluorine compounds evaporated from the melt attack heater elements. Typical heater elements have a surface load of 12.2 W/cm^2 for the normal channel operating power of 5 kW, and the maximum operating surface load is 20 W/cm^2 (88 kW/channel). If desired and for short time periods, it can be raised to 23.5 W/cm^2 (106 kW/channel).

Forebays

In small installations, a better quality of glass and better working conditions for the glassblower are achieved when electric forehearths are used instead of gas-fired ones.

Electrodes

Electrodes are used for melting and heating glass. Preferably, they should not be allowed to dip down into the glass melt from above because of corrosion hazard at the boundary between melt and air. However, a feeder channel thus equipped can be emptied without removing the electrodes, as these are not damaged in air. Glass experts consider that uniformity of temperature in feeder channels can only be obtained with electrode heating of the melt combined with indirect heating of the surface.

Tunnel Furnaces for Sintering Ferrites

Electric heating in tunnel furnaces is superior to conventional because the manufacturing process is very dependent on composition of the atmosphere and accurate temperature control, both possible with this electric furnace.

Tunnel Furnaces for Sintering Sanitary Ware, Steatite, Electroporcelain

Conversion from metallic heater elements to molybdenum disilicide ores increased production markedly in 100-meter-long tunnels with 700 x 700 mm internal dimensions for sanitary ware. Fast firing of steatite has been done since 1961 with molybdenum disilicide elements. Electroporcelain furnaces have a nominal power up to 200 kW. High-duty fire-bricks are treated in furnaces that allow intermittent operation (2-3 firings per week) with an element consumption of 0.5 to 1 set/yr.

Laboratory Furnaces

Horizontal and vertical tube furnaces often with ceramic tubes of sillimanite, and chamber furnaces have been manufactured in many arrangements. Normal operating temperature is 1500°C. To react a temperature of 1600°C in horizontal tube furnaces, Kanthal Super elements must be installed so that they are vertically suspended. In vertical tube furnaces, elements with bent terminals are used. Large chamber furnaces (up to 200 mm x 290 mm x 290 mm) can be operated up to 1600°C with 8.2 kW and 10.2 V and 137 ampere per element.

Vacuum Furnaces

An important development in electric furnace equipment is the vacuum furnace, which provides heat treatment in a partial pressure environment. Vacuum furnaces are all manufactured for electric heating.

The advantages of vacuum heat treatments over conventional salt bath or atmosphere treatments include:

- safety, simplicity, repeatability, economy;
- no hazardous atmospheres or chemicals to prepare, control, eliminate or purchase;
- absence of salt in blind holes, screw threads, etc., or stratification of atmosphere resulting in soft spots, etc.;
- absence of either carburization or decarburization;

- minimizing of grinding, polishing and burnishing after hardening–particularly on complex dies and molds, since the metal finish will be as bright as it was when the work was introduced into the furnace;
- high Rockwells for ultimate metallurgical properties, giving longer life to tools treated in a vacuum;
- repeatable processes automatically achieved through time-temperature-transformation curves by instruments and circuitry, not dependent upon operator skill or attention;
- economical operation achieved as vacuum furnaces require utilities only when operated, unlike atmosphere or salt baths which must be almost continuously in operation. Maintenance of graphite is negligible and conventional equipment requires daily attention;
- control of distortion by programming preheat cycles prior to heating to austenitizing temperature. Another method of minimizing distortion can be programmed into the cooling cycle and result in reduced distortion.[6]

When required, a liquid quench can be incorporated into the vacuum vessel (the secret is, of course, a specially developed oil with very low vacuum pressure). This permits the parts to be rapidly conveyed from the heat to the oil quench. The material can also be quenched with gas or water. The quench system is fast enough in its heat removing capacity to produce hardened parts, and the equipment design prevents any contamination that could affect recycling of the heat treating process.

A graph of the savings in time and energy obtained with vacuum carburizing (where prepurified natural gas is admitted after heat treatment in the vacuum) compared with conventional endothermic atmosphere carburizing is presented in Figure 4-1.[7] The use of graphitized cloth and graphite felt insulation has further extended the range of vacuum furnace temperatures, which are now routinely operated at 2200-2300°C, to harden molybdenum and tungsten series tool steels. For one particular application (applying thermal shocks to moon rocks), a graphilizer cloth vacuum furnace is routinely used at temperatures of 2000°C, and the heater elements withstand rock pieces bouncing off them. At the limit, graphitizer cloth goes to 2780°C if carefully shielded from oxygen.

A comparison of vacuum furnaces with nonvacuum electric furnaces shows various advantages. Since costs of commodities and labor vary widely between localities, a specific set of baseline figures has been selected:

Electric Power	\$00.01/kWh
Water	01.18/1000 ft^3
Nitrogen	00.60/1000 ft^3
Quench Oil	00.66/U.S. gal
Labor	10.00/hr

Assuming a 200-lb load and a 2-hr floor-to-floor cycle for an oil-quenched load of 0.5 x 3.0 in. shafts made of carbon steel:

Electric Power (50 kW for 2 hr)	\$01.00
Quench Oil	00.10
Nitrogen Backfill Gas	03.00
Water at 20-25 gal/min (1 U.S. gal = 0.1336 ft^3)	00.57
Direct Labor (15 min/cycle)	02.50
Maintenance (including labor)	04.00
	11.17
or (per lb)	00.055

The following atmospheric, sealed-oil quench furnace is offered for cost comparison, even though it has a limited range of heat treating capability:

Electric Power	\$00.01/kWh
Water	01.18/1000 ft^3
Endothermic Gas	01.00/1000 ft^3
Quench Oil	00.60/U.S. gal
Labor	10.00/hr
Maintenance	–

Assuming a 100-lb/hr production, this works out as:

Electric Power	\$00.40
Water	00.05
Endothermic Gas	00.15
Quench Oil	00.15
Labor	02.50
Maintenance (including labor)	04.00
	7.25
or (per lb)	00.0725[6]

Processing costs are particularly low in vacuum furnaces.[8] Computer data on vacuum-oil quench furnaces indicate that total maintenance cost was \$0.70 per operating hour in 1968, and this dropped to \$0.38 per operating hour in 1970. Utilizing the highest maintenance cost plus all of the fixed costs, our total cost of heat treating in a vacuum oil quench furnace is \$0.05/lb in 1975.[9]

Another factor lowers processing costs for vacuum carburizing: processing floor-to-floor time (the time it takes from the point where the work is put into the furnace to the time when the work is removed) is less than for conventional atmosphere carburizing. Thus, the total time the work is in the heat treatment department is cut greatly, reducing the total inventory maintained in a plant at any one time. This is a hidden saving not often recognized by those trying to justify new capital equipment.

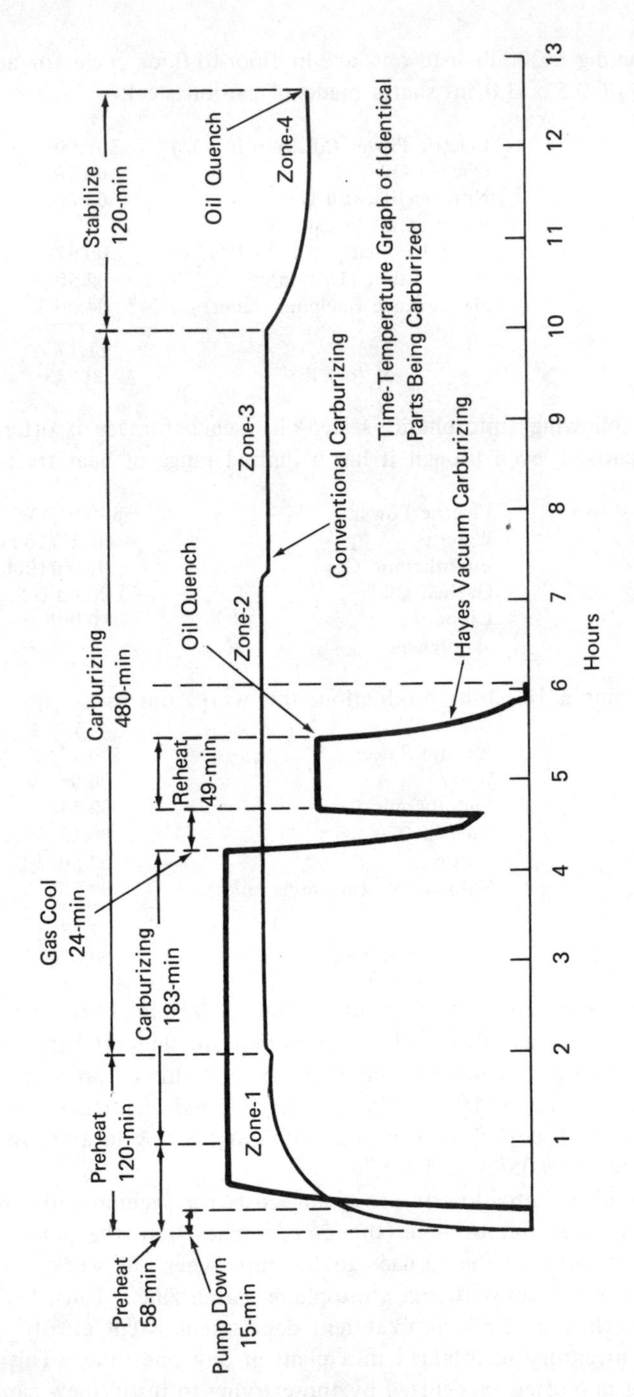

Figure 4.1. Comparison of a conventional endothermic atmosphere-carburizing method vs Hayes vacuum carburizing.[7]

Sintering powdered metals is another area where vacuum furnaces have found a broad application. The cost advantages are evident from the following example:

To produce one kilo of cemented carbide by the three most common methods:

1. Conventional atmosphere tunnel furnace to dewax, presinter and final sinter costs $1.50.
2. Conventional atmosphere tunnel furnace for dewaxing and presintering, with final sintering in vacuum, costs $0.704.
3. Continuous process using hydrogen dewax and vacuum presinter and final sinter costs $0.40.

The above costs do not include equipment amortization. However, equipment costs for the three methods are roughly the same but floor space requirements for method 3 is from one-third to one-half that needed for methods 1 and 2.[10]

The price range for vacuum furnaces is $1000-2000 per installed kW. As an example of a large vacuum furnace, Lindberg cites one of 120 kW with a chamber volume of 24 x 36 x 18 in. and costing $150,000.[11] The temperature range, power rating and cost of various types of vacuum furnaces are given in Table 4-6.

Table 4-6 Characteristics of Vacuum Furnaces[7]

	Maximum Temperature (°C)	Power Rating	Cost ($ thousands)
Oil Quench VF	1315	20- 40 60-225	35- -106
Vacuum-Sealed Quench F	982	13- 18	-126
Oil and Gas Water Quench		75-180	
Gas Quench Vacuum Furnace	1315	60- 99 (dual (chamber) 75-300 (single)	- 87
Conveyor Vacuum Furnace		9-210	-225
Vacuum Tempering Furnace		15- 16	38
Sintering			
Aluminum	595- 650	50- 75	
Bronze	760- 900	91-225	
Iron carbon steel	1120-1150	91-228	55-60
Iron	1200-1345	63-202	87
Stainless steel	1100-1150	45-207	55-60
	1205-1350	31-107	87
U oxide	1595-1705	45-105	-275
W-Ti carbide	1565-1650	26- 99	-275

POTENTIAL APPLICATIONS

In his report on the industrial heating equipment industry, Ricciuti gives an outline of how the applications for electric heat treatment are constantly growing.[1]

> "The art of heat application as an industrial manufacturing technique has advanced so rapidly that today's innovators are tomorrow's basic processes. After a long history of technological evolution, there have been in the last few years a sudden and widespread acceptance of new techniques and a highly accelerated replacement of traditional methods by new or alternate processes.
>
> "In particular, for heat processing metal treating equipment, more and more attention is being paid to getting the most out of metals by heating treating, and continuing demands are being placed on equipment builders for sophisticated equipment. Self-generating atmosphere furnaces for deep drawn steels, rapid plate quenching, and high-speed heating for galvanizing lines are examples of some of the recent developments in heat treating equipment.
>
> "Vacuum treating of steel is becoming more acceptable to industry. Once used solely for heat treating titanium and other exotic metals as the only method suitable, new developments have made it more economical for use in heat treating stainless and tool steels. While this method is still more expensive than heat treating in controlled atmospheres, certain advantages such as better control of carburization, the elimination of surface hydrogen to improve ductility, and better elimination of oxygen bearing contaminants, tend to reduce costs by eliminating other operations and prolonging the life of end products."

The industry is reportedly working actively on the development of a continuous vacuum furnace suitable for its needs. So far, however, no continuous vacuum furnaces for annealing steel coils have been put into operation in the United States, although conveyor vacuum furnaces for treating semiconductors or brazing copper exist.

GENERAL ASSESSMENT OF THE FUTURE

Basic Technical Motivation for Planning of Industrial Furnaces With Electric Resistance Elements

According to Kanthal, a great deal of development has taken place within the furnace-building field, which has been characterized by the growing importance of the electric resistance furnace for many different industrial applications. This progress has to a large extent depended on the fact that newly developed resistance alloys permit a considerable increase of the maximum temperature in resistance furnaces. Since the highest quality products permit element temperatures of up to 1800°C, it has been possible to build furnaces for operating temperatures of up to 1700°C.[5] In addition, the introduction of graphitized cloth as a heater element has permitted routine operation of vacuum furnaces at 2000-2200°C.

Kanthal also feels another reason for the rapid development is that the advantages of electric heat have become ever more obvious, such as:

- accurate temperature control through automatic regulation combined with great uniformity of temperature;
- possibility of local power distribution to suit requirements;
- great adaptability for production runs;
- improvement in quality of the products;
- improved working hygiene due to the absence of combustion products.[5]

Market Considerations

The activity in the electric heat treatment equipment industry is largely dependent upon the segments of the economy that manufacture durable goods, particularly those produced by the primary metals, transportation and machinery industries. Expenditures for industrial heating equipment as a whole tended to parallel expenditures for plant and equipment by durable goods manufacturers in past years (Figure 4-2).[1] The major investment programs in the late 1960s for modernization in the metal processing and fabrication group, particularly ferrous metals, have given a stimulus to industrial heating equipment manufacture. Improvements in technologies applied to manufacturing processes have strongly affected the economies of utilization and production of industrial heating equipment, shifting the demand from one type of equipment to the other. Resulting variations in growth rates for different product lines have had significant consequences for individual firms manufacturing this equipment where firms tend to specialize in rather narrow product lines. As a result, substantial efforts have been made to update and extend product lines, and develop exports which have profited from the development of metal industries in other nations.

Market for Industrial Heating Equipment

Sales statistics were obtained for the years 1958, 1960, 1965, 1972 and 1974 for the whole industrial heating equipment industry, including not only heat process and metal treatment furnaces, on which this report is focused, but also for metal melting, refining and holding furnaces, and hot-forming, forging, rolling and extrusion furnaces and a variety of related and ancillary equipment.[12]

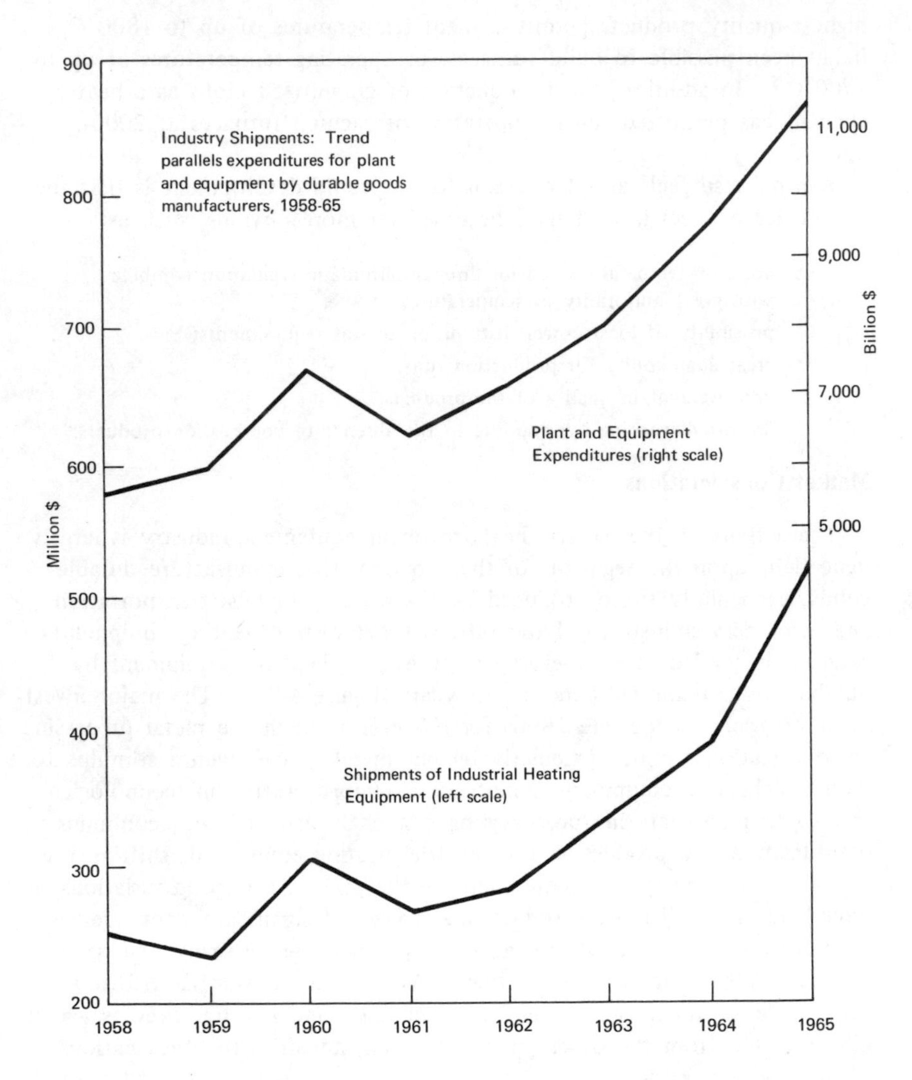

Figure 4.2 Shipments of industrial heating equipment and plant and equipment expenditures by durable goods manufacturers, 1958-1965.[1]

To show the proportion of the electric heat treatment equipment to the whole range of industrial heating equipment, comprehensive statistics on the sales of all the various types of equipments are given in Tables 4-7 and 4-8.

The Industrial Heating Equipment Association, which groups the manufacturers of process heating equipment types used in the manufacturing

Table 4-7. Industry Shipments, by Categories, 1958, 1960, 1965, 1972, 1974 ($ thousands)[1,12]

	1975	1974[a,b]	1972[a]	1965	1960	1958
Grand Total	548,000	439,000	307,077	514,078	203,390	238,955
Molten metal melting, refining, and holding furnaces (excluding blast furnaces and cupolas), total		72,900	51,488	58,924	21,113	22,909
Electric (except induction)		8,050	6,940	15,222	5,628	10,205
Ferrous		NSA	4,277	12,557	4,446	9,507
Nonferrous . . .		NSA	2,663	2,665	1,182	698
Fuel-fired		9,300	7,364	15,003	5,497	6,069
Ferrous		NSA	4,864	5,710	NSA	NSA
Nonferrous . . .		NSA	2,500	9,293	NSA	NSA
Induction		55,550	37,184	24,713	7,478	4,446
Ferrous		NSA	25,884	17,875	NSA	NSA
Nonferrous . . .		NSA	12,300	6,838	NSA	NSA
Vacuum		NSA	NSA	3,986	2,510	2,189
Ferrous		NSA	NSA	3,092	1,774	NSA
Nonferrous . . .		NSA	NSA	894	736	NSA
Hot-forming, forging, rolling, piercing, and extruding heating equipment total .		17,800	12,702	33,108	17,668	18,804
Electric (except induction)		2,670	1,985	876	NSA	NSA
Ferrous						
Nonferrous . . .						
Fuel-fired		7,830	5,779	28,570	NSA	NSA
Ferrous		NSA	NSA	23,742	14,433	15,859
Nonferrous . . .		NSA	NSA	4,828	NSA	NSA
Induction		7,300	4,938	3,662	2,596	2,407
Ferrous		NSA	NSA	2,301	1,609	1,752
Nonferrous . . .		NSA	NSA	1,361	987	655
Heat processing and metal treating equipment, total		215,550	153,031	159,118	131,871	95,877

Table 4-7, continued

	1974[a,b]	1972[a]	1965	1960	1958
Electric (except induction)	38,800	27,800	24,687	22,915	12,257
Ferrous	NSA	19,386	18,381	19,873	10,354
Nonferrous . . .	NSA	8,414	6,306	3,042	1,903
Fuel-fired	73,300	50,700	72,947	65,166	56,656
Ferrous	NSA	46,469	62,480	58,220	47,090
Nonferrous . . .	NSA	4,231	10,467	6,946	9,566
Induction	29,100	19,661	15,780	13,580	10,436
Ferrous	NSA	17,203	13,290	9,682	7,410
Nonferrous . . .	NSA	2,458	2,490	3,898	3,026
Vacuum	NSA	12,186	9,641	2,674	1,262
Laboratory furnaces	NSA	7,793	3,521	3,763	NSA
Bath-type furnaces	NSA	3,732	3,175	1,908	1,384
Heating equipment without furnace chambers	NSA	NSA	665	1,087	142
Glass and ceramic equipment	NSA	25,567	17,383	17,355	9,771
Cellulose processing equipment	NSA	NSA	1,415	198	NSA
Miscellaneous metal processing and heating equipment	NSA	NSA	9,904	3,221	3,969
Industrial process ovens, total	53,700	46,949	59,179	35,968	32,576
Electric	NSA	18,744	11,940	8,245	6,736
Fuel-fired	NSA	28,205	45,735	25,853	24,664
Steam	NSA	NSA	1,504	1,870	1,180
Dielectric heating equipment	NSA	5,600	7,268	3,005	2,521
Auxiliary industrial heating equipment	NSA	NSA	33,826	NSA	NSA
Components					
Combustion					
Equipment . . .	52,700	27,380	40,330	21,721	15,277
Gas	NSA	22,649	NSA	NSA	NSA
Oil	NSA	4,731	NSA	NSA	NSA
Atmosphere gas					
Generators and control					
Equipment	11,000	7,044	8,345	5,366	4,115
Gas generators . .	NSA	NSA	6,980	NSA	NSA
Control equipment	NSA	NSA	1,365	NSA	NSA

Table 4-7, continued

	1974[a,b]	1972[a]	1965	1960	1958
Regenerative and recuperative heat exchangers	15,400	8,483	3,978	NSA	NSA
Regenerative . .	NSA	5,343	NSA	NSA	NSA
Recuperative . .	NSA	3,140	NSA	NSA	NSA
Replacement and repair parts sold separately.	NSA	NSA	43,453	45,209	24,283
Cost of subcontracting	NSA	NSA	26,808	NSA	NSA

[a]Does not include engineering firms.
[b]Estimated.

Table 4-8. Comprehensive U.S. Industrial Heating Equipment Industry Shipment ($ millions)[1,12]

Year	Value
1956	258.3
1957	307.8
1958	252.9
1959	236.6
1960	302.4
1961	263.4
1962	278.5
1963	333.9
1964	382.1
1965	514.0
1969	367
1970	310
1971	265
1972	307
1973	338
1974	439
1975	548

industry is another source of sales data. The figures for the sales of fossil-fired and electric heating equipment by their members from 1970 through 1975 are given in Table 4-9.[13]

Table 4-9. Orders Reported by Industrial Heating Equipment Association Members ($ thousands)[13]

	Fuel-fired Processing Equipment	Electric Processing Equipment	Other Equipment	Total
1975	52,429	43,562	50,375	146,366
1974	90,397	23,765	39,326	153,488
1973	75,785	18,892	32,966	128,643
1972	41,274	12,788	25,257	79,319
1971	30,271	7,465	25,942	63,678
1970	43,944	11,020	33,510	88,474

Fuel-fired equipment sales decreased considerably from 1974 to 1975, whereas sales of electric equipment nearly doubled, because many industries had trouble in 1974 finding fuel to keep their operations going and chose electricity as the energy source for new equipment. The short supply of natural gas, which is used as fuel in most heat treating furnaces, was also a reason to go electric for new equipment.

Estimates of Installed Electric Heat Treatment Capacity in the United States

No comprehensive records of installed capacity of electric heat treatment exist in this country. Sales as recorded by the U.S. Department of Commerce are for selected years. For the missing years, figures have been obtained by using linear interpolation. Consequently, one has to derive capacity estimates from shipments under the assumption of some $/kW figure for the various categories of equipment. At the same time, one has to take inflation into account by applying the gross national product implicit price deflator for the total private sector as given in the economic report of the President for 1975 (Table 4-10).

The value of cumulative units installed since 1958 in constant 1975 dollars is shown in Figure 4-3 for electric heat processing equipment (without induction), electric industrial process ovens, vacuum and laboratory furnaces. To these we apply the $/kW value given in Table 4-11, which are mid-range values for medium-sized equipment, and find installed capacities as listed in Table 4-11.

For comparison purposes, the cost data obtained from Lindberg,[2] manufacturers of large-sized, carbon-made equipment, is as follows:

- electric furnaces (not induction), of the large production-type equipment, continuous (funnel furnaces) with some degree of automation, such as used by the automotive industry–$/400-500/kW

Table 4-10. Gross National Product Implicit Price Deflator for the Total Private Sector[5]

Year	Total
1955	91.57
1956	94.53
1957	97.92
1958	100.00
1959	101.41
1960	102.76
1961	103.73
1962	104.73
1963	105.80
1964	107.05
1965	108.83
1966	111.56
1967	114.79
1968	118.90
1969	124.30
1970	130.32
1971	135.70
1972	139.61
1973	147.56
1974	163.27

- bath or box-type furnaces, of the middle-sized class of equipment, unautomated–$325-375/kW;
- medium-sized industrial process ovens (temperatures below 700°F), made of light sheet metal and light insulation, of the type used in the paint, automotive, furniture, appliances and food can industries–$200/kW;
- vacuum furnaces–$1000-1250/kW; and
- laboratory furnaces, in the kW range–$750-850.

These values roughly confirm those that have been used in Table 4-2. They are at most 30% lower.

Larger-sized equipment units come in much smaller numbers but with much higher power rating that the average units. It is considered (by Lindberg) that of the electric heat processing equipment, 65% is composed of large units (on the order of $600,000-$1,000,000) and 35% as small units (on the order of $45,000-55,000).[11] The larger units would then be in the size range of 1500-2500 kW. Thus, for $38 million in sales in 1974, $25 million would go for about 30 large units and $13 million for about 280 smaller units. The installed capacity would be 60,000 kW in large units and 28,000 kW in smaller units. Due to the lack of accurate

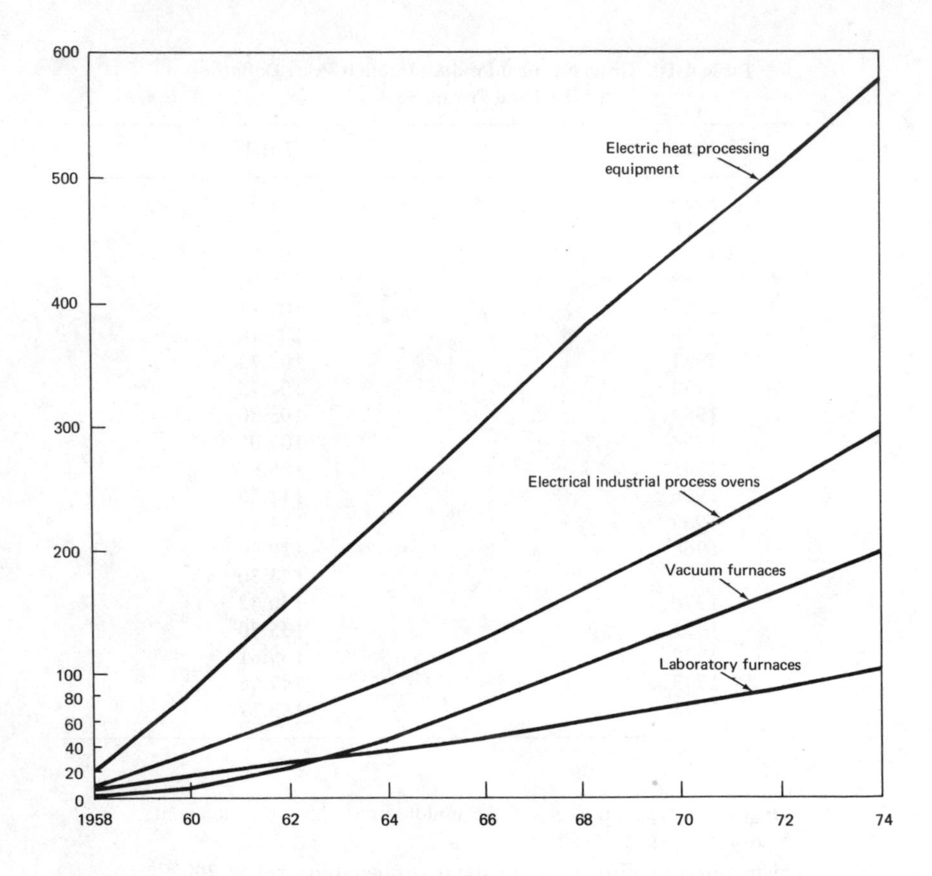

Figure 4-3. Value of cumulative units installed since 1958 in constant 1974 dollars ($ millions).

Table 4-11. Estimated Capacity of Electric Heat Treatment Equipment Installed in the U.S. Since 1958

Type of Equipment	Cost per Power Unit ($/kW)	Capacity (MW)
Electric Heat Processing Equipment	550	1060
Electric Industrial Process Ovens	250	1200
Vacuum Furnaces	1500	130
Laboratory Furnaces	1250	80

records and the broad price ranges, these figures have to be considered as indicative only.

Trends of Sales of Electric Heat Treatment Equipment

The interpolated sales statistics of the U.S. Department of Commerce clearly indicate the trends for the various types of equipment. Figure 4-4 shows the annual sales in current (not constant) dollars, as taken directly from the statistics for electric heat processing equipment (excluding induction), electrical industrial process ovens, vacuum and laboratory furnaces.

As previously mentioned, many industries have had trouble in 1974 finding fuel and keeping their operations going, and have chosen to go electric for new equipment. However, this trend might not last, and the electric equipment sales might top out sometime, when electric power becomes scarce in this country. The cost of money prevents, to some extent, the utilities building more power plants at this time. Then the choice of the users breaks down to economics. For those cases or parts of the country where fuel is not expected to be available or is expensive, one will go electric. In other cases, if there is a choice between fuel, of gas and electricity, the present economic situation will dictate the option and the less expensive method will be used. The fear of a shortage of natural gas is an additional factor that is difficult to estimate.

Economic Comparison of Electric and Conventional Equipment

Small- and Medium-Sized Equipment

An economic comparison between electric heat treatment and fuel-fired equipment is complex because some equipment may not be available or comparable, both technically and cost-wise, in both categories. For medium-sized furnaces where special atmospheres are required, however, comparable equipment exists in both categories. Gas-fired equipment requires a muffler to prevent the combustion products from entering the chamber where the workpieces are placed. Gas-type continuous furnaces are an example. Some companies that manufacture both fuel-fired and electric equipments for the same purpose sell them at the same price to avoid competition between their own products, and let the customer make the choice, based on local conditions of use, availability and price of natural gas, other fossil fuels and electricity.

Cost and energy data on Lindberg furnaces available for both gas and electric are given in Table 4-12.[11] For each furnace type, the ratio of energy consumption between the electric and the gas versions has been computed in Btus using the physical energy equivalent of 3413 Btu/kWh

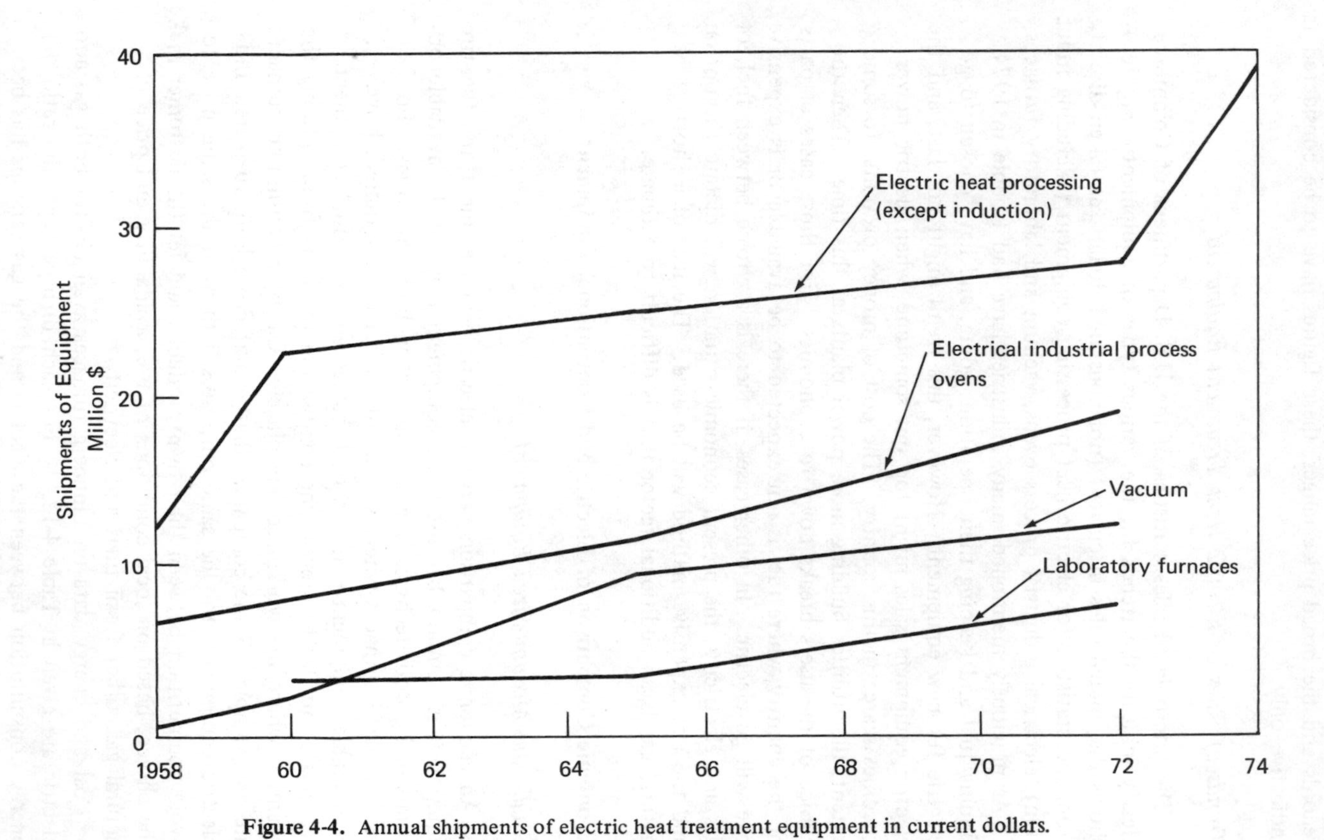

Figure 4-4. Annual shipments of electric heat treatment equipment in current dollars.

Table 4-12. Comparison of Lindberg Gas and Electric Furnaces[11]

		Electric			Gas			Electric to Gas Btu Ratio			
Type of Furnace and Chamber Dimensions	Temperature Max. or Range (°C)	Price ($)	Input (kW)	Gross Heating Rate to Furnace Temperature (lb/hr)	Price ($)	Input (Btu/hr)	Gross Heating Rate to Furnace Temperature (lb/hr)	With 1 kWh = 3413 (Btu)	With 1 kWh = 10,000 (Btu)	Specific Price for Electric Models ($/kW)	Specific Price for Gas Models ($/Btu/hr)
Integral Quench (in-out)	875										
24, 36, 18 in.		53,000	65	560	52,500	600,000	560	0.37	1.08	815	0.09
30, 48, 30 in.		64,000	145	1,500	63,000	1,500,000	1,500	0.33	0.97	441	0.04
Integral Quench (two chambers)	875										
30, 48, 30 in.		132,000	290	3,000	132,000	3,000,000	3,000	0.33	0.97	455	0.04
Tool Room Cyclone (no controlled atm)	190-675	9,700									
24, 36, 18 in.		9,700	32		10,000	175,000		0.62	303	0.06	0.06
Production-Box Cyclone (no controlled atm)	190-675										
24, 36, 18 in.		13,700	34		14,100	240,000		0.48	1.42	403	0.06
Contherm Element (atm)	1010										
24, 36, 18 in.		22,000	65		22,700	600,000		0.37	1.08	338	0.04
Rod Element (atm)	1095										
24, 36, 18 in.		12,200	440		13,500	500,000[a]		0.27	0.80	305	0.03
Draw Furnace (no controlled atm)	675										
30, 48, 30 in.		19,000	95	700	22,000	500,000	820	0.65	1.90	200	0.04

[a]Gas semimuffle furnace, does not correspond exactly to electric model.

(which is what occurs at the furnace). However, since it takes approximately 10,000 Btus to produce one kWh at the electric generating plant, the ratio of energy consumption for both furnaces has also been calculated using this value, which is what happens nationally, on the average. Locally, electric furnaces use one- to two-thirds of the energy of gas furnaces; however, reduced at the input to the power plant, the energy consumption of electric furnaces is from 1 to 2 times that of gas furnaces. Table 4-13 lists typical energy losses and efficiencies for gas and electric furnaces. The main difference is that in gas furnaces half the energy goes to the stack in the flue gas.

Table 4-13 Energy Losses and Efficiencies of Furnaces, Stabilized at Constant Temperature[2,11]

Type of Loss	Gas-Fired (%)	Electric (%)
To stack	45-50	0
Thermal to brick work, opening door, black body loss to black chamber of quench furnace, energy to heat atmosphere	15-20	20-25
Total losses (%)	60-70	20-25
Efficiency	30-40	75-80

In Table 4-14, a comparison of capital, installation, maintenance, energy and labor cost factors are given for electric and conventional heat treatment equipment. When the combustion products can mix in the chamber with the workpieces, and no muffler is required for separation, gas-fired equipment is decidedly cheaper in terms of capital cost than electric equipment.

Electric equipment will be preferred for special equipment; it is easy to control, and the electric heaters can be placed near the pieces that need heating. This also makes the electric equipment more energy-effective. For very large units (on the order of 2000-3000 kW), electric equipment is at a disadvantage, as the large supply of electrical energy might require a substation, negating any cost advantage.

In general, the larger portion (85%) of furnace cost is in materials; the rest (15%) in labor. Furnace prices are thus very dependent on the price of raw materials.

For the cost of operations, electric heat treatment furnaces are considered to be 95% efficient, *i.e.*, the heat loss is 5%. Gas furnaces are

Table 4-14 Cost Factors for Gas vs Electric Controlled-Atmosphere Quench Furnaces[11]

Type of Costs	Gas	Electric
Capital Costs	Approx. same as other	Approx. same as other
Installation Costs	Higher, because of renting and exhaust systems	
Maintenance Costs	Replacement of radiant tubes costing 2-3% of furnace in 1-1.5 yr (approx. 2%/yr)	Replacement of heaters costing 5-7% of the furnace in 3-5 yr (approx. 1.4-1.7%/yr)
Input Energy Costs per Btu	Cheaper	More expensive
Input Energy Costs per lb/hr	Approx. same as other	Approx. same as other
Labor Costs	Approx. same as other	Approx. same as other

considered to have 50% efficiency, meaning that half the heat from a fuel goes up the stack. However, the price of a Btu of fuel is currently about one half a Btu of electricity. Therefore, fuel costs nearly break even, with a small advantage towards fuel-fired. Typically, the energy required for the heat treating of 1000 lb/hr of parts is 80/90 kWh. However, the real cost of operating a furnace is manpower, not fuel.

A summary of the energy and cost comparison for gas and electric furnaces in the small- to medium-sized range is given in Table 4-15, listing the advantages and disadvantages of equipment using both types of energy supply where models for both exist.

Table 4-15 Advantages and Disadvantages of Gas and Electric Furnaces of Small to Medium Size[2,11]

	Gas	Electric
Advantages	Input energy slightly cheaper (depending on region)	Electricity more readily available, environmentally clean
	Capital costs less when flue gas does not harm workpieces	Only solution for vacuum furnaces
Disadvantages	Gas in short supply	Input energy slightly more expensive (depending on region)
	Environmental pollution by combustion products	
Approx. share of market in 1975 for Lindberg furnaces where the two options (gas and electric) exist	20%	80%

Large-Sized Equipment

In some very large-size applications where the flue gases can be used directly to heat the workpieces, gas becomes more advantageous. Lindberg cites a furnace 66 ft in diameter (used for railroad car wheels) operating at 1300°C with a heat input of 44 x 10^6 Btu/hr that cost about $1 million. By contrast, a vacuum furnace, which is electric, in a model with a 24 x 36 x 18-in. chamber operating at 1300°C with a power rating of 120 kW, would cost $150,000.

For very large-sized equipment, say for thermal metallurgy used by ferrous and nonferrous metal producers, the trend toward electric heating is not as pronounced as it is for smaller-sized equipment. One manufacturer of such thermal metallurgy equipment claims:

- more electric furnaces are purchased than fuel-fired, but only of necessity;
- capital cost of electric furnace is almost always higher;
- maintenance cost (if operated per vendor's instructions) is almost always lower;
- energy costs (where natural gas is still available) are higher for electrically heated furnaces, but depending on heat treating, this can be offset by thermal efficiency if proper equipment is installed;
- within the scope of design parameters there is little difference on the operation, regardless of energy source. Either it can be used or not; for some operations you may not be able to substitute electricity where the products of combustion become an integral part of the process.

Typical of this kind of equipment are Drever's vertical continuous annealing lines for strip steel. The one installed at the large steel plant in Hamilton, Ontario, has a production of 18.5 tons/hr and a speed of 600 ft/min with a 3000-kW electrically heated line and an operating temperature of 732°C.

In this category, much of the equipment is custom-made, and it is difficult to generalize cost or energy data. In particular, consumption data vary greatly with the size of the unit, the material being treated and the process itself. Consumption may range from 600,000-4,000,000 Btu and from 175-600 kWh per ton processed. In the past, selection of type of energy to be used was not dictated by consumption data or energy costs, but primarily by what was best for the process (best quality product) and most reliable for continuous operation. Today, it is more likely to be dictated by what energy is available at a specific geographic location.[3]

One economic feature peculiar to the electric heat treatment equipment manufacturing industry is that sales follow the cycles of the steel and metal manufacturing industries.

Another consideration in industrial electric heating equipment is the control equipment. U.S. sales of control equipment for heating of all types of furnaces is currently indicated to be $25 million per year.[14] This equipment is estimated to be about 10% of the total heat treatment equipment. This would bring us to $250 million, which is close to the $215 million estimated for 1974. This control equipment has had an annual growth rate of 20% recently, whereas the growth rate of the total heat equipment is estimated to be 5% over the long range.

ASSESSMENT OF THE FUTURE

Electric heat treatment equipment manufacture is an apparently healthy industry, growing steadily and increasing its growth rate since the fuel crisis. New high-temperature resistance heaters have been developed that have introduced electric heat treatment equipment of fairly large size in the steel-making industry. Also, electric vacuum furnaces have broadened the scope of heat treatment Trends for induction equipment, which are similar to those for equipment described in this report, have been documented.[5]

Sales of all equipment slowly increased up to 1972; however, there was an abrupt increase in the sales of global electric heat processing equipment in 1974, clearly due to the fossil-fuel crisis of 1973, and as of 1975, sales were still increasing (Table 4-9). The main reasons for this are: (1) improvement of electric heat treatment techniques and equipment by the introduction of high-temperature heaters and of vacuum furnaces; (2) lack of domestic natural gas and a fuel price increase after the 1973 crisis; and (3) enhanced Environmental Protection Agency requirements regarding air purity, gaseous effluents, etc.

Accurate projections to 1980 and 1985 could not be obtained. However, the industry reports continuous sales increases, although there is some doubt whether electric utilities will be able to expand the supply of electricity needed by the manufacturing industry in the future at a sufficient rate, if sales of such equipment continue as at present.

TECHNICAL DATA ON RESISTANCE HEATERS

The materials used for resistance heaters vary with the manufacturer. The following materials have been typically identified and are used in the temperature range indicated in Table 4-16. Ultrapure graphite cloth, consisting of practically 100% graphitic carbon, is very flexible, pliable and maintains its characteristics after repeated cycling. The strength of graphite doubles as temperature rises from room temperatures to 2480°C. It

Table 4-16 Materials Used for Resistance Heaters

Manufacturer	Material or Trade Name	Temperature Range (°C)
Lindberg	Silicium carbide	500 - 1500
	Molybdenum disilicide	1800 - 1900
	Nichrome (0.8 Ni, 0.2 Cr)	675 - 1200
	Inconel (0.72 Ni, 0.17 Cr, Fe)	980 - 1100
Leeds Northrup	Graphitized cloth	up to 1315
	Molybdenum	
	Inconel	790 - 1000
Kanthal	Molybdenum disilicide	1700 - 1800
	Fe Cr Al Co - Alloys	1375-1280
	Ni Cr - Alloys	1050-1200
	Fe Cr Al - Alloy	1050
C. I. Hayes	Graphite cloth rolled into hollow tube	1000[a]- 2480[b]
	Molybdenum foil, rod or wire	1100
	Nichrome bars	<1050

[a]Threshold oxidation temperature in water vapor.
[b]Temperature at which strength doubles.

sublimes at 3650°C. The main advantage of graphite cloth is that a vacuum furnace using heating elements made of it is not bothered by water vapor and oxygen because graphite burns with oxygen to form CO and CO_2. Threshold oxidation temperature (where graphite loses 1% of its weight in 24 hr) is 520-569°C in oxygen, 700°C in water vapor and 900°C in CO_2.

In addition to heating elements, graphite is also used for insulation purposes in the form of felt. Typical properties of graphite cloth and graphite felt are given in Tables 4-17 and 4-18.

There are advantages to using graphite cloth elements since the graphite:

- is flexible;
- is a pure form of carbon;
- does not melt or deform at high temperature;
- doubles its strength from room temperature to 4500°F (2482°C);
- is uniform, dimensionally stable and free of hot spots;
- does not contain oxidation-promoting catalyst ordinarily present in manufactured graphite;

- consists entirely of flexible graphite filaments measuring 8 μ in diameter (0.0003 in) made by graphitizing rayon cloth at 5000°F (2760°C); and
- has an average braking strength of 7000-24,000 lb/in.[2,6]

Table 4-17 Typical Properties of Graphite Cloth[10]

Physical	
Tensile Strength	
Warp (lb/in.^{2})	25
Fill (lb/in.^{2})	23
Chemical	
Carbon content (%)	99.96
Ash content (%) heated to 4500°F	0.04 max.
Electrical	
Resistance, 1-in.-wide strip (ohms/in. of strength)	
at 70°F	0.47
at 1000°F	0.38
at 3000°F	0.20
Construction	
Weight (oz/yd^2)	7.3
Thickness (in.)	0.023
Count (yd/in.)	
Warp	27
Fill	24
Filament diameter (approx. inches)	0.0003

The advantages of graphite felt for insulation include:

- superior insulating values–much better than metallic shields;
- low density–furnace heating chamber heats and cools faster;
- excellent stability–uniform density–no voids after many cycles, strength increases with temperature;
- no outgassing–no gas evolution from graphite felt in vacuum furnaces;
- easy fabrication–semirigid, cuts with scissors; and
- economical–much lower in cost than metallic shields.[6]

The basic material, molybdenum disilicide elements, is manufactured by extrusion. The rods obtained, with different diameters for the terminals and heating zones, are then dried, sintered and cut into suitable lengths. The heating zones are bent under heat to the desired shape and welded to the terminals, one end of which is ground down to the same diameter as that of the heating zone. The other end of the terminal is aluminized to ensure good electrical contact. Table 4-19 includes data on the physical properties of this and other resistance materials.[5]

Table 4-18 Typical Properties of Graphite Felt[10]

Density	5.3 lb/ft^3
Thickness	0.206 in. nominal
Shrinkage to 5000°F	nil
Water Absorption in 90% RH	nil
Specific Heat	0.17 Btu/lb at 70°F
Specific Heat	0.40 Btu/lb at 2500°F
Emissivity	0.99
Melting Point	Sublimes at 6600°F
Vapor Pressure	at 4120°F - 1 μ
	at 4420°F - 10 μ
	at 4750°F - 100 μ

Kanthal A-1 is economically applied at temperatures (up to 1330°C) attainable previously only with nonferrous heating elements or those of precious metals. It is thus particularly useful in high-grade steel furnaces, ceramic kilns and large air-heating installations in the chemical industry. It is manufactured as large cold-rolled strip and as wire above 1 mm in diameter, and is particularly suited to the manufacture of heavily loaded resistance coils for industrial furnaces.

Because of its considerable resistance to oxidation, Kanthal A is often used for medium furnace temperatures to ensure a longer element life than obtainable with nickel-chromium alloys.[5] Greater durability can be obtained by increasing the chrome content; thus, a new Nikrothal alloy with 70% nickel and 30% chrome was developed, characterized by a higher maximum operating temperature and a better service life in air and inert gases than all other Nikrothal alloys. Its use is particularly suitable in cases where operation in oxidizing and reducing atmosphere occurs alternately.

Nikrothal 20 is mainly used for heating elements for temperatures up to 1050°C. It may also be used as hot-rolled construction materials in industrial furnaces.

The Alkrothal resistance material is intended for temperatures up to 1050°C and is used in heating apparatus where, apart from the life of the element, economy plays a decisive role.

Resistance of Heater Materials to Chemical Attack in Gaseous Media

The durability of resistance alloys in air at high temperatures is due to the formation of a protecting oxide layer formed by reaction with the

Table 4-19. Analysis and Properties of the Kanthal Alloys[5]

	Alloy							
	Alkrothal	Nikrothal 80	Nikrothal 70	Nikrothal 20	Super 33	Super N	A-1	A
Form		Wire, strip, tape	Wire, strip	Wire, strip	Preshaped heating elements	Preshaped heating elements	Wire over 1 mm ϕ strip	Wire tape
Max. continuous operating temperature (element temperature) (°C)	1050	1200	1250	1050	1800	1700	1375	1330
Principal analysis: % Ni Cr Fe	Cr 15 Al 4.5 Fe 80.5	Ni 80 Cr 20 Fe –	Ni 70 Cr 30 Fe –	Ni 20 Cr 25 Fe 55	37 63	Si 37 Mo 63	Cr 22 Al 5.5 Co 0.5 Fe 72	Cr 22 Al 5.5 Co 0.5 Fe 72
Density, g/cm^{-3}, $wt/in.^3$, lb	7.28	8.3	8.1	7.8	5.6	5.6	7.1	7.1
Resistivity at 20°C, $ohm \cdot mm^2 \cdot m^{-1}$	1.25	1.09	1.17	0.95	0.3	0.3	1.45	1.39
Resistivity at 1800°C, $ohm \cdot mm^2 \cdot m^{-1}$						4	4	
Coefficient of linear expansion, $°C^{-1}$ average value for:								
20-250°C 68-482°F		$15 \cdot 10^{-6}$	$13.8 \cdot 10^{-6}$	$16 \cdot 10^{-6}$			$11.0 \cdot 10^{-6}$	
20-500°C 68-932°F		$16 \cdot 10$	$14.9 \cdot 10^{-6}$	$17 \cdot 10^{-6}$			$12.5 \cdot 10^{-6}$	
20-750°C 68-1382°F		$17 \cdot 10^{-6}$	$15.9 \cdot 10^{-6}$	$18 \cdot 10^{-6}$			$14.0 \cdot 10^{-6}$	
20-1000°C 68-1832°F	$15.0 \cdot 10^{-6}$	$18 \cdot 10$	$17.1 \cdot 10^{-6}$	$19 \cdot 10^{-6}$			$15.0 \cdot 10^{-6}$	
20-1500°C					$7.1\text{-}8.8 \cdot 10^{-6}$	$7.1\text{-}8.8 \cdot 10^{-6}$		
Heat conductivity at 20°C (68°F), $W m^{-1} °C^{-1}$	17	11	11	12			17	
Specific heat at 20°C (68°F), $kJ kg °C^{-1}$	0.64	0.45	0.45	0.45	0.48		0.46	
Melting point, approx. °C(°F)	1510	1440	1380	1380			1510	
Mechanical strength $(\sigma_B)^a$ $N mm^{-2}$	700	750	900	750			750	
$(\sigma_{o,2})$ $N mm^{-2}$	550	450	500	450			580	
Hardness (H_V)	220	180	200	180			230	
Elongation in % on 200 mm (8-in.) length	18	30	30	30			18	

[a] $1 N \cdot mm^{-2} = 0.102 kp \cdot mm^{-2}$.

Table 4-20. Maximum Heater Temperatures in °C in Various Furnace Atmospheres for Kanthal Elements[5]

Material Protective Layer	A-1 Al_2O_3	D Al_2O_3	
Atmosphere	**Temperature**	**Temperature**	**Remarks**
Air, Dry	1375	1280	Formation of protective oxide layer, mainly Al_2O_3 above 1000°C
Air, Moist	1300	1100	Steam reduces life as it disturbs oxide layer
Hydrogen (H_2)	1350	1250	Does not harm
Nitrogen (N_2)	950	900	Lifetime reduced as nitrogen reduces protective layer
Cracked Ammonia (NH_2)	1200	1100	Lifetime reduced as nitrogen reduces protective layer
Carbon Monoxide (CO)	1100	1000	
Carbon Dioxide (CO_2)	1100	1000	
Sulfur Dioxide (SO_2)	1150	1050	Higher life than nickel-containing materials
Hydrogen Sulfide (H_2S)	1050	950	Higher life than nickel-containing materials
Reducing Gas A[a]	1150	1000	
Reducing Gas B[a]	1050	1000	
Reducing Gas C[a]	1050	1000	
Hydrocarbons			
Halogens (Cl, F)			Harmful

	Nikrothal CR_2O_3		Molybdenum Disilicide SiO_2	
Atmosphere	Temperature	Remarks	Temperature	Remarks
Air, Dry	1250		1700	Favorable
Air, Moist	1250		1700	Favorable
Hydrogen (H_2)			1350(dry)-1460 (moist)	Depends on water
Nitrogen (N_2)			1600	Inert
Cracked Ammonia (NH_2)			1400	
Carbon Monoxide (CO)			1500	
Carbon Dioxide (CO_2)			1600	Favorable
Sulfur Dioxide (SO_2)		Not resistant, use	1600	b
Hydrogen Sulfide (H_2S)		Nikorothal 40		
Hydrocarbons	900-1000	Intergranular corrosion (green rot) use Nikrothal 80		Carbon deposits must be removed[a]
Halogens (Cl, F)		Usually harmful		Not suitable
High Vacuum				

[a]Approximate constituents of the gases:

	CO	CO_2	H_2	N_2
A	10	–	15	balance
B	5	29	5	balance
C	20	–	40	balance

[b]Elements used in reducing atmospheres should be reoxidized by allowing them to operate in air for 1 hr at 1500°C.

oxygen in the air. The protective nature of this oxide layer is dependent on the structure and nature of the crystals. Foreign materials usually interfere with the formation of the oxide layer, causing a reduction in life.

Table 4-20 summarizes the properties of various Kanthal alloys with respect to chemical attack in various atmospheres. The nature of the material and of the protective layer is indicated, with the maximum reliable operating temperatures. As a rule, oxidizing atmospheres are favorable; reducing atmospheres presents hazards, except with the A-1 and D alloys where hydrogen does not harm. The highest-temperature material, molybdenum disilicide, whose protective coating is fused silica, resists CO, CO_2, SO_2 and hydrocarbons (these at a lower temperature) provided the elements are reoxidized from time to time by allowing them to operate in air for about one hour at 1500°C.

Element Lifetime

According to Kanthal, if an acceptable length of service life is taken to be a period of 2-6 months in continuous operation and under otherwise normal operating conditions, the following maximum element temperatures for the different Kanthal Super qualities can be specified:

Kanthal Super ST	-	1700°C
Kanthal Super N	-	1700°C
Kanthal Super 33	-	1800°C

The length of service life applies at the maximum temperature concerned. However, if the element temperature is lowered by only an insignificant amount, the elements will have a considerably longer service life. There are examples of Kanthal Super elements still operating satisfactorily after 10 years in continual use. However, it has been found that most customers order replacement elements after 1-3 years.

In continuous operation and at an element temperature of 1800°C, the service life of Kanthal Super 33 has been found to amount to at least eight weeks. In this connection, continuous operation in Kanthal Super furnaces means that the element temperature never falls below 800°C. In intermittent operation, when a temperature of 1800°C is maintained for a few hours in each period, the service life will amount to 50-60 periods or more depending on the length of the period, the shape of the temperature curve and the lower temperature limit. The service life in intermittent operation thus depends on the individual circumstances regarding the heating cycle, element dimensions, furnace atmosphere, construction of the furnace, etc.[5]

Surface Load

Kanthal states that the correct choice of surface load regarding the construction of the furnace walls, atmosphere and temperature is an essential factor to arrive at a satisfactory service life for the elements. The curves shown in Figure 4-5, which apply to furnaces in which the elements have unobstructed radiation into the furnace compartment, show the approximate element temperatures at varying furnace temperatures and element loads. For example, at an element load of 12.5 W/cm² and a furnace temperature of 1400°C, an element temperature of about 1500°C is obtained. With the same load at a furnace temperature of 1600°C, the element temperature obtained exceeds 1700°C, which means that Kanthal Super ST or Kanthal Super N elements are overloaded. In such cases, the load must be reduced to about 8 W/cm² to give the elements an acceptable service life. Figure 4-6 shows the corresponding surface load graph for Kanthal Super 33.

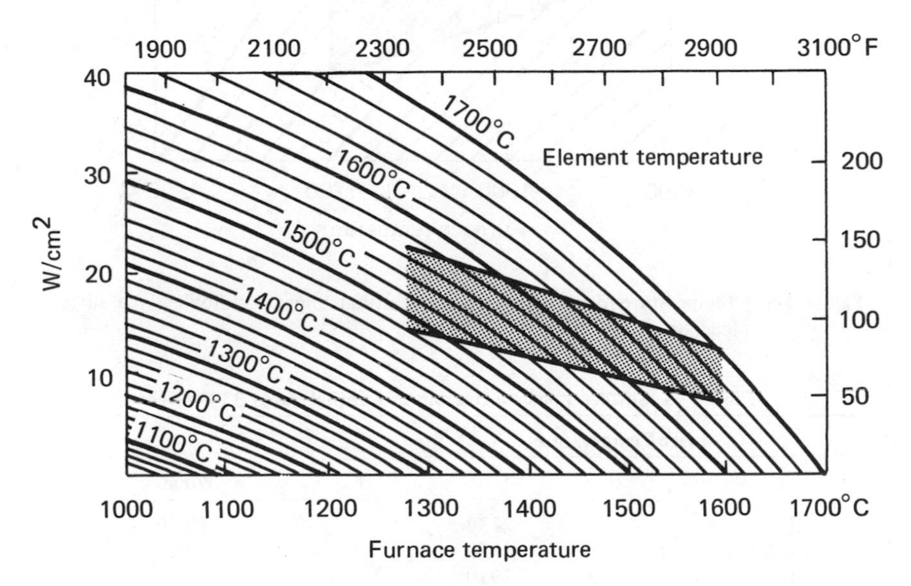

Figure 4-5. Temperature-loading diagram for Kanthal Super ST and N showing the surface load on the elements.[5]

According to Kanthal, because of the difference in temperature between the surface layer and core of the heating zone on Kanthal Super 33 elements, the surface load should not exceed 5-10 W/cm² at the maximum element temperature of 1800°C. Guiding values for the surface load on Kanthal Super 33 elements at different furnace temperature are shown in

Table 4-21, which is applicable providing the elements are able to radiate freely into the furnace chamber.[5]

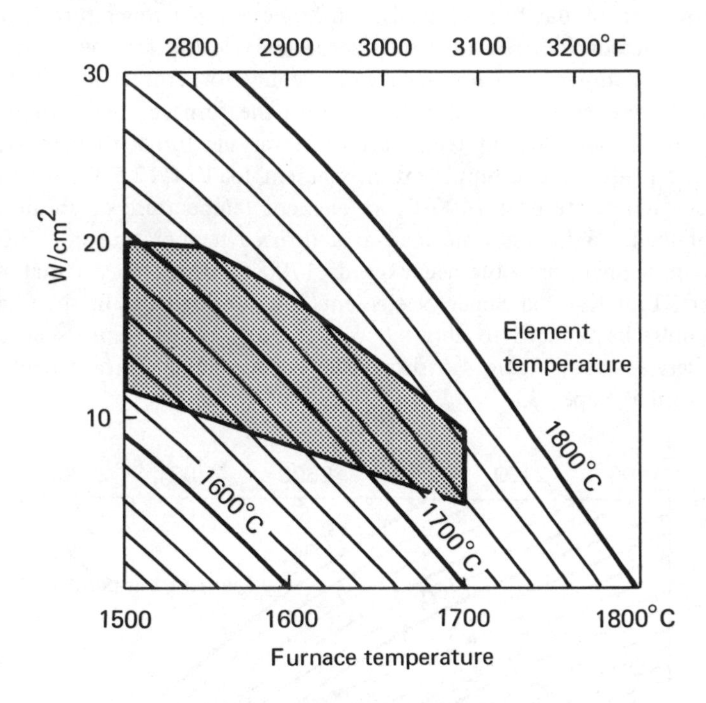

Figure 4-6. Temperature-loading diagram for Kanthal Super 33 showing the surface load on the elements.[5]

Table 4-21 Surface Load on Heating of Kanthal Super 33 Element

Furnace Temperature		
°C	°F	W/cm^2
1500	2730	12-20
1550	2820	10-20
1600	2910	8-19
1650	3000	6-12
1700	3090	5-10

REFERENCES

1. Ricciuti, R. A. "The Industrial Heating Equipment Industry," U.S. Department of Commerce (1969).

2. Burkhalter, Samuel. Private communication.
3. Johnson, F., Drever Company. Private communication.
4. Leeds-Northrup Corporation. Private communication.
5. "The Kanthal Super Handbook," Bulten-Kanthal AB, S-73401 Hallstahammar-Sweden, Kanthal Division.
6. Brennan, H. F. "An Account of Development 1925-1969: Vacuum Furnace Story," available through C. I. Hayes, Inc.
7. Hayes, C. I., Inc., Cranston, Rhode Island. Sales literature.
8. Diman, W. C. "Five Decades of Vacuum Furnace Progress," *Metal Prog.* (September and November, 1975), available through C. I. Hayes, Inc.
9. Diman, W. C. Private communication.
10. Brennan, H. F. "Improved Vacuum Furnace for Sintering Cemented Carbides," reprinted from *Prec. Metal* (1973), available through C. I. Hayes, Inc.
11. Herring, D., Lindberg Division, Sola Basic Industries. Private communication.
12. Ricciuti, R. A., U.S. Department of Commerce. Private communication.
13. Industrial Heating Equipment Association, Arlington, Virginia.
14. Smith, Merle, Eurotherm, 1100 Brush Hill Lane, Lake Zurich, Illinois. Private communication.
15. "Economic Report of the President (transmitted to Congress) and Annual Report of the Council of Economic Advisors" (Washington, D. C.: U.S. Government Printing Office, 1975).

5

SINTERING OF POWDERED METALS*

INTRODUCTION

The powder metallurgy industry is divided into two segments: the production of metal powders, and the fabrication of parts from the powders. The companies in the metal powder and the metal powder parts segments of the industry are generally independent of each other, although some companies are active in both areas. The metal powder industry is a commodity-type bulk marketing business with repetitive sales of a limited number of materials; the powder metal parts industry, on the other hand, makes parts almost entirely on a custom design basis for component and equipment manufacturers. Powder metallurgy also represents one of the markets for mechanical and process equipment. Their equipment suppliers manufacture specialized process equipment for use in the production and fabrication of these powders.

The metal powder industry segment encompasses the manufacture of powders of elemental metals or alloys. The raw materials range from metallic ores to metal scrap and refined virgin metal. Powder manufacturing methods vary with the producer. For an individual powder, any of the several physical and/or chemical methods can be employed, including mechanical comminution, spray atomization, chemical reduction, electrolysis, or precipitation. The objective is to produce powders with appropriate characteristics for operations such as molding or sintering.

The powder metal parts industry processes metal powders into shaped parts. Again, the most common processing treatment consists of cold pressing a blended powder composition to a "green" shape (compact) and sintering the compact under appropriate conditions of temperature, time

*By: W. E. Jacobsen, B. Bovarnick

and pressure so that part dimensions are controlled and the desired physical properties achieved. Post-sintering treatments (such as coining, repressing and resintering, impregnation, or surface coating) enhance the physical properties and service performance of the finished part.

Powder metallurgy makes it possible with little or no machining to fabricate metal powders to geometrically complex parts having essentially finished dimensions. Within the broad limits of the process, any metal powder material can be fabricated to a part shape. The process especially permits the production of parts with characteristics or compositions difficult or even impossible to produce by any other process, for example, cobalt-bonded tungsten carbide materials, porous metal bearings (subsequently impregnated with oil), iron-copper compositions, and tungsten-silver contact alloys. The particulate nature of the powders also makes them suitable for chemical reactants or for incorporation into nonmetallic materials such as plastic or concrete.

There are five metal powders produced in significant quantities–iron, aluminum, copper, nickel and tungsten (primarily for tungsten carbide). Each can be blended with alloy additions to obtain a wide range of grades and properties. Each metal serves a distinct segment of the powder-using and parts-production industries. As shown in Table 5-1, U.S. shipments of powder for the five metals amounted to $300 million in 1975. The value added by the structural parts-producing industry in 1975 was estimated as $210 million or on the same order of magnitude as the value of the metal powder shipments. Including the value of carbide cutting tools, estimated at $500 million, total industry sales appear to be over $800 million.

Table 5-1. Estimated U.S. Shipments of Important Metal Powders for 1975

Metal	Quantity (thousand short tons)	Change from 1974 (%)	Value of Shipments[a] ($ millions)
Iron	140	-22	50
Aluminum	45	-47	52
Copper	17	-38	30
Nickel	10	+3	44
Tungsten	6	-20	120
Metal	2.7		
Carbide	3.5		

[a]Including the value of the base price of the metal.

A related facet of powder technology, but outside the normal scope of this industry, is the production of many specialty metals whose extraction process yields a fine powder or sponge particle as the primary metal product. The more important among these metals are the refractory metals and the reactive metals. Except for tungsten, most of which is converted to carbides, and tantalum, most of which is converted to sintered capacitor anodes, the powders of these relatively exotic metals are normally converted into billets for subsequent metalworking into mill shapes.

Poor economic conditions from 1974-76 interrupted the historic pattern of rapid commercial growth of powder metallurgy. Despite the poor economic environment, technical development of the art has continued, and the newer areas of powder forging and hot isostatic pressing are gaining acceptance for industrial utilization of the process. The trend to increased shipments and sales in 1976 and 1977 indicates potential return to their historic growth patterns as economic conditions improve.

In order of quantity of powder shipments in 1975, five metals of major commercial importance as powder or for subsequent sintering to engineering components were iron, aluminum, copper, nickel and tungsten as the metal and the carbide. Each metal powder is important for different types of applications. Other powders growing in importance are stainless steels, molybdenum and, more recently, superalloys.

Iron powder had the greatest quantity of shipments in 1975–140,000 tons. There are only five important iron powder producers who serve the North American markets; three in the U.S. two in Canada. The major consumption was for sintering of molded parts for use in structural components and friction devices in mechanical equipment, and reached an estimated level of 105,000 tons or about 75% of 1975 iron powder shipments. This quantity was a substantial decline from the 1974 levels of 136,000 tons or 76% of iron powder shipments, which had also declined from the preceding record year of 1973. The principal end use of iron powder parts occurs in the automotive industry, which normally takes about two thirds. The number of powder parts fabricators continues to increase, now numbering about 180 plants of which 25% are custom fabricators providing parts to customer order, and the balance are captive fabricators producing for parent company consumption, as well as for external sales.

The metal powder having the next largest quantity of shipments during 1975 was aluminum, with estimated shipments of 45,000 tons, the lowest in 10 years. Aluminum, in the form of atomized or granular powder, usually constitutes about 80% of aluminum powder and is primarily used as a chemical reactant. The previous year's shipments were 80,000 tons total and 64,000 tons of atomized powder. Consumption of aluminum powder had reached the record level of 138,000 tons in both 1968 and 1969. Aluminum powders are produced by the primary aluminum companies as well as by several aluminum secondary remelters. Its consumption in conventional powder metallurgy for sintered aluminum parts is believed to have approached 300 tons, a substantial gain. Several metal powder parts fabricators, working in conjunction with Alcoa, have been developing the methodology, and a number of parts have been selected for production.

Copper powder for bearings was the first metal powder used commercially in sintered powder parts, about 50 years ago. In conjunction with parallel developments for other metals, the copper powder metallurgy industry grew to a record in 1966 of 33,000 tons which was approached in 1968 and again in 1973. Since then, copper powder shipments have declined to 27,000 tons in 1974 and to about 17,000 tons in 1975, the lowest level in 20 years. There are three general types of copper powder: (1) granular powder of unalloyed copper for powder metallurgy structural parts, bearings and friction materials; (2) copper-base alloy powders for similar applications; and (3) flake, pigment and paste grades of powder in copper and copper-base alloys. These classes of copper powder typically amount to 60, 30 and 10% of total copper powder. Price increases and price fluctuations of copper-base metal have caused considerable uncertainty in the future outlook for gorwth of consumption of copper powders. Among its several suppliers, AMAX is the most prominent.

The principal end uses for copper powder in sintered parts are: bearings, which consume about 75% of the granular powder, and structural parts and friction materials, which are all predominantly consumed in the automotive and domestic appliance markets.

Within the past decade, nickel powder has gained importance in the U.S. for powder metallurgy. Consumption increased from less than 3,000 tons in the mid-1960s to just over 10,000 tons in 1975. Essentially all the nickel powder is imported from Canadian plants of the International Nickel Company or Sherritt-Gordon Mines. The former produces the nickel powder by thermal decomposition of the carbonyl while the latter uses its proprietary reduction of ammoniacal sulfate solutions. The recent growth of nickel powder can be attributed to growing interest in a wide variety of specialty nickel-bearing products, such as nickel and nickel alloy rolled powder strip, nickel-cadmium battery porous electrodes, nickel-alloyed steel powder parts, stainless steel powder for porous filter media and, most recently, hot forged preforms and the hot isostatic pressing of nickel base and, for jet engine applications, nickel-alloyed super-alloys.

The shipments of tungsten metal powder products in 1975 amounted to 5,500 out of 6,300 tons for all uses of tungsten. About 1,300 tons were metallic powder, and 4,200 were carbide powder, of which one sixth was crushed-cast or crystalline powders and the balance carburized from the metal powder. The consumption of tungsten in 1975 declined almost 40% from record consumption of 10,000 tons in 1974 for all uses, consumption of carburized powders declined only about 29% from 4,900 to 3,500 tons. The dominant factor in the U.S. tungsten supply is the government-operated stockpile, which is used to balance out fluctuations of supply to meet domestic needs and, frequently, for export as well. The tungsten supply in 1975 was composed of 3,500 tons of imports with Canada supplying about 20% and domestic mine production of about 4,300 tons. A tungsten mine at Bishop, California, operated by Union Carbide Corp. and a molybdenum mine at Climax, Colorado, operated by AMAX Inc., are the principal domestic sources of tungsten. The principal use of tungsten is in cutting tools and wear-resistant materials which have been taking an increasing proportion of total tungsten, reaching a level of 64% in 1975. About 40 companies are active in the production of tungsten metal and carbide powder products, the most prominent being Kennametal, with current annual sales greater than $100 million, and General Electric, with somewhat fewer sales in carbide products but with wire production in addition. The combined value of sintered tungsten metal and tungsten carbide products for 1975 were estimated to be about $500 million. The cumulative market value of sintered metal powder parts shipments for 1975, excluding tungsten powder products, amounted to approximately $210 million, a decline of about 20% from the preceding year. The value of sintered tungsten products, particularly tungsten-carbide cutting tools, is estimated to be about $500 million, a decline of about 20%. The cumulative value of the metal powder industry shipments, including both powders and sintered parts, is estimated to be in excess of $800 million. With reasonable projections for the rate of economic recovery, the shipments of this industry should have a cumulative market value in excess of $1 billion dollars in the near future.

The growth of powder metallurgy in Japan has followed a similar pattern. Consumption of powder from 1960-1975 increased at differing rates for diverse products. Total powder consumption increased at an average annual rate of 15% to about 23,000 tons, while sintered structural parts increased at a rate of 20% to 16,000 tons. Sintered bearings increased at an average rate of 9% to about 4,400 tons. Electrical contacts had a slow growth to 160 tons in 1975 for a rate of 7% annually, well below Japan's average GNP growth rate. Tungsten carbides had the slowest growth with an average rate of 6% annual growth to a level of 900 tons in 1975.

Nation-wide energy requirements for sintering the various metal powders are summarized in Table 5-2. Energy requirements for producing powder metal parts depend strongly on sintering temperatures. Another significant

Table 5-2. National Energy Requirements for Sintering Metal Powders

Metal	Gross Energy (10^9 Btu)
Iron	4200
Aluminum	24
Copper	450
Nickel and Nickel Base	320
Tungsten Metal	306 (electric[a])
Tungsten Powder Parts	24
Hydrogen-Sintered Tungsten Carbide	102 (electric[a])
Vacuum-Sintered Tungsten Carbide	24 (electric[a])
Total Energy for Sintering Powder Parts	5420

[a]1 kWh equivalent to 3400 Btu (electric).

factor, however, is sintering furnace production throughput. A typical example–fabricating a piston ring from iron powder–indicates that energy costs (including operating and idling duty cycles) are on the order of one cent per unit produced in a gas-fired sintering furnace. The enrgy cost is about 6% of the total production cost of each piece. Corresponding energy costs per unit sintered in an electric furnace would be higher by a factor of about six. Higher temperature sintering of tungsten is generally accomplished with supplemental electric resistance heating.

Although at present, gas furnace operations show favorable energy costs, the higher-temperature sintering operations (approximately 1500-2800°C) require electric vacuum techniques.

GENERAL PRINCIPLES

The powder metallurgy industry has two main functions: production of metal powders, and their conversion to engineering applications. (See uncited references for general works on powder metallurgy).

Metal powders are produced with particles having controlled characteristics suitable for use in subsequent processing. In the conversion, the principal concern is consolidation of the loose powder to solid materials with controlled porosity and density, as the processed powder metals approximate those of the corresponding compositions which have been conventionally cast or wrought.

Metals in powder form permit preparation of unique materials with extraordinary properties that cannot be attained in any other technical art. In addition, metal powders facilitate the preparation of known materials in physical forms, which cannot be made by any other fabricating methods, or in complex shapes that are difficult or expensive to make by conventional methods. The preparation of known materials to simple shapes can also be accomplished at lower cost with powdered metals than from the better known and more common manufacturing technologies. As a result, the field of powder metallurgy has attracted a high level of scientific, engineering and industrial interest. Industy has acted in consort to stimulate the development of the requisite technology of rapid growth and continuing commercial success. From beginnings in diverse areas that were essentially insignificant about 50 years ago, powder metallurgy in the United States is rapidly approaching a gross annual market on the order of $1 billion.[1]

Powder metallurgy begins with the production of the powders themselves. A large number of methods have been developed for this, most of which can be grouped into three general classes: mechanical, physical and chemical. The method chosen in a given case depends on the metal chemistry, the nature of the raw material, the physical properties of the powder, and the degree of control that can be obtained for particle sizing of the product powder.

Among the mechanical methods are machining solid stock to chips by conventional metal-cutting processes; grinding or milling coarse-sized feed to finer particle sizes by stamping or milling; and mechanical shock. The physical processes employ melting to form the powder particles by granulation during solidification; shooting the metalthrough a screen or rotating shotting disc into a shot tower; or spray atomization of the melt with a controlled jet. Another class of physical processes is the condensation of metallic powder from the vapor state for those elements that have high pressures or in which high vapor pressures can be induced by subjecting them to very high temperatures, as in an electric arc. Condensation methods usually allow powders to be obtained with very low impurity levels and very fine particle sizes. Thermal decomposition of an unstable compound such as a metallic carbonyl, or a quasi-compound such as a metallic hydride, is also used to physically produce powders. The physical processes were originally developed for lower melting metal and alloys and have since been extended to those with melting points as high as iron.

Among the chemical methods are electrolysis, with solution and reprecipitation of the metal as a friable electrodeposit; precipitation from solution by an exchange reaction or hydrogen reduction of the solute metal ions; ionic reactions between the metal salts; and the reduction of the solid metallic oxide by an active reductant such as hydrogen or carbon. This is the most common and readily accomplished process, if the metal is easily reduced without losing its particulate characteristics. Another chemical approach that has been employed for some alloy metal compositions is the attack of the metallic grain boundaries, but this method has never had serious commercial acceptance.

With the wide variety of powder production methods available, there are virtually no solid metals that could not be made into powder. A listing of the resulting powder form types is presented in Table 5-3 for the metals of commercial importance in the American powder metallurgy industry. The powder particles produced by any one of the powder manufacturing methods will have external configurations and internal structures that are characteristic of the specific method. The powders are produced with a distribution of particle sizes that can be controlled to some degree by adjustment of the powder manufacturing conditions.

Table 5-3. Metals of Commercial Importance in U.S. Powder Metallurgy[1]

Metals	Source	Types of Powder by Process
Aluminum	Domestic	Atomized, ground, flake, paste
Brass	Domestic	Atomized, ground, flake, paste
Bronze	Domestic	Atomized, ground, flake
Cobalt	Imported	Precipitated, gas reduced
Copper	Domestic	Precipitated, electrodeposited, atomized, gas reduced, ground, flake, paste
Iron	Domestic	Sponge, gas reduced, atomized, electrodeposited, ground, carbonyl
Lead	Domestic	Atomized
Molybdenum	Domestic	Gas reduced
Nickel	Imported	Gas reduced, carbonyl
Stainless	Domestic	Atomized, ground
Superalloys	Domestic	Atomized
Tantalum	Domestic	Precipitated
Titanium	Domestic and Imported	Sponge
Tungsten	Domestic	Gas reduced
Tungsten Carbide	Domestic	Carburized, ground
Zinc	Domestic	Atomized

The next step in powder metal processing, except for those powders used in the powder form, is to convert the loose powder particles to a coherent solid body having a selected final density and residual porosity. This has conventionally been accomplished by a two-stage process: (1) consolidating the loose particles of the powder to the desired geometric shape; and (2) treating the consolidated shape at elevated temperatures to enable sintering between the particles. New practices will include accomplishing both functions in a single step with the development of hot-pressing processes, or adding a third step, the hot forging of a sintered shape. Both developments are designed to obtain a sintered powder part which has better metallurgical and therefore mechanical properties than can be achieved by conventional practice.

Consolidating the powders is variously called briquetting, molding, compacting, or pressing; the pressed shape is then known as a briquette or compact. The powders are prepared by mechanically screening and blending selected compositions, size ranges and powder types. The prepared powder is put into the cavity of a molding or compacting die specially designed to produce the desired part shape and dimensions. The actual consolidation takes place by applying pressure to the powder through top and bottom punches of the dies. The required pressure is a function of the powder being pressed and the desired density of the pressed part. The range of compacting pressure is usually between 10,000 and 200,000 psi and, most frequently in commercial practice, between 40,000 and 100,000 psi for structural parts. The powder compact is described as "green" when it is in the as-pressed condition. It has sufficient strength in the green state to maintain its dimensions during subsequent handling, but it is weak and friable.

The pressing loads and forces, and in some cases the rate of load application applied to the powder during compaction, influence the subsequent sintering treatment. Sintering is the most important factor in obtaining bonding between the particles and controlling the final properties of the material. The sintering treatment is a time-temperature combination cycle of heating, soaking temperature and holding time, and cooling. During sintering, mass transfer occurs between the particles by surface and volume diffusional processes to form and enlarge metallurgical bonds between adjacent particles. In some cases, liquid phases are present so that mass flow mechanisms can also be involved. Each metal powder has its own nominal sintering temperature regime, depending upon the basic powder in the mix and any alloying additions. For aluminum-base powders, the temperature should be about 600°C, for copper about 950°C, for iron or nickel 1150-1250°C, for tungsten carbides about 1450°C, and for metallic tungsten about 2800°C. The time at temperature for sintering is a less critical parameter than the temperature itself. Times will typically range from 20 minutes up to one hour in commercial production, but more extended times could be used in developmental programs.

Compared to conventional practice, the developments of hot pressing, such as hot isostatic pressing or hot forging, use somewhat higher temperatures so that the metallurgical phenomena can be accelerated and the processing completed in the shortest possible time. The acceleration is essential to gain requisite process economics through high productivity, because capital equipment costs tend to be high.

EXISTING APPLICATIONS

Use of Powdered Iron

More iron powder was consumed than any other in 1975, making it a mainstay of the powder metallurgy industry. Shipments were reported at 140,000 tons in 1975, and are expected to increase with the growth of manufacturing in the economic recovery, to the high shipment levels of previous years.[2]

Shipments of metal parts made from iron powder were reported at 105,000 tons in 1975, a decrease of 24% from 1974.[2] The largest market for iron powder is in the manufacture of molded parts for use as structural and friction components in mechanical equipment (Table 5-4).

Table 5-4. Iron Powder Consumption by Application

Application	Iron Powder Consumption (%)
Molded Parts	70-80
Welding Electrodes	15-20
Flame Cutting and Scarfing Operations	5-10
Electronic and Magnetic Materials and Chemical Use	balance

The consumption of iron powder in molded parts had grown at an annual compounded rate of 13%/yr for the past 20 years, but less than 10% annually for the past 10 years. Iron powder consumption in non-molding applications (welding electrodes and flame cutting and scarfing) by comparison, have grown at a compounded rate of only 4% per year. The estimated shipments of iron powder are shown in Table 5-5.[3]

Parts molded from iron powder compete with structural parts made by other part-shaping operations including (primarily) machining, forged parts, cast parts and fabricated assemblies. The higher and gradually increasing costs of competitive processes, the improved powder metallurgy equipment, and the development of improved manufacturing technology to take advantage of potential cost benefits are factors that will contribute to the growth of demand for iron powder parts. In addition, serious efforts are being devoted to the intensive development of single-step forging of powder preforms, which may be partially or fully sintered.[4] As the cost of powder is reduced and properties are improved, hot-forged preforms are expected to be able to compete more broadly with small shapes forged from solid steel stock. There are now about 20 hot-forged parts in standard production.[2]

Table 5-5. Estimates of Iron Powder Shipments, 1965-1975[3]

Year	Total (1000 tons)	Molding Powders (1000 tons)	Share (%)
1965	86.8	54.6	63
1966	100.0	69.0	69
1967	94.0	63.0	67
1968	112.5	78.0	69
1969	126.9	89.1	70
1970	114.6	79.1	69
1971	127.9	93.6	73
1972	154.4	118.2	76
1973	194.5	151.5	78
1974	178.9	136.0	76
1975	140.0	104.7	75

The industry expects that annual consumption of iron powder in parts will increase to about 150,000 tons in the late 1970s and could reach 250,000 tons by the mid-1980s. Two important, related reasons for this growth will be increased consumption of molded parts in the automotive industry, and lower comparative costs for iron powder parts. The results are expected to stimulate expanded utilization in the production of mechanical components.

Existing capacity for production of iron powder is about 250,000 tons/yr; no new capacity is now committed. Over-capacity combined with the decrease in the 1975 market is believed to have caused expansions plants to be suspended. Lowered pricing by new suppliers is anticipated as part of their strategy to gain market entry when conditions would be appropriate.

The major producers of molding-grade iron powders and their present capacities are listed in Table 5-6. Hoeganaes Sponge Iron Corporation, Riverton, N.J., with a capacity of 140,000 tons, dominates this market. The company, now a subsidiary of Interlake Steel, was founded by Hoeganaes Corporation of Hoeganaes, Sweden, an important supplier of iron powder to world markets outside the United States and also of basic iron ore raw material used by the Riverton plant. The U.S. demand for imported iron powder declined after the Hoeganaes Riverton plant was started in 1954. The initial capacity of 20,000 tons/yr was increased to 40,000 tons/yr about 1960, to 90,000 in the mid-1960s, and to 140,000 in the early 1970s. Hoeganaes' existing capacity employs the carbon reduction of iron ore to sponge iron; its new capacity employs the atomization process.

The Metal Powder Division of Domtar Chemicals of Canada has plants in Canada and at Ridgway, Pennsylvania; Domtar supplies atomized and reduced powders. A. O. Smith Company, Milwaukee, Wisconsin, began commercial iron powder production in the early 1960s, and has formed a

Table 5-6. Major Iron Powder Suppliers[1]

Present Suppliers	Capacity (1000 tons/yr)	Manufacturing Process
Hoeganaes Sponge Iron Corp.	90	Sponge iron
	50	Atomization of melted steel scrap
Pyron Division of AMAX, Inc.	20	Hydrogen reduction of mill-scale and iron ore
Metal Powder Division, Domtar Chemical	25	Atomization of melted steel scrap
A. O. Smith Company	25	Atomization of melted steel scrap
Quebec Metal Powders, subs. Quebec Iron & Titanium Corp.	40	Atomization of molten pig iron
Total	250	

joint venture with Inland Steel for expanded production. The company introduced and has promoted the use of high-compressibility powders made by atomization rather than electrolysis. These powders have been an important factor in allowing production of larger and more complex shapes. The company's capacity is intended primarily for the production of its molding grade powders. Captive production of iron powder is also being done on small scales for proprietary uses by a few companies.

Three Canadian companies announced in 1968 plans to install iron powder production facilities. Canadian Petrofina established Fina Metal Limited as a subsidiary, with the installation of a 15,000 ton/yr plant in the Montreal area, at a cost of $5 million. Fina Metal employed hydrogen reduction for making iron powder from nearby low-cost high-grade ores, but has since shut down. The Peace River Mining and Smelting Company undertook construction of a 50,000 ton/yr plant in the Windsor, Ontario, area at an estimated cost of $15 million and employed hydrogen reduction of ferrous chloride. The raw material for this process can be either hydrochloric acid pickle liquor from steel plants or low-grade steel scrap dissolved in hydrochloric acid. This plant, however, never operated. Quebec Metal Powders is a subsidiary of Quebec Iron and Titanium Corporation (QIT), a joint venture of Kennecott Copper and New Jersey Zinc. Quebec Metals was on-stream by the end of 1968 and began shipments of specification product. The company's reported investment cost of $7 million was lower for the amount of capacity than Fina Metal's or Peace River's because its plant is an adjunct of the existing ironmaking facilities at QIT. Quebec Metals is making powder from the pig iron produced as a by-product of titanium oxide concentration. The company achieved savings by atomization of the QIT product directly from the molten state without intermediate cooling and remelting.

The steel industry pays close attention to powder metals, which is a robust industry, and several steel companies are now becoming involved. Interlake Steel, for example, has taken a major investment position in Hoeganaes; Inland Steel and A. O. Smith have announced joint participation in the existing powder operations and plans for a new 100,000 ton/yr plant; Republic Steel is active in fabrication of basic steel shapes directly from powder; and U.S. Steel is known to be studying the industry.[1]

In the fabrication of parts from iron powder, about 180 companies are actively engaged, ranging in size from the largest industrial firms to small independent shops employing as few as 20 men.[2] An estimated three fourths of the fabricators are captive facilities; the balance are independent shops that manufacture parts as custom vendors on a contract basis. Many of the captive plants also produce parts as custom vendors for outside customers. The capabilities of the fabricators vary considerably and depend primarily on equipment capacity rather than technological proficiency, which is generally quite good throughout the industry. Skill and capability in the production of iron powder parts is usually associated with corresponding skill and proficiency in the production of copper powder parts; the same firms are active in both areas.

The major iron powder parts producers are the captive operations of General Motors (Delco-Moraine Division), Chrysler (Amplex Division), and Ford (General Parts Division). A number of independent automotive parts suppliers also have captive iron powder parts manufacturing operations, including Federal-Mogul, Clevite and Ametek.

Several relatively important contract fabricators have been absorbed by larger corporations in recent years; among them are Clezite, acquired by Gould; Burgess-Norton, recently acquired by Ametek; International Powder Metallurgy, now a division of Allegheny-Ludlum; Merriman, Inc., acquired by UTD Corporation (formerly Union Twist Drill), which in turn was merged into Litton Industries; and Presmet and Bound Brook Bearing both acquired by GKN of England.

Most of the independent fabricating shops are privately owned and have sales of less than $5 million a year; none has sales as high as $20 million per year. The organization by a few individuals of a small independent company for the contract fabrication of powder parts is still a relatively frequent occurrence.

Use of Powdered Aluminum

Industrial usage of aluminum has shown modest growth in recent years. The oldest applications of aluminum powder are its use as a flake and paste pigment in metallic paint, and as a chemical reactant in the reduction of reactive metals. It is also used for roll-bonding coatings onto steel sheet and other flat mill products for corrosion protection.

In the past few years, two chemical reactant applications have become increasingly important. Increased use of aluminum alkyl catalysts has resulted from pressure to change detergents to more biodegradable compositions. Precise data is not available, but it is estimated that consumption of aluminum alkyl has reached 20 million pounds in recent years. Similarly, the development of slurried mining explosives with a 5-20% aluminum powder content has given rise to a market that did not exist years ago. Aluminum consumption in this application is estimated at 40 million pounds currently but should increase with mining activity. Another dramatic growth of aluminum powder shipments in past years was due primarily to the substantial increases in military consumption. In the military market, aluminum powder is used in ordinance pyrotechnic devices and as an important constituent in solid rocket fuels. Military use grew faster than 100%/yr in 1965-1967, and by 1968 consumed more than 100,000 ton/yr. This market, however, has now declined to insignificant levels.[1]

The primary domestic suppliers of aluminum powder are Alcoa, Reynolds Metals Company, Alcan Metal Powders, Inc. and U.S. Bronze Powders, Inc. The standard powder is granular and is made by atomization from the melt. Particle shape and size are controllable to broad limits by the atomization and subsequent treatment conditions. A constant problem encountered by this process is the threat of explosion due to the energetic oxidation activity of fresh aluminum powder fines upon exposure to ambient atmosphere. The producers take elaborate precautions to minimize this threat. Estimates of aluminum powder shipments are shown in Table 5-7. Data of recent years indicates that atomized powder represents about 80% of total powder.

Use of Powdered Copper

Over the long term, shipments of copper powder have grown in line with the gross national product; from 1950 to 1968, copper powder shipments grew 6.5%/yr, from 10,000 to 31,000 tons. Over the short term, however, copper powder shipments fluctuate in line with copper metal prices, which in turn reflect important fluctuations in the copper market supply and demand situation. Copper powder pricing is led by AMAX, the leading supplier, and essentially amounts to a toll price premium above the base price for copper. This premium was about 13¢/lb in the 1950s, rose during the 1960s, and has increased to about 23.5¢/lb delivered.

Copper powder consumption is divided among unalloyed granular copper powder (60%) brass- or bronze-alloyed granular powders (30%), and flake powder (10%). The market shares of unalloyed and alloyed granular copper

Table 5-7. Estimates of Aluminum Powder Shipments, 1965-1975[3,5]

Year	Total (1000 tons)	Atomized
1965	29.4	
1966	55.7	
1967	117.2	
1968	138.2	
1969	138.0	
1970	102.3	
1971	83.8	
1972	113.8	
1973	128.7	104.7
1974	80.0	64.1
1975[a]	45.0	34.0

[a]Estimated.

powders, relative to each other, have been consistent over the years, within narrow limits. The consumption of flake powder, used mainly for ink, paints and decorative purposes, has been steady in a narrow range of 200-300 ton/yr since 1960.

From the mid-1960s until 1975, shipments of copper powder had been within 10% of 30,000 tons and shipments of parts made from copper powder within 10% of 27,000 tons as shown in Table 5-8. Shipments in 1975 were about 17,000 tons of powder and 15,000 tons of parts. Industrial consumption of copper powder became significant after World War I, with the commercialization of porous bronze bearings made by pressing and sintering copper or copper-alloyed powders. Subsequently, the application of copper powder was expanded by the development of electric motor brushes based on copper or copper-alloyed powders. Today, the copper powder parts market is divided into three segments–porous bearings, structural parts, and friction materials for brake and clutch components bearings, accounting for three-quarters of the market for copper powder in parts.[6]

Among suppliers now active in the copper powder market, the principal producer is United States Metals Refining Co., a subsidiary of AMAX INC., which reportedly converts to powder copper anode from electrolytic refining. Greenback Industries, a subsidiary of Handy & Harman, Inc., has established its position as the second largest supplier by careful attention to the requirements of the automotive industry, the principal market for copper powder. The third largest supplier is Glidden, whose copper powder facilities are located in Hammond, Indiana. Alcan Powders (formerly Metals Disintegrating Corp.), a subsidiary of Aluminum Limited of Canada, markets

Table 5-8. Estimates of Copper Powder Shipments, 1965-1975[3]

Year	Total (1000 tons)	Granular (1000 tons)
1965	31.0	
1966	33.0	
1967	28.0	25.0
1968	31.1	28.0
1969	30.3	28.2
1970	23.8	21.8
1971	26.3	24.1
1972	27.0	24.6
1973	32.3	29.4
1974	27.3	24.0
1975	17.0	15.0

a wide variety of metal powders, including copper, only some of which it manufactures itself.

Copper powder parts are produced by two groups of fabricators. Friction materials are generally produced for captive use in the manufacture of brake and clutch components and assemblies. With metallic friction materials, the compositions used and process treatments employed are closely guarded secrets, and to a large degree they determine the effectiveness of the devices in which they are used. Consequently, primary powder suppliers do not supply the blended powder mixtures, and outside vendors seldom are asked to manufacture the components.

Copper powder porous bearings and structural parts are produced primarily by original equipment manufacturers (OEMs) for both captive consumption and external sales, and by contract fabricators to customer order. In addition, some captive and independent powder parts fabricators produce porous bearings to standard specifications as catalog items. The major OEM producers of copper powder parts and bearings are Amplex Division of Chrysler Corp. and Delco-Moraine Division of General Motors Corp., both of which produce parts and bearings for internal use and external sale. Generally, the contract producers of sintered copper powder parts are the same companies that produce iron powder parts; the technology used is essentially the same. The major independent producers of porous bearings are automotive parts suppliers, such as Federal-Mogul and Clevite. Smaller, independent producers include Bound Brook Bearing, and Wakefield Bearing Co., Wakefield, Massachusetts.

Use of Powdered Metal

Of the several metallic powders having important commercial significance, nickel powder is the most recent addition to the group. In the early 1960s, nickel powder was still considered a specialty metal powder and consumption was under 3000 tons. Principal uses of nickel powder at that time were in areas other than powder metallurgy for sintering to parts. The principal use was in preparation of nickel alloy billets of specialty alloys for electronics and instrumentation.[7] The alloy billets were subsequently converted to strip, rod and wire by conventional hot and cold working metallurgical operations.

The advantages offered by nickel powders compared to conventional melting and ingot casting practice was that the nickel powders were available in high purity. The alloy compositions could be prepared to precise chemical specifications, and the contamination that was introduced into a melt from refractories, fluxes, slags and melting additives could be avoided. The nickel powder billets offered the opportunity of producing alloys to stringent specifications for electron tube construction, resistance alloys, controlled expansion alloys and magnetic alloys. In addition, nickel powders were also used in chemical processes such as electroplating and catalysis.

In the mid-1960s, Canada undertook the production of coinage from nickel strip rolled from nickel powder obtained by Sheritt-Gordon from its proprietary process of hydrogen reduction of ammoniacal sulfate solutions.[8] At about the same time, the potential of nickel-cadmium batteries as a power source for cordless appliances was recognized. A parallel event in powder metallurgy was the development of atomization processes for production of nickel-bearing, high-strength alloy steel powders with high compressibility. These powders offered the compacting advantages of high-priced electrolytic iron powders at the price range of reduced sponge iron or mill-scale powders. These developments stimulated strong interest in nickel powder. In anticipation of future demand for nickel powder, International Nickel installed a new powder production facility in Canada with substantial capacity for producing carbonyl nickel powders.[9] The expectations have been realized by the rapid growth in consumption of nickel powder since 1970 (Table 5-9).

The demonstrated improvement of mechanical properties in sintered nickel steel alloys coupled with the prospective availability of high-quality nickel powders in large quantities led to extensive development efforts in powder metallurgy for nickel superalloys.[10] These efforts demonstrated that superalloys made from powders subsequently forged to full density could meet the alloy property specifications and be economically

Table 5-9. Estimates of Nickel Powder Shipments, 1969-1975[3]

Year	Quantity (tons)
1969	2,800
1970	3,100
1971	2,700
1972	4,800
1973	7,600
1974	9,800
1975	10,100

competitive. With this approach, one jet engine part now going into production requires 6.5 pounds of forged powder alloy to produce 1 pound of finished shape, compared with 19 pounds of cast alloy ingot.[11]

Use of Powdered Tungsten

Primary metallic tungsten is initially produced in the form of powder from its ore by chemical extraction process. The bulk of the tungsten is retained in the form of powder and used in the fabrication of tungsten carbide and heavy metal alloys. In addition, some powder is converted directly into billet and made into solid bodies or mill shapes in extended hot and cold metalworking operations; these are considered part of the tungsten metal industry rather than a part of the powder metallurgy industry and will not be discussed further. Tungsten powder shipments totaled 7700 tons in 1974 and 6200 tons in 1975.[2] About 60-70% went into the production of tungsten carbide for use in either cemented carbides or special tungsten-base alloy parts, and the balance into heavy metal alloys, tungsten wire, rod, dense forging blanks and mill shapes.

Cemented carbides are a specialty metal product in which very hard carbide grains (primarily tungsten carbide) are alloyed into a two-phase structure by sintering with a relatively ductile binder phase of metallic cobalt. Products fabricated from sintered tungsten carbide provide high abrasion resistance and high strength. Shipments of cemented carbides totaled about 3500 tons in 1975.[2]

The origin of cemented carbides goes back to attempts at the end of the nineteenth century to develop hard abrasive materials as substitutes for diamonds. Efforts by Moissan to synthesize diamonds by heating carbon with refractory metals in an electric arc furnace resulted in the production of extremely hard particles, which were subsequently discovered to be refractory metal carbides rather than the diamonds sought.[12]

As a result of these experiments, tungsten carbide was recognized as offering significant potential, and its development was undertaken and pursued by Osam Geselleschaft in Europe.

The earliest (and still significant) applications for tungsten carbide were dies for drawing fine wires, and cutting tools for cast iron and nonferrous metals. Cemented carbides for these applications were composed of tungsten carbide only with a cobalt binder.[13] The discovery that adding titanium carbide[14] or tantalum carbide[15] to the tungsten carbide substantially reduced degradation of the cutting edges allowed for the machining of steel. Since that discovery, the quality levels have improved so that cutting performance has been enhanced, and the machining of steel has become the largest use for tungsten carbide.

There has been a growing demand for sintered carbide for rock- and well-drilling tools in preference to the steel tools predominantly used in the past.[16] The use of tungsten carbide in wear- and abrasion-resistant parts in expensive equipment, to increase its lifetime and reduce downtime, has also been increasing. Another innovation in the use of tungsten carbide–the snow tire stud–had gained wide acceptance in the United States for some time, but the use of studded tires has begun to decline in the past few years. The most recent innovation in the use of the sintered tungsten carbide has been in structural applications, which can benefit from the metal's exceptional modulus of elasticity of 90 million psi, and compressive strength of greater than 500,000 psi. Applications of these properties are especially important in high-precision machine tools and steel rolling mills where stiffness and high rigidity under dynamic loads can be essential to attaining close tolerances.

There is no immediate prospect for replacement of cemented carbides in present uses, although the machining industry has made a continuing effort to utilize other materials for cutting tools.[16] Examples are the development of ceramic cutting tools based on aluminum oxide and titanium carbide tools. This effort is at least 10 years old and has met with limited commercial success in special machining problems, in some cases completely displacing tungsten carbide.

Tungsten carbide applications can be divided into the following broad categories according to service conditions to be satisfied: chip removal, wear and abrasion resistance, and high compressive strength. The chip-removal category includes all metal cutting operations: single-point, multipoint and continuous cutting such as drilling. Chip-removal applications are segregated into steel cutting and cast iron and nonferrous metal cutting. The wear- and abrasion-resistant uses encompass dies for the drawing of rod and wire, wear parts of mechanical equipment, snow tire studs and soft stone cutting tools. The high-compressive-strength category includes press operations requiring punch faces and die body liners, hard-rock drilling and mining tools, and armor-piercing anti-tank ammunition. The production of carbides can be roughly apportioned to the major areas of application, as given in Table 5-10.

Table 5-10. Major Applications of Carbides[1]

Application Category	% of Total
Chip Removal–Steel Cutting	60
Cast Iron and Nonferrous Metal Cutting	10
Wear- and Abrasion-Resistant	15
High Compressive Strength	15

The high-compressive-strength category has been the fastest growing segment of the industry; the growth from the transition and use of carbide in wear parts and mining tools is still in process, but the acceptance of carbide-studded tires should moderate in a few years. The initial step in the manufacture of cemented tungsten carbides is the production of metallic tungsten powder via the preparation of ammonium paratungstate (APT) as the first product of purification and extraction from tungsten ore concentrates. Tungsten metal powder is obtained from ammonium paratungstate by thermal decomposition of the APT to tungstic oxide, followed by hydrogen or carbon reduction of the oxide to metal powder. Throughout these thermal treatments, the time-temperature schedule is carefully controlled to attain desired particle sizes. Hydrogen-reduced tungsten powder is converted to the monocarbide by reaction with carbon in a graphite tube furnace at about 1600-1700°C. The carburized fine powder product is controlled according to the process treatment to yield selected sizes for various grades and applications.

In the production of cemented carbides, tungsten carbide (with additions of titanium and/or tantalum carbide, in the case of steel-cutting tool grades) is blended with cobalt, which is used in proportions ranging from 3-25% by weight. The blended powder is then pressed and sintered at 1400-1450°C in a hydrogen atmosphere or vacuum, depending upon the grade of carbide or the producer. During the sintering treatment, the cobalt becomes liquid, and the tungsten carbide powder particles are pulled together by the liquid phase, forming a tungsten carbide skeleton. After sintering, the carbide piece is at full hardness and is finished to precision dimensions by diamond grinding (the finishing of tungsten carbides is the greatest single use of diamond-grinding wheels).

The various grades of carbides have been classified by the Cemented Carbide Producers Association according to the properties they offer or the service conditions to be satisfied. Each manufacturer identifies the nominal service category for each grade of his material; a manufacturer may have more than one grade in a category or may recommend a grade for inclusion in more than one category. The grades are governed by the proportions of carbides, cobalt content, the carbide nominal grain size and the treatments employed in producing the finished carbides.

The major carbide producers have undertaken extensive efforts in product development to evaluate the performance characteristics and limitations of carbide materials. They have also made extensive efforts to improve the grades of carbides available. In recent years the trends have been toward more complex carbide proportions, finer carbide grain size, and more precise control of process operations to obtain improved product quality. A range of grades is now available to meet a wide variety of metal cutting and other service conditions.

The heavy metal alloys are tungsten-base compositions with specific gravities of 16 to 17. They contain 90% or more of tungsten and additions of nickel, iron and/or copper. The heavy metal alloy compositions have developed from the consolidation of binder systems. Parts made from the heavy metal alloys are custom fabricated essentially to shape by the standard powder metallurgy processes of blending, cold pressing and liquid-phase sintering.

Their significant strength levels permit the heavy metal alloy parts to be used in dynamically loaded, space-limited applications. They are used extensively as hinge and crank counterweights in jet aircraft flight control surfaces and rotating machinery and also are used in gyro rotors and shielding. Depleted uranium competes with tungsten-base heavy alloys in some applications and is available from NL Industries.

The domestic tungsten supply situation is dominated by Union Carbide Corp., which owns the most important mine in the country (at Bishop, California) devoted primarily to the production of tungsten ore. Union Carbide also has extensive facilities for beneficiation of tungsten concentrates and their conversion to APT and tungsten powder. The second largest domestic supplier of tungsten ore concentrates is Climax Molybdenum of AMAX which extracts and sells tungsten ore as a by-product of its molybdenum mining operations at Climax, Colorado.

Among significant Canadian producers is Canada Tungsten Mining Corp., Ltd., of which AMAX owns a substantial portion. It has been producing tungsten concentrate at a rate of about 4-5 million pounds a year. AMAX also owns the Mactung mine in Canada.

Historically, the price of tungsten has fluctuated violently with surges in supply and demand. The federal government made stockpile purchases until 1958. After the termination of the stockpiling purchases, world prices for tungsten concentrate fell to a low of about $8 a short ton unit (stu). One short ton unit of tungsten concentrate is equivalent to 1% of a short ton or 20 lb of contained tungsten trioxide (WO_2), which contains 15.86 lb of tungsten. A metric ton unit (mtu) of concentrate would contain 10 kilos of WO_3 or 7.93 kilos of elemental tungsten. In the early 1960s, they rose to around $20 a unit and reached record levels of $103/stu in September 1974.[17] Sales from the stockpile (which has been reduced to about 100 million pounds) by the U.S. General Services Administration have been set at shelf-item prices. The stockpile is increasingly important both as a source of supply for domestic consumption and for export to foreign consumers, as well as for price stabilization at times of unstable market conditions.

The industrial growth of cemented tungsten carbides led a number of producers to enter this business; today, the cemented tungsten carbides market is divided as follows:[1]

Company	% of Market
General Electric	20-25
Kennametal	20-25
Valeron	10-15
Fansteel	10-15
Allegheny-Ludlum	5-10
Teledyne	5-10
Adamas Carbide Corp.	5-10

There are also about 30-40 smaller companies. Carbide producers have been substantially retaining their shares of the market, so all have benefited by the growth in demand.

The major carbide producers generally purchase ammonium paratungstate (principally from Union Carbide), but periodically the larger producers undertake the captive conversion of purchased concentrate to APT. General Electric, Kennametal, Fansteel and Sylvania produce not only tungsten carbide but also other tungsten products, such as powder, heavy metal alloys, mill shapes, or incandescent light bulb filaments, for sale or captive use in their own product lines. Producers of tungsten products other than cemented tungsten carbides include P. R. Mallory, Westinghouse and Union Carbide. They produce such products as mill shapes, wire and filament, heavy alloys, contacts and hard coatings.

POTENTIAL APPLICATIONS

Aluminum

Substantial development efforts have been underway for several years in two areas of aluminum powder metallurgy: (1) the direct fabrication of parts such as bearings and light-duty structural components by the usual powder metallurgy processes of cold press and sinter; and (2) the production of flat-rolled sheet and strip from powder feed.[18,19] Production of aluminum powder parts in both Japan and Europe has been reported; the parts are being produced both from independently developed processes and under license from Alloys Research. The principal U.S. licensee is Sinteral Corp.

U.S. Bronze has also published promising results of its work on aluminum powder alloy blends containing modest additions of copper.[20] Alcoa has been working in this area intensively and now prepares and sells a preblended powder. Alcoa presents detailed discussion of its progress at the annual Design Show and has extensively promoted aluminum powder parts fabrication and utilization.[19]

A number of powder parts makers are cooperating with the powder producers by conducting development programs for aluminum and have been achieving promising results. Current press equipment is adequate, but new furnaces could be required. The industrial acceptance and use of parts made from aluminum powders will depend on their cost and performance in comparison to that of parts made from alternative materials. Reynolds had also worked on the production of flat-rolled sheet and strip from powder feed. The details of its work showed that the technical characteristics of the products appear comparable to those of conventional stock. Since the program was a pilot-plant effort, the economics of the process are difficult to assess.

Nickel

One improvement in the conventional three-step processing of powdered metals was the development of hot isostatic pressing (HIP) to accomplish the same result in one step.[11] In hot isostatic pressing, the powder is in a semifinished shape, which is heated and, while hot, subjected to uniform pressure sufficient to accomplish full densification. As with forged powder material, alloy property requirements are satisfied. The advantage of HIP is that the same part which needed 6.5 lb of material for forging only needs 3.5 lb. The potential benefits of HIP are expected to lead to selection of this process for a number of jet engine parts in nickel powder-base superalloys. At this time, it is too early to designate the major participants in HIP processing of superalloys. However, it is likely that they would include United Technologies and General Electric (Pratt & Whitney Division), the primary builders of jet engines.

Tungsten

For the long term, growth in tungsten carbide demand will be dictated primarily by requirements of metal machining and shaping technology and mining and rock drilling operations, supplemented by structural uses requiring extreme stiffness. Barring major changes in technology or in engineering materials, a modest growth rate of about 9%/yr for fabricated tungsten carbide product value, currently at $500 million, is anticipated.

ASSESSMENT OF THE FUTURE

Economics of Powder Metallurgy

Among the principal motivations for the growth of sintered powder metal parts is their relatively attractive economics, although former lack of recognition of the true benefits offered by powder metallurgy may have retarded its acceptance. There are several distinctive cost elements in powder metallurgy that are more or less common to all parts: the raw material cost, the compacting cost, the sintering cost and the finishing cost. The raw material cost, however, must be excluded from a general consideration of part production, for each part is made of a designated composition whose basic powder can range from iron at 18¢/lb to tungsten at $1015/lb (1975).

In general, the process technology for powder metallurgy is quite similar for most metals. The powders of desired composition are purchased as elemental powders except for some limited composition classes, such as the stainless steels. The powders are mechanically blended to the appropriate composition. The mixed powders have incorporated into them pressing binder or lubricant which is volatilized on subsequent sintering. The lubricated powders are shaped in high-speed presses with compacting pressures ranging from 40,000-100,000 psi, depending upon the powder mixture, the die cavity configuration and the properties desired in the finished part. The "green" or pressed compacts of the powder are then sintered in continuous furnaces at temperatures appropriate to the specific composition. The temperatures are about 1800°F for copper-base compositions, and 2500-3000°F or even higher for more refractory metal compositions. The metals are sintered in reducing atmospheres or vacuum.

The technological level for production of sintered iron powder parts is moderately sophisticated and quite broadly disseminated throughout the whole industry. Among the most sophisticated activity is that being done by manufacturers of business machines, such as National Cash Register, who are concerned with the production of parts of densities approaching the theoretical density of iron in order to obtain improved performance, corrosion resistance and service life in their equipment. The higher performance parts are produced by starting with higher quality powders, such as electrolytic iron powder, and subjecting the part after sintering to a second pressing and second sintering operation with perhaps even a third and final pressing or coining operation.

A cost analysis for production by powder metallurgy techniques is best described by example. From the range of available sizes, a representative part is a ring shape with dimensions of 4-in. outside diameter x 3/16 in. wall x 1/8 in. thick. These dimensions result in a ring weight of 0.075 lb for an expected sintered density of 85% (6.7 g/cm^3). A typical 100- to 150-ton compacting press would easily be capable of producing 500 rings per hour for an annual output of 1 x 10^6 rings per single shift operation. A standard belt sintering furnace with a hearth space of 15 ft^2 would be able to treat about 300 rings per hour in a single layer loading. Since furnaces would be operated on a three-shift bases, an annual throughput of 2 x 10^6 pieces per year could be achieved by each furnace using single-layer loading on the belts. The maximum output of most recent press designs are considered to be about 1500 units per hour for minimum die fill. A press of this type operated on a three-shift basis would have a productivity approaching 10 x 10^6.

Thus, within moderate ranges, a single press and furnace combination can be a reasonably close match for an integrated production line. The capital cost of the units described are in the ranges of $150-175,000 each, so that a basic press and furnace installation would have a capital cost of $350,000. Additional equipment requirements would depend on the specific requirements for post-sintering finishing treatments such as machining, grinding, heat treating, coining or heat shaping.

The direct operating costs of a sintered powder ring plant would entail charges for raw materials, labor, maintenance, utilities and other processes materials. Since all the powders can be purchased preblended, there is no charge for pressing lubricants. The basic cost of iron powder in 1975 was about 18¢/lb delivered in truckload quantities, plus the value of any additional elements such as copper at about 90¢/lb, nickel at $2.20/lb or molybdenum at $5.50/lb. Iron powder prices were expected to increase to approximately 20¢/lb, and with nominal allowance for producer charges for preblending of lubricant and alloy additions, final powder cost would then be 26¢/lb or about 2¢ a ring. The yield of powder to rings should approach 99%. The labor requirements would call for one man to oversee operation of the press and one man to load the furnace. The rings would now be ready for finishing operations.

Maintenance requirements are moderate and the annual cost could be taken at 4% of the equipment cost for round-the-clock operation. Utility requirements are primarily for the gas for heating the furnace. If we allot a rate of $1.80/1000 ft³, the furnace feed demand of 1500 ft³/hr would have an operating cost of $2.70/hr and weekend idling cost of $1.80/hr. The only other material charge would be for ammonia at a rate of 2000 ft³/hr with a typical charge of $3.50/1000 ft³ or $7.00/hr during operation and possibly only $2.70/hr when idling on weekends. One additional cost that should be anticipated is for the compacting die to be used in the press. Die life will be governed by the erosion due to powder abrasion, but a life of 1,000,000 units could be expected with cemented carbide liners. The die cost could be on the order of $3000 or 3 cents per unit. The direct operating costs are summarized in Table 5-11 and are compared in Table 5-12 with a similar ring made from steel tubing.

The preliminary cost for this example, which was for piston rings, indicates processing costs for the pressed and sintered rings to be about 17.5 each at 1,000,000 units per year throughput, including costs of capital recovery. The costs can be compared with costs for cutting similar-sized rings from tubing. The results show that on direct processing costs alone, powder appears only marginally competitive. However, when total costs are compared, the powder rings offer a potential cost savings of 15%. While this would not be sufficient to assume the risk of technical development, the proven development of sintered iron piston rings has eliminated much of the uncertainty and risk in the technical feasibility of this processing concept. The engines in the small 2- and 4-cycle class are now using these rings. Among them are the Muskegon rings for nonautomotive engines and the Brico rings in England being made for the Volkswagen, Rover 2000 and Triumph 2000 engines. These rings are smaller than sizes used in American automotive engines, so that the performance of larger rings still has to be established. Some modification of composition from small engine rings might be required, but it presents a recognized starting point.

Energy Consumption in Sintering Powder Metals

To produce metal parts by sintering powder compacts, examination of energy consumption must consider several factors, the most important being the specific sintering temperatures for each of the individual metals;

Table 5-11. Estimated Production Costs of a Sintered Powder Metal Ring[1]

Cost Element	Costs/Million Parts			
	Unit Price	Annual Units	Unit Cost (¢/part)	Annual Cost ($/yr)
Material–Powder	26¢/lb	75,000 lb	1.95	19,500
Direct Labor				
Pressman	$7.50/hr	2,000 hr	1.50	15,000
1½ Furnacemen	$7.50/hr	3,000 hr	2.25	22,500
Helper	$5.50/hr	2,000 hr	1.10	11,000
Plant Overhead @ 50% DL			2.42	24,200
Maintenance	4%/6,000 hr	varied	0.58	5,800
Fuel @ $1.80/MCF				
Operating	15 MCF/hr[a]	3,000 hr	0.81	8,100
Idling	10 MCF/hr	1,400 hr	0.25	2,500
Ammonia				
Operating	$7.00/hr	3,000 hr	2.10	21,000
Idling	$2.50/hr	1,400 hr	0.35	3,500
Compacting Dies	$3,000 each	1,000,000 parts	0.30	3,000
Combined Direct Costs above Material			11.66	116,600
Depreciation	10%/yr	varied	1.45	14,500
Capital Recovery	16.7%/yr	varied	2.42	24,200
Total Cost Including Material			17.48	174,800

[a]MCF/hr = thousand cubic feet of natural gas per hour.

Table 5-12. Estimated Production Costs of a Conventional Ring[1]

	Costs/Million Parts			
Cost Element	Unit Price	Annual Units	Unit Cost (¢/part)	Annual Cost ($/yr)
Material–Tubing	$480/ton			
60% Cut-off Yield	$800/ton	75,000 lb	3.00	30,000
Direct Labor				
2-cut-off Operators	$11/hr	4,000 hr	4.40	44,000
1 cut-off Helper	$9/hr	2,000 hr	1.80	18,000
Plant Overhead @ 50% DL			3.10	31,000
Maintenance	4%/6,000 hr		1.000	10,000
Fuel	–		–	–
Ammonia	–		–	–
Jigs & Fixtures	$500/yr/cut-off		0.25	2,500
Combined Direct Costs Above Material			10.55	105,500
Process Losses–5%			11.10	111,000
Depreciation	10%/yr		2.50	25,000
Capital Recovery	16.7%/yr		4.23	42,300
Total Costs Including Material			20.83	208,300

second is the shape factor and the mass volume ratio of a given part, and the third is the sintering times and furnace duty schedules (continuous or intermittent).

In most powder sintering, the individual pieces are raised to temperature by heat transfer from heated furnace walls. For reasons of efficiency in energy utilization, as well as for manufacturing productivity, continuous-duty furnaces are preferred for most production sintering operations. The combinations of the shape and mass factors and the times and schedule factors make up the governing parameters for individual production of furnace loading. Obviously, high furnace loadings result in lower energy consumption per unit part.

In most cases, although the consumption of energy in sintering is small, it is still a significant share of the cost for production of a part. The example given previously has shown energy costs for an iron part with a relatively low mass-volume shape factor. The low shape factor results in a low furnace loading and, therefore, a relatively high energy consumption and high cost of energy per part or per unit weight. By comparison, a part with the configuration of a spur gear of the same approximate outer diameter and thickness as the ring of the previous example would have had a nominal weight of 0.3-0.35 lb, or slightly more than four times that of the ring. The production rate for the spur gear would not be significantly less than that of the ring so the energy consumption per unit would be decreased proportionately, although energy consumption per unit part would be about the samc. Thus, energy consumption of sintered metal parts must be defined precisely to avoid confusion or misunderstanding.

The comparative consumption of energy for sintering parts of various metals, all other factors being constant, is related to the sintering temperature required for each specific metal composition. In the previous discussions on the various metals, their sintering temperatures were indicated to be nominally: aluminum 600°C, copper 950°C, iron 1200°C, nickel 1250°C, tungsten carbide 1450°C and tungsten 2800°C. The energy required to attain and maintain these temperatures in a sintering furnace will be proportional to the temperature as long as the heat losses are primarily due to conduction, but will increase to fourth-power dependence on absolute temperature if radiation becomes the dominant heat loss mechanism. Temperatures for the sintering of parts of tungsten carbide or tungsten are such that generally radiation predominates; for the other metals, the conduction mechanism is predominant. Either type of loss in sintering can be minimized by either proper design of refractory insulation for conduction, or by radiation shielding where radiation is predominant.

The heat losses in sintering the iron powder rings of the previous example were shown to consume about 1500 ft^3/hr of natural gas during production, and about 1000 ft^3/hr of natural gas for idlings during weekends and nonproductive periods. For the production rates, which were projected to be 300 pieces per hour, energy consumption would amount to 1.5 million Btu/hr during production, for a direct energy demand of 5000 Btu per piece or about 70,000 Btu/lb. Energy losses during furnace idling periods would add about 30% to the energy consumption for a combined energy demand of nearly 90,000 Btu/lb. If the parts being sintered had the spur gear configuration instead of the ring, the direct energy consumption demand would decrease to about 15,000 Btu/lb, and the combined energy with idling time would total about 20,000 Btu/lb.

By analogy, based on the sintering temperatures of the other metals, similar part shapes in aluminum would have energy requirements of half those of iron; copper would have about three-quarters the energy requirement; nickel about the same requirement. These energy estimates must be adjusted, however, especially aluminum, for density of the elemental metal, as seen in Table 5-13.

Table 5-13. Energy Requirements for Sintering Powder Parts

Metal	Sintering Temperature (°C)	Direct Sintering Energy by Shape Factor (Btu/lb)	
		Low Shape Factor	High Shape Factor
Aluminum	600	100,000	25,000
Copper	950	65,000	10,000
Iron	1200	70,000*	15,000
Nickel	1250	5,000	14,000

*An equivalent production energy supplied for this operation in an electric furnace would be 70,000 Btu/lb (2.93×10^{-4}/Btu) = 20.5 kWh/lb. If the industrial energy rate is 3¢/kWh, the cost is 20.5 kWh (3¢/kWh) (.075 lb/piece) = 4.6¢/piece versus 0.81¢/piece using natural gas.

By comparison, similar-shaped parts manufactured by more conventional chip form methods from solid metals consume very little energy, except for postmanufacturing heat treatment, which could also be required for sintering powder parts. In common mills, the energy consumption in machining operations is usually considered insignificant as a direct cost and is accounted for only as an element of indirect or overhead cost of production.

The most severe energy demand in powder metallurgy is in sintering tungsten powders to dense bodies without the presence of a liquid phase.

The required sintering temperature, about 2800°C, causes radiation power losses from the outer surfaces to approximately 3 kW/in.^{2} or 0.5 kW/ cm.^{2} Excessive losses incurred from heating a furnace chamber, excluding considerations of furnace design, are avoided by high-temperature sintering through internal electrical resistance heating of the tungsten itself. For this reason, tungsten is usually sintered as a billet or preform shape and subsequently hot-forged and worked to attain desired density. The consequence of these requirements is that production of metallic tungsten parts and stripped rod and wire is limited to a few primary producers such as General Electric, GTE-Sylvania Westinghouse and Fansteel.

The commercial production of tungsten powder to billet for rod wire and forging blanks, which is about 20% of the total tungsten consumption, has probably the most complex sintering treatment of all metal powders. The energy consumption for each stage of the process is shown in Table 5-14. The sintering of tungsten carbide pieces also follows a complex schedule because of the unique metal of the carbide compact. The

Table 5-14. Direct Energy Consumption for Sintering Tungsten Powder to Billets[1]

Treatment Step	Temperature (°C)	Time	Energy (kWh/lb)
Presinter	1200	1 hr	3
First-State Sinter	2500	5 min	2
Final Sinter	2800	20 min	15
Total Direct Energy Consumption			20
Estimated Additional Furnace Power Losses			10
Total Energy Consumption			30

conventional sintering practice employs a three-stage treatment in electrical resistance heated furnaces with a protective reducing atmosphere of dry hydrogen. The first stage is to de-wax the green pressed carbide compact by heat in an intermediate temperature zone of 400°C. This step is required to volatilize the wax incorporated as a pressing binder for the hard carbide particles. After sufficient time has elapsed for the wax to be eliminated, the carbide enters an intermediate temperature zone of 800-1000°C for presintering and temperatures for homogenization to occur. The carbide then passes into the high temperature zone at 1400-1500°C for final sintering, depending upon the grade of carbide. The furnace is designed so that the residence time would be about one hour in the hot zone before the carbide would be moved into a cooling zone.

Trends in the production of carbide are to replace the continuous hydrogen sintering furnaces with vacuum furnaces, either batch or semicontinuous, as required by the production demand. In the vacuum furnaces, the sequence of the heating treatments is similar to the practice in the hydrogen furnaces. However, the vacuum furnaces offer the advantage of having lower residual concentrations of oxygen than can be obtained in a hydrogen atmosphere furnace. Two benefits are obtained: the refractory carbide of titanium or tantalum are sintered without oxygen contamination, and the reproducible vacuum atmosphere provides more uniform sintering conditions, which result in more consistent properties in the sintering pieces. The benefits of improved sintering properties have stimulated a growing trend to vacuum sintering of carbides.

The energy consumed by the conventional hydrogen atmosphere furnaces and the newer vacuum furnaces, as with sintering of other metal parts, is dependent in large measure on the shapes of the pieces and the furnace loadings. In general, the packing density tends to be as high as practical. For sintering carbides, energy consumption is in the range of 5-10 kWh/lb in the conventional hydrogen furnaces; it is decreased to the range of 2-5 kWh/lb for vacuum furnaces, with the larger, high-production systems having the more efficient energy utilization.

The examination of energy consumption for sintering metal powder parts allows an estimate to be made of a total energy consumed by this industry as a function of total production. This has been shown for each metal powder in Table 5-15, it amounted to a total of 5450 x 10^9 Btu for sintering metal powder parts in 1975.

Evaluation of the Potential of Sintering Powder Metals

Based on past and recent trends in U.S. shipments of powder metals and on prospective future applications for molding and sintering powder metal parts, considerable increases in this technology are anticipated.

Among the principal metal powders, iron has by far the predominant utilization. About three-fourths of iron powder is molded into parts and sintered. Although total shipments of iron powder have undergone a downward trend–due primarily to unfavorable economics throughout the metals industry–reversal of this trend is certain. Forming structural parts based on iron and steel powder metallurgy techniques, with their favorable economics, will become an increasingly robust industrial process. Projecting an average long-term increase of about 9000 short tons per year in

Table 5-15. Energy Consumption for Sintering Metal Powders in the U.S. (1975)[1]

Metal	Sintered Parts Short Tons	Unit Energy[a]	Gross Energy (10^9 Btu)
Iron	105,000	20,000 Btu/lb	4,200
Aluminum	300	40,000 Btu/lb	24
Copper	15,000	15,000 Btu/lb	450
Nickel and Nickel-Base	4,000	40,000 Btu/lb	320
Tungsten Metal	1,500	30 kWh/lb[b]	306 (electric)
Tungsten Powder Parts	1,200	10,000 Btu/lb	24
Hydrogen Sintered Tungsten Carbide	2,500[a]	6 kWh/lb[b]	102 (electric)
Vacuum Sintered Tungsten Carbide	1,200[a]	2 kWh/lb[b]	24 (electric)
Total Energy for Sintering Powder Parts			5,450

[a]Computed for assumed average production practice.
[b]One kWh equivalent to 3400 Btu electric.

powder shipments results in growth to a level of nearly 200,000 short tons in 1980. Average increases in iron powder utilization during the past decade have been nearly 10%/yr, compounded.

Total shipments of aluminum powder have been declining for more than six years. Of the total material produced, sintering powders (primarily for bearings and light-duty structural components), have represented a very small fraction. The primary uses of aluminum powder are currently as an additive to paints and in blasting agents for mining. Should the decline in shipments of aluminum powder be soon arrested and stabilized, the anticipated shipment of this product would remain in the neighborhood of 40,000 short tons in 1980.

Utilization of copper powder has remained nearly stable throughout the past decade, with even a slow, steady decline experienced more recently as worldwide copper prices have risen. Three-quarters of the used powder has gone into the fabrication of porous bearings, with the remainder applied to the manufacture of structural components and friction parts such as brakes and clutches. The total domestic shipment of copper powder will probably not exceed 25,000 short tons per year by 1980.

Nickel powder, introduced into broader commercial applications in the mid-1960s, has undergone an upward trend in its nationwide usage for more than four years. This increased market is expected to continue as more high-temperature superalloy structures are produced (*e.g.*, turbine engine parts) and as the market for nickel-cadmium batteries and coinage expands. In 1980, about 20,000 tons of nickel powder will be required in the U.S.

Tungsten powder is similarly projected to increase appreciably in sintering applications, particularly for tungsten carbide manufacture as well as for use in heavy metal alloys. The former requirement is primarily in metal-machining and shaping, plus mining and rock drilling; the latter is for parts of high stiffness and/or high-temperature resistance. With a 9% growth anticipated for this valuable commodity, an expanded tungsten powder shipping requirement would be achieved in 1980, peaking at about 10,000 short tons.

POWDER METALLURGY IN JAPAN

Activity on powder metallurgy research and development in Japan began early in the twentieth century. A minimal commercial activity was undertaken after World War II, and by 1960 consumption of sintered metal powder parts in Japan had reached an estimated level of 2000 tons annually.[21] This included approximately 35% structural parts, 40% bearings, 10% in friction materials, 10% in cemented tungsten carbide cutting tools and a small share in sintered electrical contacts. By the mid-1960s, consumption of sintered parts had increased by about 50% to approximately 4500 tons. The consumption of structural parts, which increased by almost 150%, was the major factor in this increase.

Powder metallurgy in Japan enjoyed a dramatic increase by 1960 to total consumption of 15,600 tons of iron and copper-base structural parts. Consumption of tungsten carbide materials and contacts amounted to another 1700 tons. Reported consumption of sintered metal powder parts in 1975 was 22,000 tons.[22] Extrapolation of historic growth patterns suggest that iron would supply almost 90% and copper the balance of this demand. Structural parts would probably have consumed about 75%, bearings about 20%, and sintered friction materials only about 5%. In addition, consumption of 900 tons of tungsten carbide in 1975 was reported, a substantial decline from 1500 tons of carbide in 1970 and presumably reflecting the depressed state of Japan's manufacturing industries during the reporting year. The consumption of electrical contact materials in 1975 was reported to be 160 tons, essentially unchanged from consumption of 157 tons reported for 1970. This pattern of powder metallurgy is summarized in Table 5-16.

Prior to 1965, most of the molding-grade iron powders for sintering of structural parts and bearings were imported from Sweden, as were

Table 5-16. Estimate Shipments of Sintered Powder Parts in Japan[21,22]

	Quantities by Types of Parts (thousands of tons)			
	1960	1965	1970	1975
Molded Parts				
Structural Parts	1.0	2.4	11.0	16.5
Bearings	1.2	1.3	3.7	4.4
Friction Materials	0.3	0.2	0.8	1.1
Annual Totals	2.5	3.9	15.5	22.0
Tungsten Carbide	0.4	0.6	1.5	0.9
Electrical Contacts	0.06	0.08	0.16	0.16

most the copper powders. By the early 1970s, a variety of iron powder types were being produced. These powders were made by reduction of steel mill oxide scale, electrolytic deposition, pulverizing iron powders and atomization of steel powders. The Japanese powder producers are not able to supply the major share of Japanese requirements for coppers, tungsten, and tungsten carbide powders, as well as iron powder.

The markets served by sintered metal powder parts in Japan have been characterized as transport electrical machinery, industrial machinery and other uses. The relative shares of their usage amounts to about two-thirds for transport, which is presumed to be automotive, about 15% for electrical machinery, 15% for industrial machinery, and only about 5% for other uses.

REFERENCES

1. Bovarnick, B. Consultant. Private communication.
2. Johnson, P. Deputy Executive Director, Metal Powder Industries Federation, Princeton , New Jersey. Private communication.
3. Metal Powder Industries Federation (MPIF). Private communication
4. *Internat. J. Powder Metallurgy* 6(4):29 (1970).
5. Defense Services Administration. Private communication.
6. Taubenblat, P.W. *Internat. J. Powder Metallurgy* 10(3):169 (1974).
7. Henry Wiggen and Co., Ltd. "Nickel Alloys by Powder Metallurgy," (1965).
8. Sherritt-Gordon Mines, Ltd. "Metallurgical Products " (1963).
9. International Nickel Co., Inc. "Nickel Powders, Properties, and Users " (1973).
10. Kelley, W. *Internat. J. Powder Metallurgy* 10(1):53 (1974).
11. Matthews, P.S. Powder Metallurgy in Ordinance, Seminar, Dover, New Jersey, 1976.

12. Schwarzkopf, P. *et al. Refractory Hard Metals* (New York: MacMillan Co., 1953).
13. Schröter, K. U.S. Patent 1549615, 1925.
14. Schwarzkopf, P. U.S. Patent 1959879, 1934.
15. Comstock, G. U.S. Patent 1973428, 1934.
16. American Metal Market, Tungsten Forum, New York, 1976.
17. UNCTAD, Committee on Tungsten. *Quart. Bull.* X(1) (1976).
18. Dudac, J.G. and W.A. Dean. *Internat. J. Powder Metallurgy* 5(2):21 (1969).
19. Aluminum Company of America. "Potential Applications of Aluminum" (1971).
20. Matthews, P.S. *Internat. J. Powder Metallurgy* 4(4):39 (1968).
21. Kachi, S. *Internat. J. Powder Metallurgy* 9(3)151 (1973).
22. Matsuda, Y. President, Sumitomo Electric U.S.A., Inc., 345 Park Avenue, New York, New York. Private communication.

UNCITED REFERENCES

Bovarnick, B. *Powder Metallurgy* (Cambridge, Mass: Arthur D. Little, Inc., 1969).

Goetzel, C.G. *Treatise on Powder Metallurgy,* Vol. I, II, III (New York: Interscience Publishers, Inc. 1949, 1950, 1952).

Hirschorn, J.D. *Introduction to Powder Metallurgy,* (New York: American Powder Metal Institute, 1969).

Jones, W.D. *Fundamental Principles of Powder Metallurgy,* (London: Edward Arnold, 1960).

Powder Metallurgy Technical Assistance Mission, No. 141, Organization for European Economic Cooperation, Paris, 1955.

Symposium on Powder Metallurgy, 1954, Iron and Steel Institute, Special Report 58, London (1956).

Wulff, J., Ed. *Powder Metallurgy, Proceedings of Conferences, 1940, 1941,* Massachusetts, Institute of Technology,(Cleveland, Ohio: American Society of Metals, 1942).

6

NONCONTACTING, ELECTRICALLY ASSISTED MACHINING*

INTRODUCTION

When tools must be made of hard materials such as carbides, or when delicate parts must be machined without force being applied to the workpiece, conventional machining methods such as milling, drilling, or shaping cannot be used. Electrically assisted machining processes, however, which use an electrical current to erode the workpiece, can machine hard metals, and do so without contact between the tool and workpiece. In these machines a tool is automatically fed into the workpiece. The workpiece is mounted onto a table which has two degrees of freedom, an electrical power supply is connected between the tool and workpiece and a fluid is used to assist in the erosion process, to flush out the eroded particles, and to cool the tool/workpiece.

There are two generic types of processes: those that use an electrical breakdown, and those based on electrolysis. The principal processes using electrical breakdown are conventional electric discharge machining (EDM) and traveling wire EDM. In these processes, a voltage is applied between the tool and workpiece. This voltage is of sufficient magnitude to cause the fluid, in this case a dielectric, to break down. The current waveshape is a unipolar pulse. The charged particles that bombard the workpiece cause local intense heating which results in the disintegration of the surface. With conventional EDM, the tool must be shaped to match the shape to be machined; as the tool erodes the workpiece, the eroded area will be similar to the tool. Because of the nature of the process, there will be a gap called the overcut between the tool and workpiece. The overcut depends on the machining conditions and is predictable. Conventional EDM permits machining of hard materials and can produce punches

---

*By: Oscar Farah

and dies for cutting strips or roll stock as well as for forming. A summary of technical data on EDM is given in Table 6-1.

Table 6-1. Summary Information on EDM Machines

Power Supply	
Voltage	Up to 1000 volts for relaxation oscillator type; 20 volts and up for pulse generator type
Current	Up to 500 A for pulse generator type Capacitance to 25 μF for relaxation oscillator type
Power Input	Typical Power requirements between 5 and 10 kVA for a single-head machine
Electrode	Materials: graphite, copper, brass, copper tungsten and others
Dielectric Coolant	Kerosene, transformer oil, water, compressed air
Machine Properties	
Machining Rate	For ideal conditions to .05 in ^{3}/hr/A decreasing with increased frequency and poor flushing
Tolerance	±.0005" practical; ±.0001" possible
Example	
Finishes	30 RMS at .010 in ^{3}/hr; 400 RMS for 3 in ^{3}/hr
Overcut	Typical .0006-.022 in.
Recast layer	.0002-.005 in.
Ancillary Equipment	Tape control, arcing prevention, adaptive control for electrode drive
Current Cost	$5,000 to over $100,000 (average $25,000)

With wire EDM, the tool is replaced with a wire, which is continuously renewed. The wire is made to traverse the workpiece and the latter is driven by a computer-controlled table in a path that makes the wire trace out the periphery of the required cut. The machine acts essentially as a highly accurate, computer-driven band saw with a continuously replaceable saw. Wire EDM cannot be used to cut out blind shapes; hence, it can only be used to produce punch press tools, for use only for cutting and not for forming. Table 6-2 is a summary of data on wire EDM.

The principal electrolysis processes are electrochemical machining (ECM) and electrochemical grinding and deburring (ECG, ECD). In these processes a voltage is applied between the tool and workpiece and erosion is due to the electrolytic (reverse plating) action. The metal removal rate is governed by Faraday's Law and hence is proportional to the average current flowing between tool and workpiece. For high metal removal rates,

Table 6-2. Summary Infomation on Traveling Wire EDM Machines

Power Supply	Similar to conventional EDM
Electrode	Copper wire .002 - .010 in. in diameter
Dielectric Coolant	Deionized water
Machine Properties	
Machining Rate	.030 in./min in 1-in. thick steel
Tolerances	$\pm$.0005 in.
Finish	30 RMS
Overcut	Typical .001 in.
Recast Layer	Less than conventional EDM because of water cooling
Ancillary Equipment	Tape control (a necessity)
Current Cost	$30,000 and up

it is therefore advantageous to operate with direct current with minimum ripple at a voltage slightly lower than the voltage that would cause the tool/workpiece gap to breakdown. The fluid used in this process is an electrolyte, which supplies hydroxyl ions that combine with the removed metal ions to form a precipitate that does not plate onto the tool.

ECM is similar in operation to EDM in that a shaped electrode is used for the tool; however, the metal removal rates far exceed EDM or conventional machining rates. With ECM, under normal operating conditions (*i.e.*, with no arcing), there is no tool wear. Table 6-3 is a summary of data on EDM machines.

In ECG, the tool used is a conducting grinding wheel and, unlike the other processes, there is light contact between the wheel and workpiece. Under normal operation, 90% of the material removed with ECG is through ECM action and 10% by abrasion from the grinding wheel. The process is capable of finishes comparable to those obtained with conventional grinding. ECD is similar to ECM except that there is no relative motion between the tool and workpiece. It is used to remove hard burrs left when machining hard materials with conventional tools. A summary of technical data on ECG and ECD machines is given in Table 6-4.

Market prospects for electrically assisted machining varies for each type of machine. In general, however, the machines described can be used on hard materials which could not be machined with conventional machines; they are not replacing conventional machines at the present time. While the electrically assisted machines may be used on materials of lesser hardness as well, the cost is higher than that for conventional machines. However, if an electrically assisted machine is purchased, it should be operated

Table 6-3. Summary Information on ECM Machines

Power Supply	
Voltage	10-20 V dc
Current	Up to 20,000 A
Power Input	Typical 50-150 kVA per head
Electrode Materials	Copper, brass, stainless steel, titanium, beryllium copper, copper tungsten
Electrolyte	Sodium chloride, sodium nitrate (for ferrous materials, sulfuric acid or hydrochloric acid (for nonferrous metals)
Machining Properties	
Machining Rate	$V = \frac{\text{current}}{96{,}500} \times \frac{\text{atomic weight}}{\text{Valance}} \times \frac{1}{\text{density}} \times \text{time (cm}^3)$
Tolerances	$\pm$.002 in. practical, $\pm$.0005 in. possible
Example Finishes	50 - 125 RMS possible. Finish improves with high current and high feed rates. For cybirdrically symmetric parts which are rotated, finish to 5 μ in. RMS possible
Overcut	Typical .006-.050 in.
Recast Layer	None–intergranular attack may weaken surface
Ancillary Equipment	Tape control
Current Cost	Prices start at $100,000

Table 6-4. Summary Information on ECG and ECD Machines

	ECG	ECD
Power Supply		
Voltage	4-15 V dc	3-24 V dc
Current	Up to 1000 A	Up to 500 A/station
Power Input	4-10 kVA + 12 hp (max.) for grinding motor	5 kVA
Electrode	Metallic grinding wheel	Similar to ECM
Electolyte	Similar to ECM	
Machining Properties		
Tolerances	$\pm$ .001 in. easily held $\pm$ .0005 practical Higher accuracy possible with more use of grinding wheel	
Finish	Depends on grinding wheel	
Overcut	None	Possible
Recast Layer	None	None

continuously, even on conventional materials, to recover its cost. As labor costs increase, computer-controlled machines will become more economical. Electrically assisted machining units, which lend themselves easily to computer control, should therefore eventually replace conventional machines.

The demand for conventional EDM machines should continue at its present pace. Wire EDM, which is a new, highly computerized process, should experience increased demand for some time to come. The future for EDM, however, is not very optimistic. ECM has lost its aerospace market because of weakening of the aerospace industry as a whole, and has not been able to replace it as yet. It may survive if it can be diversified to include a number of other markets: the need for a machining process with ECM-like properties is important, particularly in expansionary times. ECG and ECD should both enjoy a moderate market with a possible future decrease. A summary of the market data for all five processes is given in Table 6-5.

In Japan, it appears that EDM machines of domestic manufacture are inferior to their U.S. and Swiss counterparts in terms of accuracy and versatility. Wire EDM is being manufactured for sale locally by Hitachi, under license by AGIE Industrial Electronics Ltd. of Switzerland. The Japanese ECM market, on the other hand, is well developed with an estimated 300 units in service and an estimated current demand of 50-60 units per year. While there is no aerospace industry in Japan, Hitachi, the major ECM manufacturer, has diversified its market to include the auto industry, tool and die makers, and the aluminum extrusion industry.

Energy consumption is not a major factor in the market penetration of these processes. Table 6-6 lists the energy consumption for those machines. The voltage requirements range from 220 V three phase to 3000/3300 V for the layer ECM machines. Although the energy consumption of an electrically assisted machining unit may be significant when compared to the energy consumption of a small village (approx. 50 kW), total energy consumed by all such units will be insignificant when compared to the consumption of a complete industry.

GENERAL PRINCIPLES

Accurate mass production of metal parts requires tools that will withstand wear. When complex shapes (other than round holes or straight cuts) are to be made in large quantities from raw materials in sheet or rod form, a punch press and/or a screw machine is used. Both machines require tools to form or cut the raw material; punches and dies are used in the punch press, and complex cutting tools are used in the screw machine. For both machines, the tools must be made of hard materials to assure long life at high operating speeds.

A complex shape usually requires a multistep punch and die set in which the shape is produced in successive operations. The number of steps depends on two functions; characteristics of the raw material, including its Young's Modulus, hardness, maleability and feasibility of constructing a tool of the required shape. Another consideration is speed of operation: a multitool operation may be selected over a single tool operation if the former will yield more output.

The conventional method of constructing tools is to machine them from soft steel and subsequently heat treat or case harden them. In heat treating, distortions could occur and final finishing with grinding tools may be required. In cases where the tools must be made of hard steels such as the carbides, conventional machining is very difficult, time consuming and tedious.

Noncontacting electrically assisted machining, however, eliminates the difficulties of machining hard materials. Electric current is used to erode the workpiece, and there is no physical contact between the workpiece and the tool. Hence, the hardness of the workpiece has very little bearing on the rate of machining. With this method, it is possible to machine intricate shapes in the hardest of materials, and since heat treatment is not required, further finishing is eliminated. The elimination of the heat treatment step also prevents stress redistributions, which can occur when parts are reheated with conventional methods. Since no mechanical forces are applied to the workpiece or tool, intricate shapes, including adjacent cavities separated by thin walls and tools for miniature parts, may be easily machined.

Two of the most widely used noncontacting, electrically assisted processes are electric discharge machining (EDM) and electrochemical processes including machining (ECM), deburring (ECD), and grinding (ECG). In EDM, the heat generated by the impact of heavy ions is used to erode the surface being machined; in the electrochemical processes, electrolysis is used to remove material, in a reverse plating action.

ELECTRIC DISCHARGE MACHINING

EDM was first developed for machining of metals by Lazarenko and others in Russia during World War II.[1] There are claims that the Russian development was based on an American design of a tap extractor sent to Russia between 1942-1943. The tap extractor had a vibrating electrode which collided with the broken tap, and used a water/oil coolant. A dc supply was connected between the electrode and the tap, and as the

Table 6-5. Summary of Market Data for the More Important Electrically Assisted Machining Processes

Process	Principal Application	Approximate Current Stock	Expected Near Term Demand (units/yr)	Market Motivation	Principal Users
EDM	Extremely accurate but slow machining of hard materials	10,000	500	Lowest cost, accurate for machining hard materials	Tool manufacturers
Wire EDM	Programmable, extremely accurate machining of hard materials	1,000	1,000	Low cost, accurate for machining hard materials excluding blind holes	Tool manufacturers Metalworking Industry
ECM	Extremely fast machining of hard materials	500	10-50 (depends on finding a market)	High cost, but quick; must be used full-time to recover capital cost	Tool manufacturers; heavy industry, forging and die casting industry (auto, aeroplanes, turbines)

Table 6-6. Summary Costs and Energy Data for the More Important Electrically Assisted Machining Processes for the United States

		Energy Consumption					
		1975			1980		
Process	Average Capital Cost per Unit ($)	Estimated Stock (#Units)	Estimated Installed MVA	Estimated Energy Consumption[a] (kWh)	Estimated Stock Units	Estimated Installed MVA	Estimated Energy Consumption[a] (mWh)
Conventional EDM	25,000	10,000	80	124,800	12,500	100	156,000
Wire EDM	30,000	1,000	5.5	8,580		33	51,480
ECM	300,000	500	50	78,000	550-750	55-75	85,000-117,000
			Total	210,000		Total	310,000

[a]Installed MVA x utilization factor x average load required/available load x hr/day x days/yr or installed MVA x .5 x .5 x 24 x 260.

electrode moved away from the tap an arc was drawn. The repetitive arcing resulted in disintegration of the tap. The machines have since evolved to very sophisticated units with computer control. Although the physics of the process is not too well understood, the units are presently state-of-the-art and available "off the shelf."

An electric discharge between an electrode and the workpiece is used to erode the workpiece. The electrode is shaped to produce the required cavity in the workpiece. As shown in Figure 6-1, both the electrode and the workpiece are immersed in an insulating dielectric which has a low breakdown voltage.[2] When a potential is applied between the electrode and the workpiece, the free electrons in the gap are swept away. They collide with the dielectric atoms as they move, resulting in ionization of some of those atoms and thereby releasing additional electrons. The heavier positive ions are then slowly accelerated towards the negative electrode. The initial current consists essentially of electrons; as the process continues, however, the ions are also added.

Under standard operating conditions, the electrode is negative and the workpiece is positive. Hence, erosion depends on the kinetic energy of the electrons and negative ions. However, both the workpiece and the tool erode, as indicated in Table 6-7. For certain electrode-workpiece material combinations a no-wear operating condition, in which the electrode wears very little, can be obtained by reversing polarity and applying relatively long pulses. The long pulses result in increased evaporation and sputtering of the workpiece material with subsequent deposition and solidification of some of the metal vapors onto the electrode. This layer of metal which is eroded out and redeposited after each pulse protects the electrode, resulting in little or no wear of the electrode. The penalty for this mode of operation is decreased metal removal rates and rough finish.

There are two types of EDM: conventional EDM in which machining is performed by use of a shaped electrode, and traveling wire EDM in which cutting is performed by use of a wire electrode directed by a tape control unit to cut the desired shape. In conventional EDM, blind shapes, as well as cuts that traverse the workpiece, may be machined. In traveling wire EDM, only cuts that traverse the workpiece may be made.

Three mechanisms have been hypothesized to explain the erosion of the workpiece and electrode; we feel all three contribute to the erosion simultaneously. All hypotheses are based on the fact that when the charged particles impinge upon the surface of the workpiece or the tool, their kinetic energy is converted to heat. The heat may then shape the workpiece by: (1) melting and vaporizing the metal; (2) forming an air bubble which, when it collapses, generates a shock wave that cavitates the surface; or (3) melting the surface of the metal; the thermal shock which

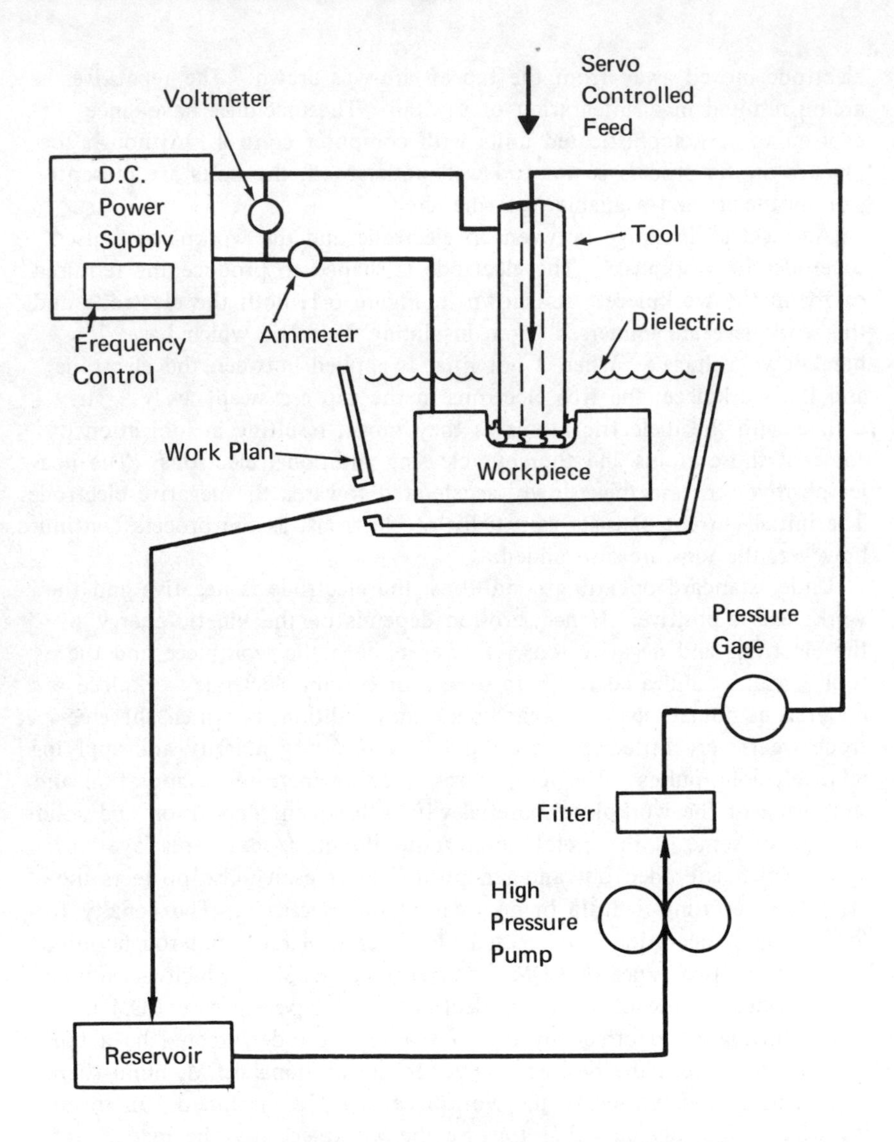

Figure 6-1. Schematic of EDM Machine.[2]

occurs when the dielectric fluid cools the molten metal then results in the disintegration and flaking of the metal.

Under normal operating conditions, the discharge will occur at different points on the surface, so large surfaces can be machined evenly; breakdown will occur where the gap is narrowest. In some cases, however, a particle

Table 6-7. Erosion Rates for Different Electrode Workpiece Combinations[3]

Electrode	Workpiece	Polarity[a]	Ratio of Electrode Wear to Work Done: Electrode	Work
Brass	Brass	Std	1	1-½
Brass	Carbide	Std	5	1
Brass	Steel	Std	1-½	1
Copper	Graphite	Std	1	3
Copper	Steel	Rev	1	2-½[b]
Graphite	Graphite	Std	1-½	1
Graphite	Copper Tungsten	Std	2	1
Graphite	Steel	Std	1	3
Graphite	Steel	Rev	1	6[b]
Graphite	Carbide	Std		
Graphite	Aluminum	Std	1	6
Steel	Steel	Rev or Std	1	1
Copper Tungsten	Copper Tungsten	Std	1	1
Copper Tungsten	Graphite	Rev	1	1
Copper Tungsten	Steel	Rev	1	4[b]
Copper Tungsten	Carbide	Std	1	2-½

[a]With standard polarity, the electrode is negative and the workpiece is positive.
[b]This combination can also be used in the no-wear mode.

may become trapped between the workpiece and tool. This will cause an arc to occur repeatedly in that area and damage the workpiece and/or tool.

The same type of damage will occur if the pulse repetition rate is too high. In this case, ionized particles will not clear before the next pulse occurs, and breakdown will occur at the same place.

Conventional Electric Discharge Machining

Components of an EDM Machine. The basic components of an EDM machine are a pulse power supply, an electrode that can move axially (usually servocontrolled), a table for mounting the workpiece, and dielectric fluid circulation and filtration apparatus. In addition, tape control units are sometimes added for more sophisticated control.

The power supply for an EDM machine must supply a train of unipolar pulses of adjustable width, height and repetition rate. The amount of material removed per unit time is proportional to the pulse repetition rate.

On the other hand, the finish of the machined surface depends on material removal per pulse. The finish becomes rougher with increased material removal per pulse. Also, for minimum tool erosion, short pulses are preferred to minimize the possibility of sputtering.

A number of designs of power supplies that basically satisfy the requirements have been used. In the relaxation oscillator, a dc source is used to charge a capacitor through a charging resistor (Figure 6-2). The capacitor is connected directly across the tool-workpiece gap. When the voltage on the capacitor reaches the gap breakdown voltage, the capacitor will discharge in the gap. When the current in the gap decreases below the holding value, conduction will terminate. To allow this, the charging resistor must be large enough so that the maximum current available is below the holding current. This places an upper limit on the pulse repetition rate. Silicon-controlled rectifiers as well as transistors have been used to disconnect the dc supply from the capacitor during the discharge cycle and immediately reconnect it when conduction stops (Figure 6-3). In this case, the probability exists of reconnecting the supply during discharge (as shown in the third pulse in the figure) resulting in possible damage.

A serious disadvantage of the relaxation oscillator type of supply is that the discharge voltage, and hence the maximum energy per pulse, are defined by the gap breakdown voltage, not by the operator. Since the ionization potential will vary from pulse to pulse due to varying conditions in the gap (such as particle content, cleanliness, gap length), so will the energy per pulse. Although the law of large numbers can predict that on the average, the amount of material removed over a long time will be the same, the finish will tend to be rougher than with power supplies which operate more uniformly. In addition, overcuts cannot be predicted accurately and corrected. Large variations are possible when biases in operating parameters occur (such as variations in the gap length and in dielectric fluid composition).

A variation on the relaxation oscillator power supply (Figure 6-2) can be made by inserting an inductor in the charging side (Figure 6-4) to make the current more uniform and speed up charging. Improved finishes with more predictable overcuts and increased metal removal result from this modification.

With a pulsed power supply, a dc source is connected in series with a resistor and switch between the tool and workpiece (Figure 6-5). The switch is a solid state switch capable of high switching rates and high currents. The resultant output is a train of evenly spaced, rectangular pulses of the same height. The current pulses, however, will vary in width between pulses due to variations in the ionization time. In some cases, where the ionization potential exceeds the applied potential, current will

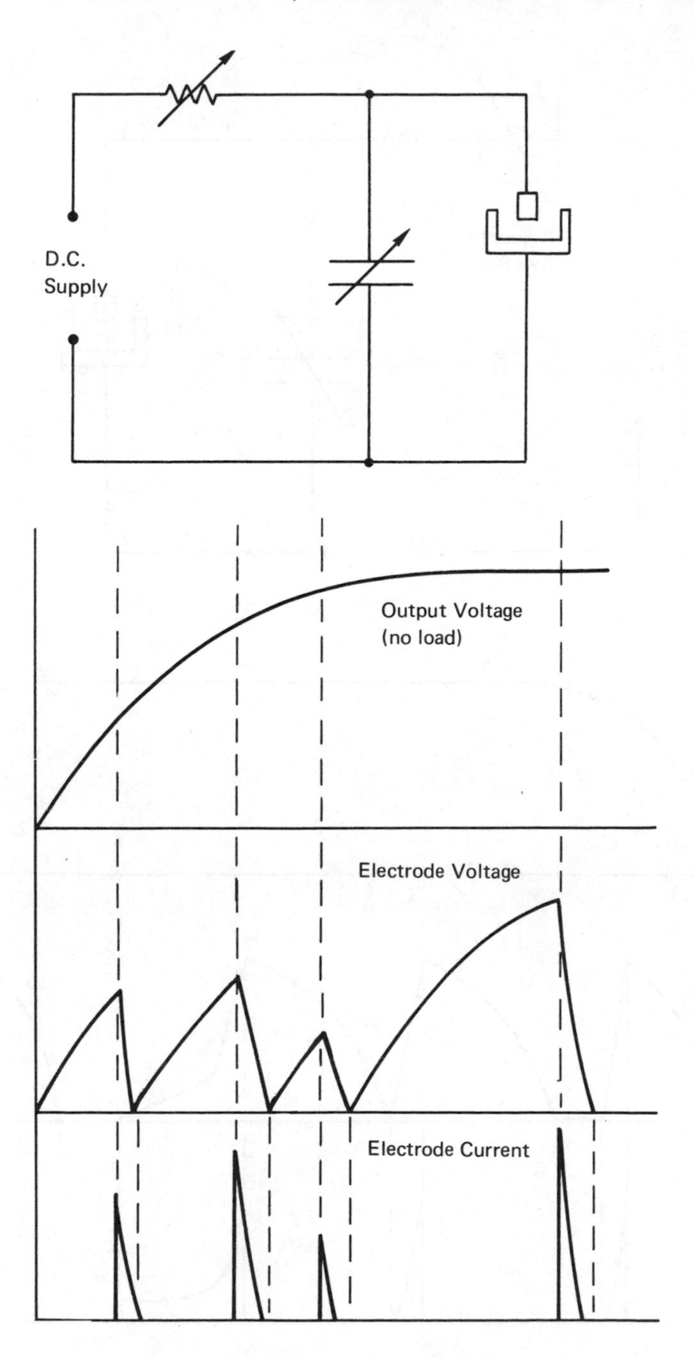

Figure 6-2. Relaxation oscillator power supply.[2]

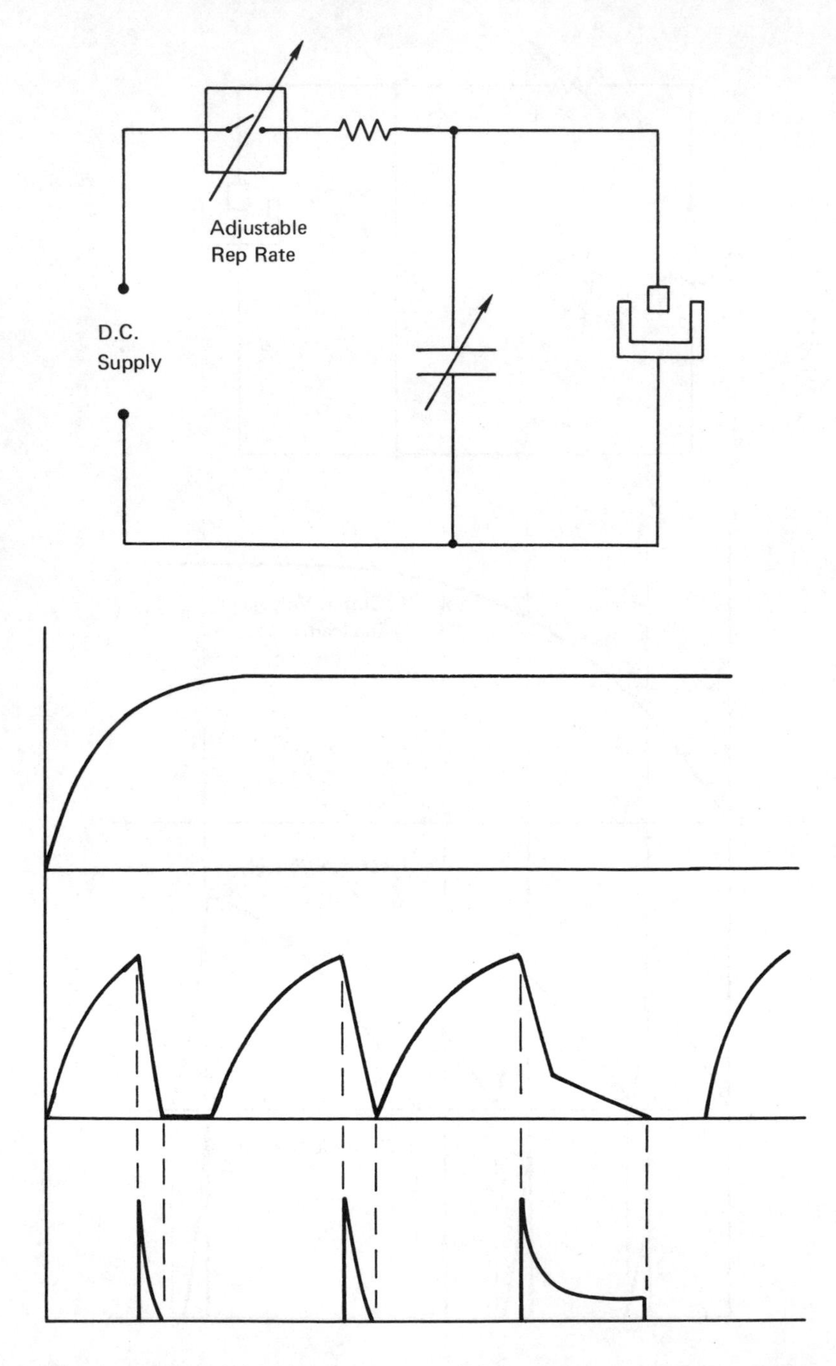

Figure 6-3. Relaxation oscillator power supply with switch in charging circuit.[2]

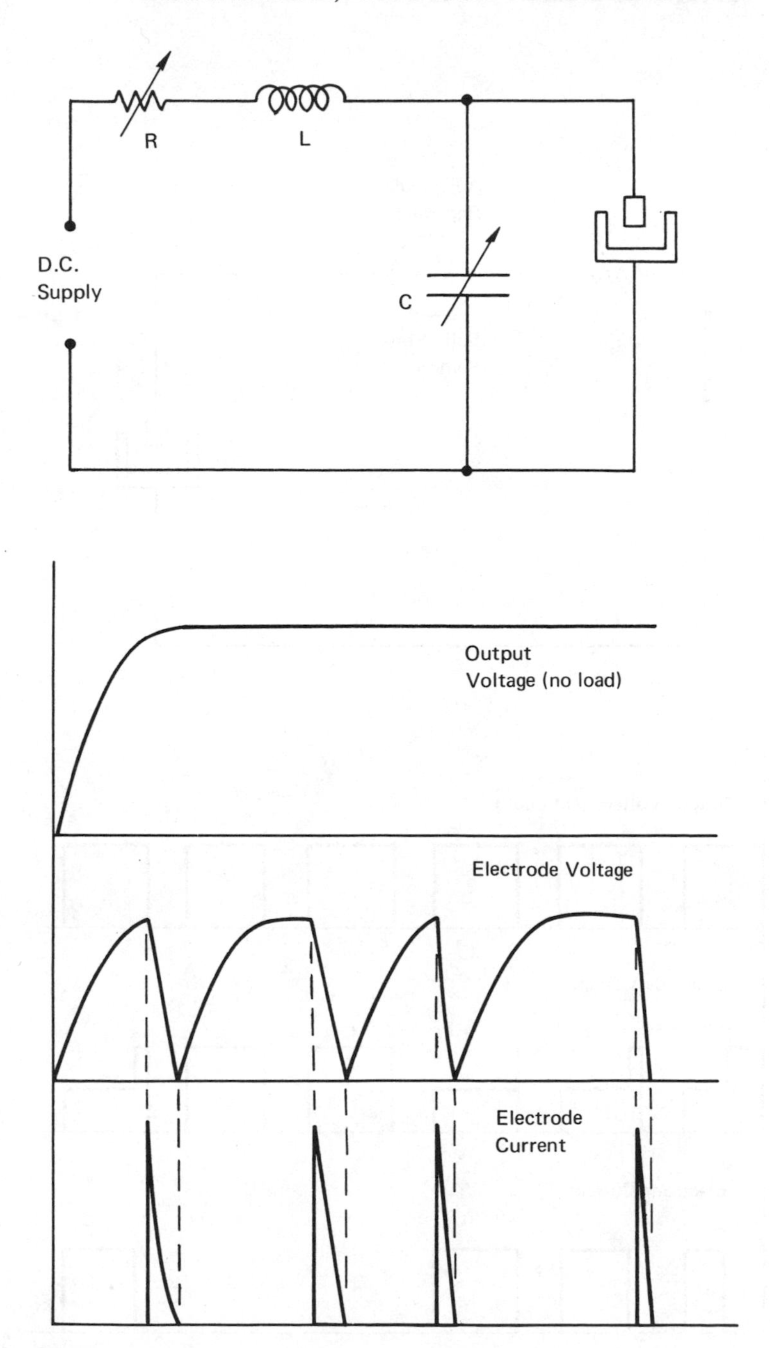

Figure 6-4. Relaxation oscillator power supply with inductor in charging circuit.[2]

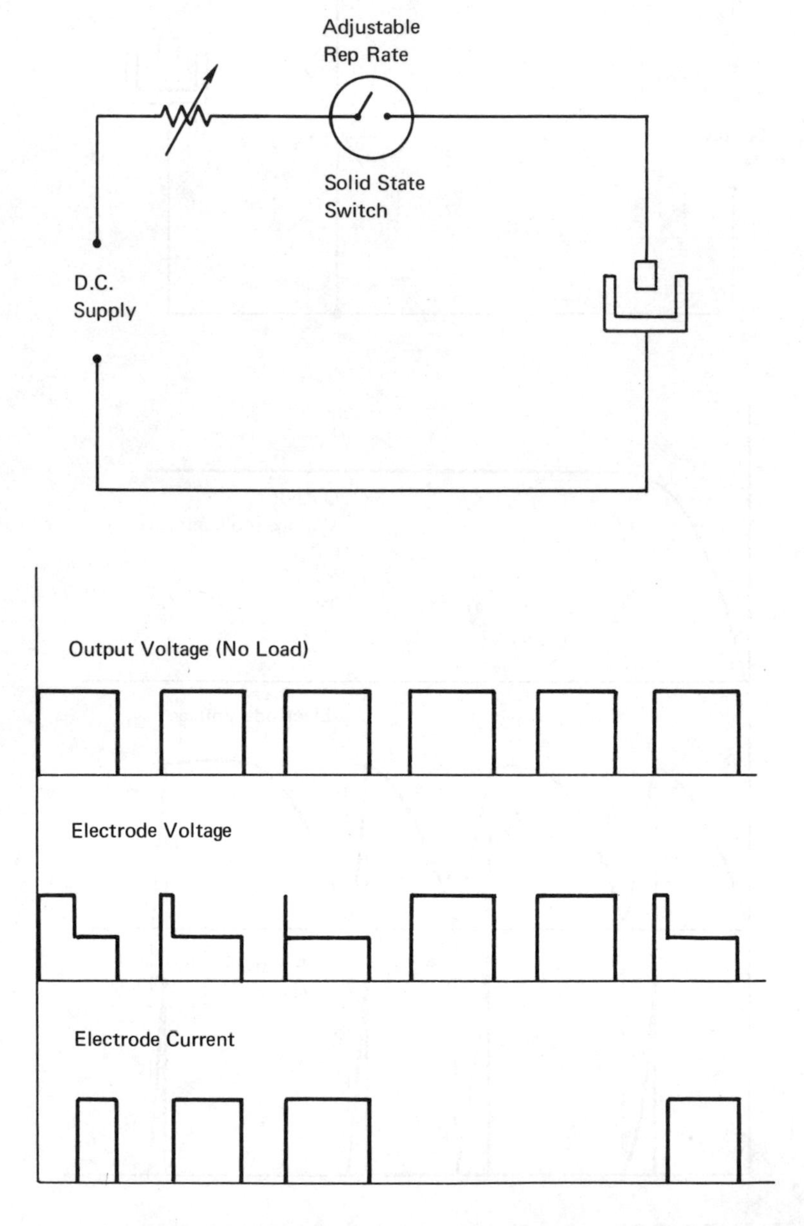

Figure 6-5. The poised power supply and its waveforms.[2]

not flow. This type of supply is capable of very high currents and, hence, high rates of metal removal with a rough finish and unpredictable overcuts. It is used primarily for rough cutting.

With a constant energy supply, a sensor detects the initiation of ionization through the current flow in the circuit (Figure 6-6). The control circuit then turns off the supply at a predetermined time after current begins to flow. The result is a uniform current pulse of constant height and width, but with variation in repetition rate. The accurate current control results in repeatable uniform finishes, overcuts, and maximum metal removal for a given surface finish.

The prevention of arcing is very important in EDM, continued arcing seriously affects the precision and surface finish, and makes it impossible to predict overcut exactly. In the relaxation oscillator type of power supplies, the charging resistor is normally large, to assure commutation of the spark gap. This results in a decrease in the machining rate. Krishnan proposed a method for sensing the occurrence of arcing and for temporarily shorting out the power supply on the charging side.[4] This would permit the use of smaller charging resistors. Results of experiments on Krishnan's circuit show an increase in machining rate coupled with improved surface finish. Because it disconnected the power without operator assistance, Krishnan's circuit essentially accomplishes adaptive arc control.

Adaptive control of EDM has also been proposed by King. In his paper, King proposes that in EDM, an adaptive control system may be used to adjust pulse amplitude, width and wave shape, and to control arcing and excess current density.[5]

For use in EDM machines, electrodes must be shaped slightly smaller than the desired cavity, to allow for overcut. They are normally made of an easily machined material such as graphite, copper, or brass. The tool is mounted onto a servomechanism which feeds it into the workpiece and maintains the correct gap between it and the workpiece. The conventional method of maintaining the correct gap is shown in Figure 6-7. Gap breakdown voltage is compared to a predetermined voltage, and the error signal is amplified and made to control the position of the hydraulic ram attached to the tool.

In the design of an EDM tool, there are two important considerations. First, flushing holes must be provided in the tool, for failure to provide correct flushing of the surface could result in arcing and damage to the workpiece. Some alternate flushing methods are shown in Figure 6-8. Second, the shape to be machined must be thoroughly studied to determine whether economics in time could be realized by multistep operations. For example, in Figure 6-9 the required cavity may be machined with a single tool as in (a), or with four different tools as in (b).

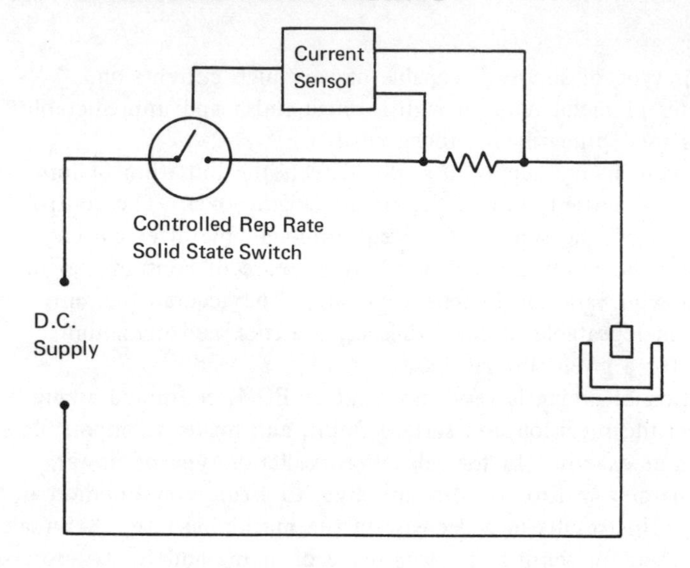

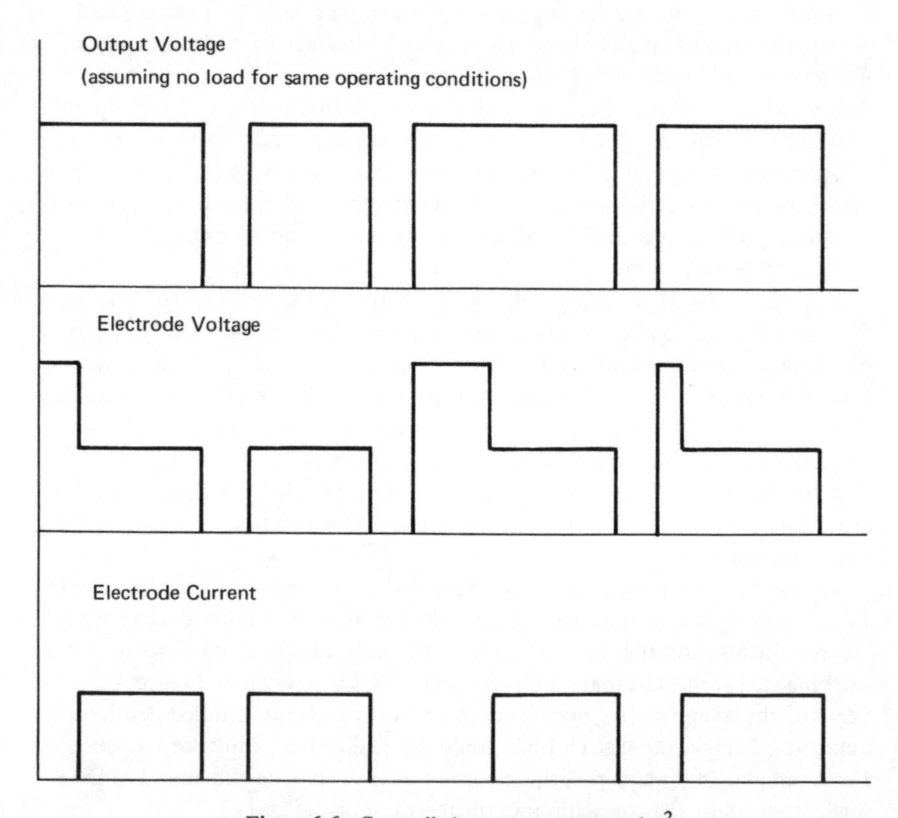

Figure 6-6. Controlled energy power supply.[2]

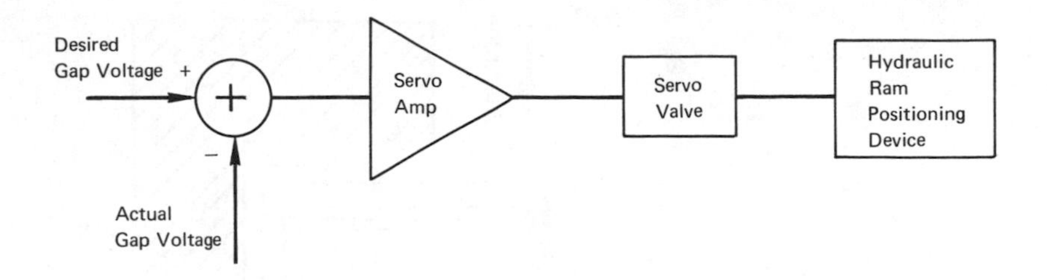

Figure 6-7. Electrode gap adjusting servosystem.[2]

The table required for EDM machines must allow for accurate X-Y translation of the workpiece. Standard tables used from conventional machines (*e.g.*, milling machines) are satisfactory. In EDM applications, however, the table and workpiece are immersed in the dielectric fluid; hence, it is important that the table be of a material that will not be corroded by the fluid. Also, the table must be insulated electrically from the tool.

In an EDM machine, the dielectric fluid fulfills four very important functions:

1. It concentrates the breakdown to a narrow region in the gap;
2. It dissipates the heat and hence contributes to the workpiece erosion;
3. It flushes out the eroded particles and hence prevents arcing; and
4. It lowers the breakdown voltage of the gap.

According to Pashen's Law, breakdown of an air gap can never occur at less than 350 V. The presence of the fluid allows EDM to be operated at voltages as low as 20.

The fluid circulation system must therefore consist of a reservoir, a pump and a filter (Figure 6-1). The fluid is forced into the tool-workpiece gap by various methods, including pressurized pumping, vacuum and tool vibration. Some of the more common fluids used are kerosene, transformer oil, distilled water, silicon oil, and compressed air.

Machining Properties of EDM. The machining properties of EDM are the following:

- The machining rate is proportional to both the energy contained in the pulse and the pulse repetition rate;
- The material removal rate under ideal conditions is .05 $in.^3$/hr/A. This rate decreases with increased repetition rate and poor flushing.
- The practical tolerances that can be held under optimum conditions are $\pm$.0005 in. with $\pm$.0001 in.;

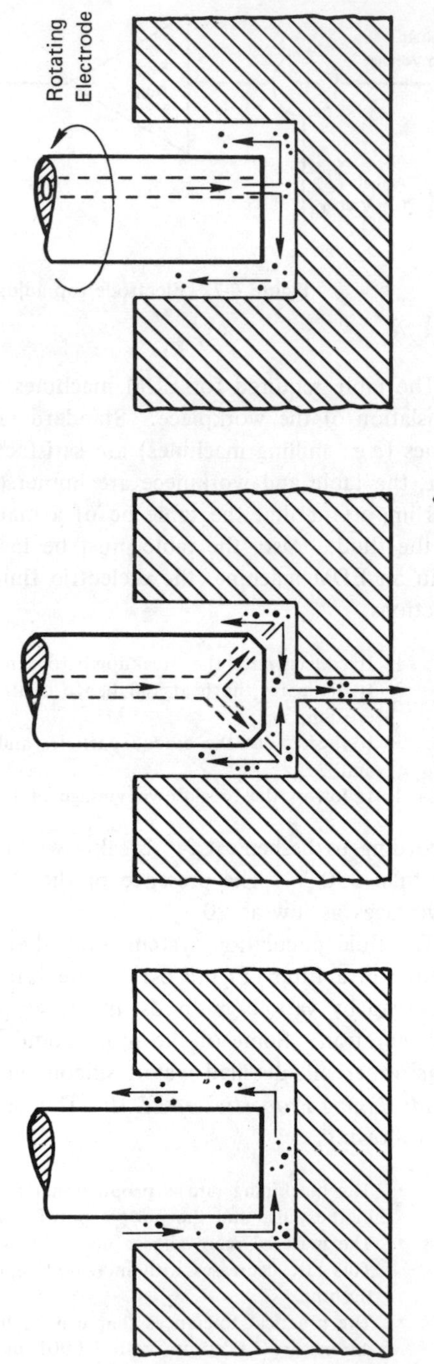

Figure 6-8. Tool design to improve flushing.[2]

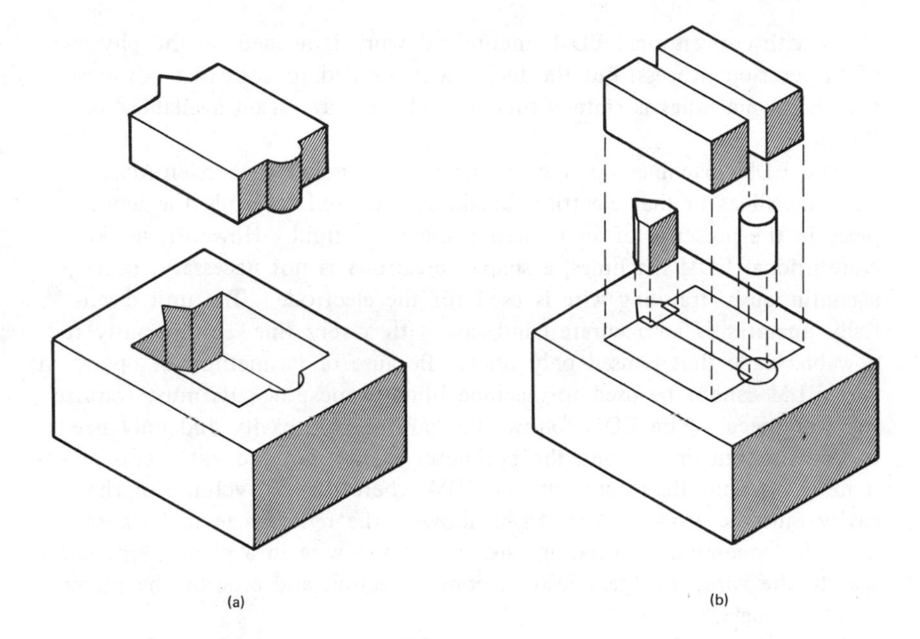

Figure 6-9. Tool design to minimize material removal.[2]

- The finish depends on the pulse repetition rate, pulse energy and the shape. Some typical finishes are 30 RMS at .010 $in.^3$/hr material removal rate and 400 RMS for 3 $in.^3$/hr ;
- Overcut depends on pulse energy and repetition rate. Typical figures are .0006 in. to .022 in. at low and high material removal rates, respectively. Generally it is possible to predict the amount of overcut for given operating conditions;
- EDM machining leaves a recast layer at the surface of the cut which may be an advantage or a disadvantage. When undesirable, it may be removed by a secondary operation. The depth of the recast layer depends on the operating conditions and and may be between .0002 in. and .005 in. thick; and
- EDM machining does not leave a burr that will require removal in a secondary operation. This is particularly significant with hard materials.[3,6-21]

Traveling Wire Electric Discharge Machining

Traveling wire EDM (or wire EDM) was first developed in Switzerland and introduced to the U.S. in 1969. It did not gain popularity until 1974 when two U.S. companies (Andrew Engineering and Elox) began manufacturing the machines.

As with conventional EDM, theoretical work is needed on the physics of the erosion process, but the technology needed to construct advanced wire EDM machines is state-of-the-art, and units are again available "off the shelf."

Wire EDM machines operate on the same principle as conventional EDM machines in that electrical breakdown is used to erode the workpiece in the presence of an insulating dielectric fluid. However, unlike conventional EDM machines, a shaped electrode is not necessary; rather, a continuously traveling wire is used for the electrode. The unit essentially operates as an accurate band saw with a very fine, continuously renewable blade that is used only once. Because of its method of operation, wire EDM cannot be used to machine blind shapes; all cuts must traverse the workpiece. Wire EDM follows the outline of a cavity and only needs to erode a thin line around the perimeter of the required cut. This makes it more efficient than conventional EDM where the full volume of the cavity must be eroded. Wire EDM allows a die relief angle to be automatically generated by rotating one end of the wire in a plane perpendicular to the wire, thus generating a conical section and eroding the material at an angle.

The dielectric fluid used with wire EDM is deionized water. Water wets the surface better than oils and hence the recast layer is thinner than with conventional EDM. Wire EDM has the same pitfalls as conventional EDM with respect to arcing and excessive pulse repetition rate.

Components of a Wire EDM Machine. The basic components of a wire EDM machine are shown in Figure 6-10. They are a pulse power supply, a wire supply and tensioning system, a servocontrolled table with two degrees of freedom for mounting the workpiece, and a table location control system. Although it is conceivable that an operator could control the table motion manually, a tape control unit is generally used to perform this function. The more sophisticated machines have computers that translate simple instructions into a stored program to drive the work table. The computer will also take into account overcut and automatically machine a punch or a die. In some of the larger machines where the table screw drives may not be linear to .0001 in., the computer can be made to correct for repeating errors in locations if those errors are programmed in ahead of time.

The different power supplies used for conventional EDM are useable for wire EDM as well. The wire used is usually copper with a diameter ranging between .002 and .020 in.; the most popular size is .008 in. Finer wires are also available for micro-EDM operations. An accurate tensioning and positioning system is required for optimum operation. The current is usually fed to the wire at equal distances on both sides of the

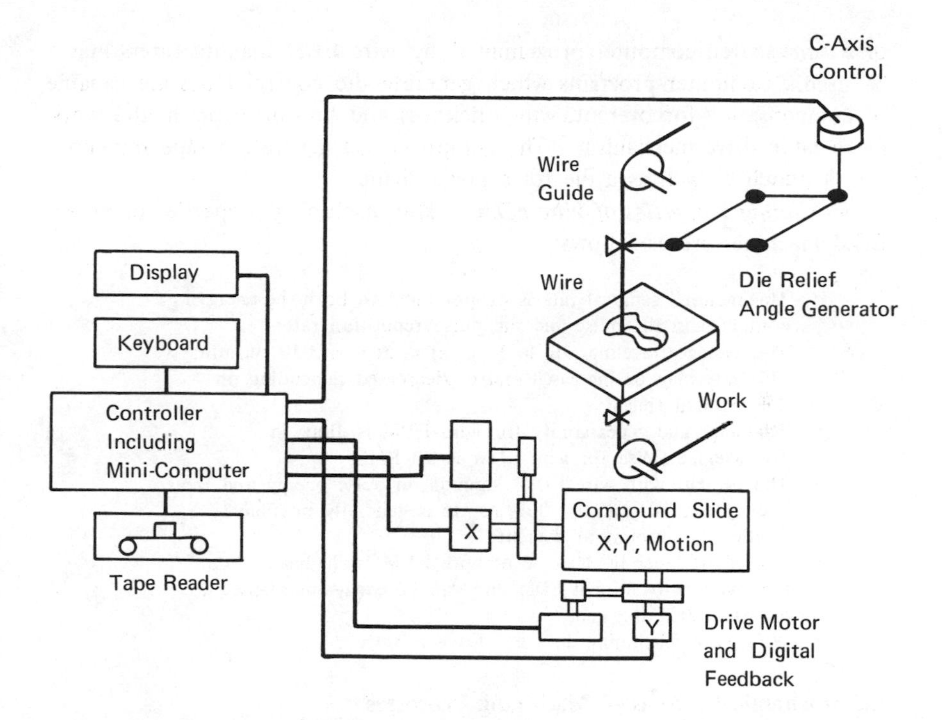

Figure 6-10. Schematic of traveling wire EDM.[10]

workpiece to balance out the voltage drop in the wire that is prevalent at high frequency.

The table required by a wire EDM machine is more elaborate than that required for conventional EDM. A table with two degrees of freedom which may be driven accurately and repeatedly with a tape drive is required. Unlike conventional EDM, the table need not be immersed in the dielectric fluid. The fluid is instead forced into the workpiece in the cavity around the wire and is collected on the other side of the workpiece. The table, however, must still be electrically isolated from the wire.

The dielectric fluid used in wire EDM fulfills the same function as in conventional EDM. The fluid is forced around the wire either by pressure, by a vacuum, or by a combination of the two.

In wire EDM, automatic control is mandatory, and is usually obtained through use of a tape control unit. The tape can be programmed directly, but this requires qualified, specially trained programmers. A properly programmed computer can also generate the tape when given simplified instructions about the desired cut. The computer can be part of the machine or a time-shared computer programmed by wire EDM manufacturers may be used. Computer programs which generate the control tapes are capable of compensating for overcut, wire thickness and known errors in the workpiece table drive mechanism. The computer can generate a tape for cutting a punch as well as a die for a given shape.

Machining Properties of Wire EDM. The machining properties of wire EDM machines are as follows:

- The material removal rate is proportional to both the energy contained in the pulse and the pulse repetition rate;
- The average traveling rate in 1-in.-thick steel is .030 in./min; This rate may be increased and/or decreased depending on the required finish;
- Tolerance and repeatability for wire EDM is .0005 in.;
- The average finish for wire EDM is 30 RMS;
- The overcut with wire EDM depends on pulse energy and frequency and is typically .001 in. It is generally possible to predict the amount of overcut;
- The recast layer in the case of wire EDM is shallower than that with conventional EDM and can be easily maintained to less than .001 in.; and
- Wire EDM machining does not leave a burr.

Electrochemically Assisted Machining Processes

Electrochemically assisted machining processes, all well-known, state-of-the-art processes, include:

- electrolytic polishing and etching, in which a current is used to electrolytically dissolve the surface of a conducting workpiece. A layer is removed evenly from the total surface of the piece;
- electroforming, in which the current is used to deposit a metallic layer over a conducting form. The form is later removed after a sufficient thickness has been deposited on it;
- electrochemical machining (ECM), which is similar to the etching operation except that the current is localized between an electrode and a portion of a workpiece resulting in much faster erosion of the workpiece next to the electrode. This process was originally patented by a Russian, W. Gusseff, in 1929, but was not put to use until the 1950s;
- electrochemical grinding (ECG) which is a variation of EDM in which the tool is a rotating, conducting, grinding wheel; and
- electrochemical deburring (ECD) which is also a variance of ECM except that it is used for removing burrs caused by other machining processes.

ECM, ECG and ECD only will be discussed in this chapter.

Electrochemical Machining

In ECM, a tool which has a shape similar to the intended shape of the cavity to be machined is immersed in an electrolyte together with the workpiece. A cd power supply is connected between the tool and workpiece (Figure 6-11). If the gap between the tool and workpiece is kept

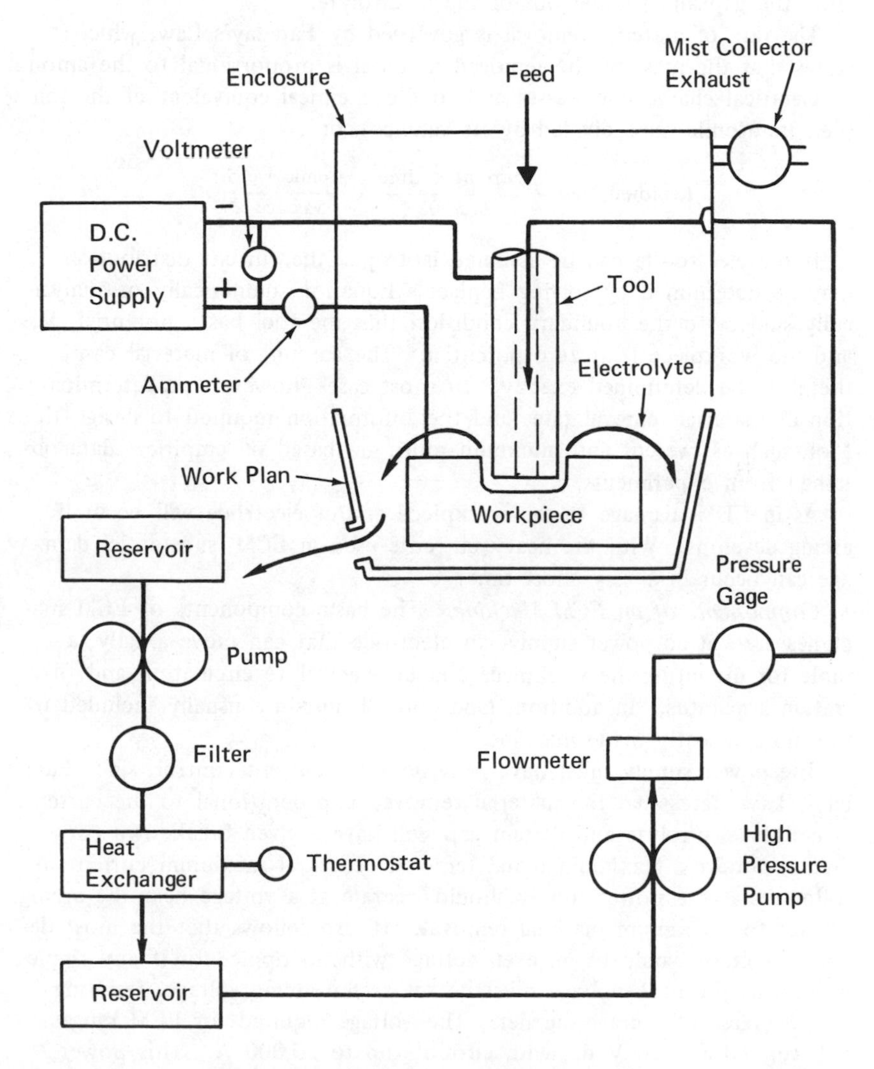

Figure 6-11. Schematic of ECM machine.[2]

small, most of the current will flow in the volume defined by the tool surface opposite the workpiece. If the tool is made negative and the workpiece positive, material will be etched away from the workpiece to be deposited on the tool. However, if the electrolyte is a salt such as sodium chloride, the metal removed from the workpiece will form metal hydroxides, which are insoluble in water. If the electrolyte is made to flow between the tool and workpiece, this precipitate can be removed from the gap and filtered out of the electrolyte.

The rate of material removal is governed by Faraday's Law, which states that the mass of the removed material is proportional to the amount of electrical charge that passes and to the chemical equivalent of the ion (*i.e.*, its atomic mass divided by its valence), or

$$\text{Removed Mass} = \frac{\text{current x time}}{96{,}500} \text{ x } \frac{\text{atomic weight}}{\text{valence}}$$

If the electrolyte can be assumed isotropic, the current distribution may be determined by solving Laplace's Equation, numerically or analytically subject to the boundary condition that the tool has a potential, V, and the workpiece is at zero potential. The amount of material can therefore be determined exactly. In most cases, however, predetermination of material removal rates and the information required to design the tool, such as overcut and minimum radii, are based on empirical data obtained from experiments.

As in EDM, damage to the workpiece and/or electrode will occur if arcing develops. With the heavy currents used in ECM, substantial damage can occur in a very short time.

Components of an ECM Machine. The basic components of ECM machines are: a dc power supply, an electrode that can move axially, a table for mounting the workpiece and an electrolyte circulation and filtration apparatus. In addition, tape control units are usually included in the more recently made machines.

The power supply must have provisions for current control, since Faraday's Law states that the material removed is proportional to the current. A certain electrolyte and certain gap will have a given breakdown (or arcing) voltage. Maximum metal removal occurs at maximum current; it follows that the power supply should operate at a voltage near the arcing voltage for maximum material removal. It also follows that the most desirable voltage would be an even voltage with no ripple, for if any ripple is present, the peak voltage must be set at the arcing voltage, resulting in a decrease of average currnet. The voltage required for ECM ranges between 10 and 20 V dc, with currents up to 20,000 A. This power cannot realistically be supplied by batteries; stepdown transformers and rectifiers must be used to convert the commercially available supply to the required dc supply.

For the required power levels, filtering is not normally used, hence the dc supply has a ripple voltage superimposed over it. To minimize this ripple, supplies are usually three phase with six-phase secondaries (using delta and wye combinations) with a ripple frequency of 720 Hz (for 60-Hz input). Also, although continuously variable output power supplies are desirable (and are available at the smaller power ratings), they are impractical for the higher power ratings. In this case, multitap transformer supplies are usually supplied for stepped adjustments of voltage. Phase delay rectifiers using silicon controlled rectifiers are also used to control the current; however, as the firing angle is decreased, the ripple increases, decreasing material removal rates. In some cases, phase delay rectifiers are combined with a multitap transformer to supply continuously variable dc power with low ripple.

It should be noted that at the relatively low voltage required and high currents, the drop in the semiconductor rectifiers becomes a significant portion of the input power. This heat must be dissipated, and air- or water-cooled heat sinks are used. In either case, because of the corrosive atmosphere in the vicinity of the machine (because of electrolyte evaporation, splashing, spillage), the power supply must be corrosion resistant.

Another important required feature is arcing detention and quick shut-off. Because of the high currents, arcing, if it occurs, can cause excessive damage in a relatively short time. Hence, rapid shut-off protection is usually used to protect the tool-workpiece. The cost of such shut-off provisions is inversely proportional to the speed required for shut-off. The larger the electrode, the more current is used and the quicker the shut-off required. Quick shut-off adds an appreciable amount to the cost of a machine.

Tools shaped smaller than the desired cavity (to allow for overcut) must be made for ECM machines. They are usually made of materials that can withstand the corrosive effect of the electrolyte. Materials often used include stainless steel, brass and copper tungsten, and must also have good machining properties. This is especially important when the shape to be machined is complicated. Good thermal and electrical conductivity is another important characteristic of the tool material. Good thermal conductivity is important to minimize damage in case of arcing. If the heat produced by the arc can be easily dissipated, there is less likelihood of melting the tool.

The sides of the tool are usually insulated to prevent further enlargement of the cavity as the tool penetrates the workpiece. Insulation materials include epoxy, porcelain, vinyl, enamel and Teflon®.

With ECM, it is particularly important to design proper electrolyte flow orifices. The electrolyte pressure is over 200 psi, and sharp corners or sudden changes in fluid velocities could adversely affect the finish of the workpiece. Also, the particle-laden electrolyte could abrade and deteriorate the insulation. For this reason, it is desirable to design the flow path so that fresh electrolyte comes in contact with the machined surface and the contaminated electrolyte is removed through the tool, as shown in Figure 6-12.

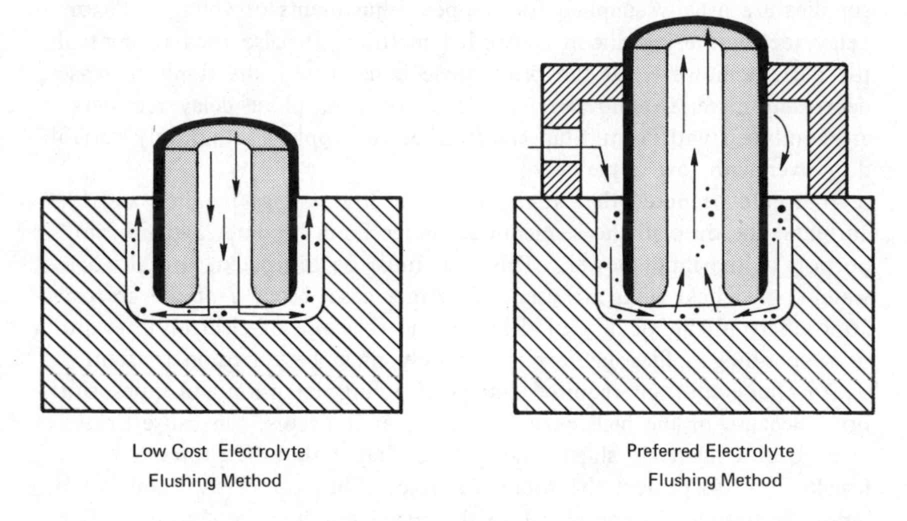

Figure 6-12. Methods of electrolyte circulation in ECM.[2]

As with all other components in an ECM system, the workpiece holding table must be corrosion resistant, as it is immersed in the electrolyte and is subject to constant sprays and splashes. Also, since the fumes from such a process are potentially hazardous, the tool and work table must be enclosed and provided with an exhaust system to remove fumes. Compliance with local pollution requirements must be observed.

In an ECM machine, the electrolyte fulfills four very important functions:

1. It provides a conductive path for the current between the tool and workpiece;
2. It provides the hydroxyl ions required to combine with the metal ions to form the metal hydroxide precipitates;
3. It carries away the hydroxide precipitates; and
4. It cools the electrode/workpiece.

It is important that the components of the electrolyte circulation system be corrosion resistant. A heat exchanger is required in electrolyte circulation loops to maintain constant electrolyte temperature, since the conductivity of the electrolyte varies with temperature and repeatability, and predictability depends on accurate control of the variables.

The size of the filtration system depends on the machine size. Three filtration systems are available:

1. settling tanks, which must be very large because of the slow settling rate of the metal hydroxides;
2. filter elements, which require frequent replacement; and
3. centrifugal separation, which although most expensive is also relatively maintenance free and requires one tenth the floor space of the settling tank.

Machining Properties of ECM. The following are the machining properties of ECM:

- The machining rate is given by

$$\text{Volume of metal removed} = \frac{\text{current x time}}{96{,}500} \text{ x } \frac{\text{atomic weight}}{\text{valence}} \text{ x } \frac{1}{\text{density}}$$

- The practical tolerances that can be held are $\pm.002$ in. with $\pm.0005$ in. possible;
- The finish depends on the metal removal rate and improves with high current and high feed rate. Typical finish is 50-125 RMS;
- Under certain conditions, intergranular attack of the machined surface may occur, resulting in weakening of the surface. This may be remedied by changes in electrolyte, tool material or a secondary ECM or EDM operation to remove the weakened layer;
- The overcut depends on current and feed rate and ranges between .006 in. and .050 in. but is predictable. Because of the relatively large overcut, sharp corners may not be machined with ECM;
- Surface striation and nonuniform finish sometimes occur if the electrolyte flow path is not properly designed; and
- ECM machining leaves no recast surface and no burrs.[21-23]

Electrochemical Grinding

ECG is very much like ECM except that a rotating, conducting grinding wheel is used to remove about 10% of the material with ECM action removing the other 90%. ECG is usually used for face-wheel grinding for small pieces where the entire surface may be ground at once or in applications in which wheel redressing is impossible or impractical. With 90% removal with ECM action, the interval between wheel dressings is lengthened.

As shown in Figure 6-13, the components of the system are similar to those for an ECM system except for several details. The power supply is smaller in ECG, with an output between 4 and 15 V dc and current up to 1000 amps, and the electrolyte circulation and filtration system is

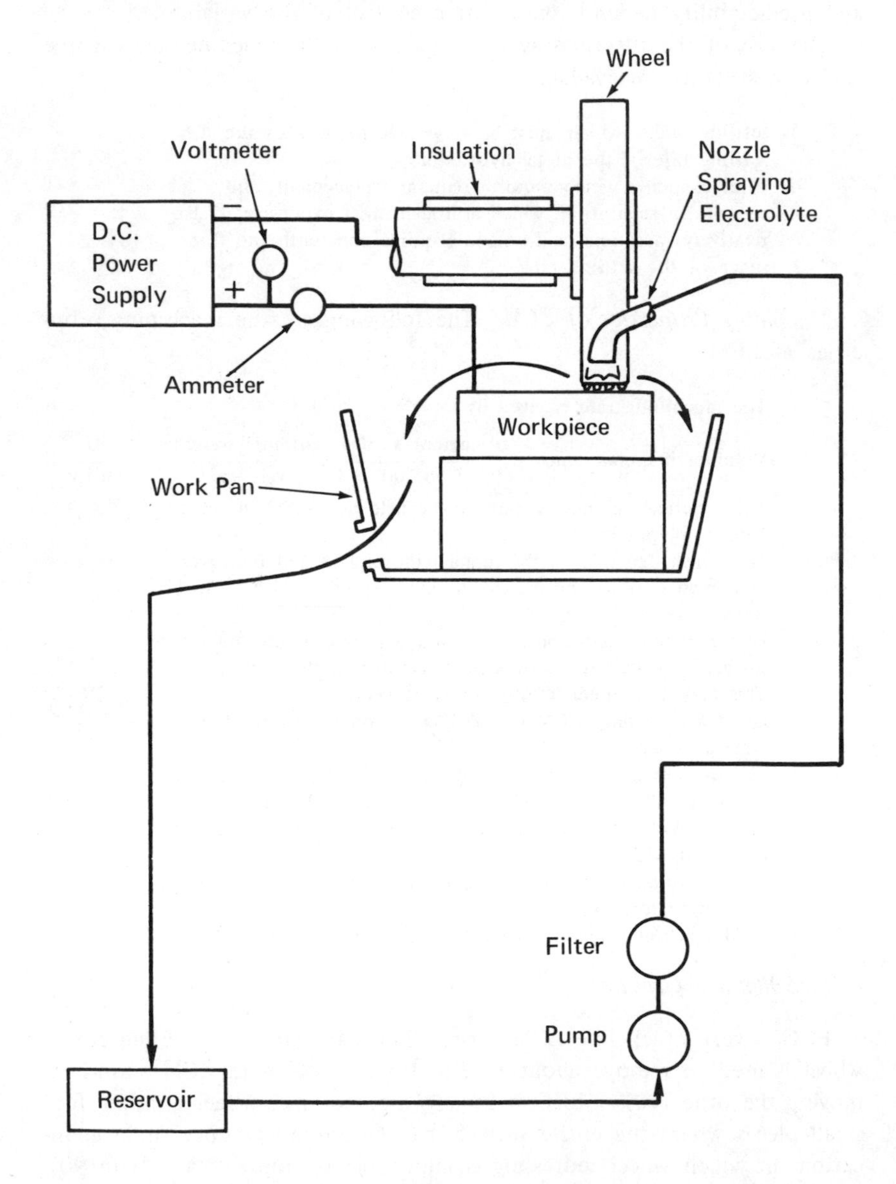

Figure 6-13. Schematic of ECG machine.[2]

smaller. Also, the tool is a rotating grinding wheel driven by a separate motor of up to 12 hp.

The machining properties of ECG are similar to those of a conventional grinder:

- Finish depends on grinding wheel surface;
- Tolerances of ±.001 in. can easily be held; ±.0005 in. is possible with 90% material removal with ECM. High accuracy is possible at the cost of more use of the grinding wheel; and
- Overcut can be eliminated with ECG.

ECG may be retrofitted to present grinders if the grinders can be made corrosion resistant.[24,25]

Electrochemical Deburring

ECD is also like ECM in that it depends on electrolysis to remove burrs left over from other machining methods. In ECD, however, the tool and workpiece are placed in a fixed relative position (*i.e.,* they do not feed toward each other) before the electrolyte is introduced and the current applied. Hence, the metal removal rate will vary with time as the gap increases between tool and workpiece.

ECD is usually used for removing burrs from parts with surfaces that must not be scratched or altered. Also, in cases where the burr is complex but predictable, such as with helical spur gears or hydraulic valve spool bodies, ECD may be the only economical method for removing the burr.

The components of an ECD machine are similar to those of an ECM machine (Figure 6-14), although the power supply is smaller, with an output between 3 and 24 V and current up to 500 A per stations. The higher voltage is required to cope with the increased gap as the burr is dissolved. An ECD machine is usually a multiple station machine, and many parts may be deburred simultaneously. The ECD tool must be specially designed for the particular part to be deburred, and it is important that the burr occur at the same place in each part.[26,27]

APPLICATIONS

Major consumers of electrically assisted machining equipment are tool and die manufacturers who, in turn, serve almost every market that depends on metal or plastic parts. Because of this widespread utilization of these machining processes, a definition of market areas would be too broad. Therefore, a description of the kinds of operations for which each process is applicable is provided.[28]

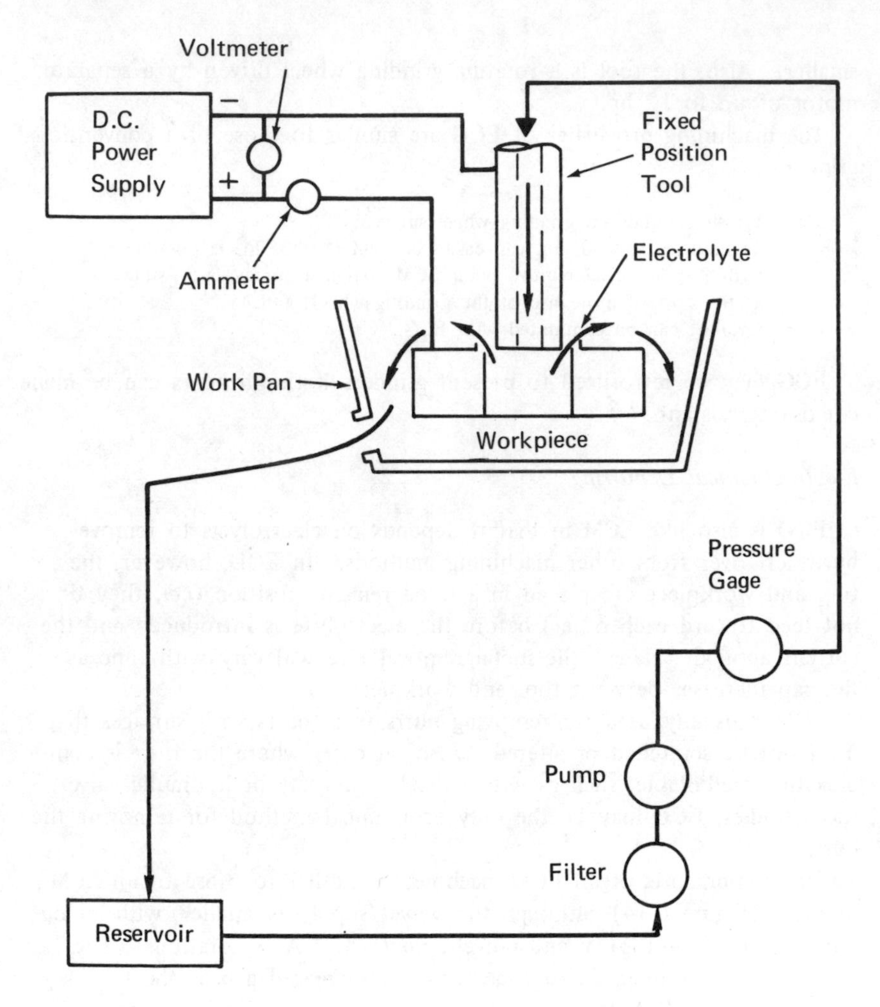

Figure 6-14. Schematic of ECD machine.[2]

Conventional Electric Discharge Machining

Conventional EDM machines are used for difficult and specialized machining of hardened steels and hard alloys that are difficult to machine by conventional methods or special operations for which conventional methods are slow or difficult. Principal cutting applications for conventional EDM are:

- dies (stamping, cold heading, forging, injection molding);
- carbide-forming tools;
- tungsten parts;

- burr-free parts;
- odd-shaped holes and cavities;
- small-diameter, deep holes;
- high-strength and high-hardness materials;
- narrow slots (0.002 in. wide); and
- honeycomb cores and assemblies and other fragile parts.

Traveling Wire Electric Discharge Machining

Like conventional EDM, wire EDM is useful for cutting hardened steels and alloys that are difficult to machine by conventional methods. Although faster than EDM because only the periphery of a cavity is cut, it cannot be used to cut blind holes. Principal machining applications are:

- punches and dies;
- carbide-forming tools;
- tungsten parts;
- burr-free parts;
- odd-shaped holes;
- small-diameter, deep holes;
- high-strength and high-hardness materials; and
- narrow slots (0.004 in. wide).

Electrochemical Machining

In ECM, metal is removed from the required areas of the work piece by deplating. Principal machining applications include:

- high-temperature alloy forgings;
- turbine wheels with integral blades;
- jet engine blade airfoils;
- jet engine blade cooling holes;
- odd-shaped holes and cavities;
- small, deep holes;
- honeycomb cores and assemblies and other fragile parts; and
- high-strength, high-hardness materials.

Electrochemical Grinding

In electrochemical grinding, 90% of the material removed from a workpiece is by ECM action, and 10% by abrasive action from the tool–an electrically conducting grinding wheel. Principal applications are:

- face wheel grinding–for smaller pieces where the entire ground surface is attacked simultaneously and maximum current used;
- peripheral or surface grinding–best for large pieces when used in conjunction with a large diameter wheel. The large diameter generally produces faster metal removal, accuracy and electrolyte flow;

- cone wheel grinding–ideal for large, flat contact areas where high current can be applied for maximum metal removal. This method is most accurate for large, flat surfaces; and
- form-wheel grinding (square grinding)–applied to the same workpieces as conventional form grinding but, because of negligible wheel wear, expensive dressing is reduced and wheel life is increased, usually 7-10 times.

Electrochemical Deburring

For parts in which selective deburring is desired and for surfaces that must not be scratched or altered, ECD is the best and often the only possible method. ECD is ideally suited for removing complex burrs from parts such as helical spur gears, hydraulic valve spool bodies and those parts with cross-drilled holes.

GENERAL ASSESSMENT OF THE FUTURE

Economic Considerations

Electrically assisted machining processes require specialized equipment, which is expensive compared to conventional equipment. In cases where hard materials are to be cut, however, they may be the only choice. Table 6-8 shows a comparison of the cost of material removed by ECM, EDM and conventional milling on 4340 steel. It is clear that even with

Table 6-8. Comparison of Costs of Machining Methods

	ECM	EDM (Conventional)	Conventional[a] Milling
Removal Rate (cm^3/sec)	2.64	1.33[b]	1.11[b]
Surface Finish (μ)	10-250	25-1200	50-500
Depth of Possible Damage (cm)	.005	.0125	.0025
Approximate Machine Cost ($)	100,000	10,000	4,000
Normalized Cost of Material Removal	10.5	2.1	1.0

[a]Stagger tooth milling of 4340 steel.
[b]The effective removal rate may be much larger depending on the method of machining (see text).

such a steel, conventional machining is at least half as costly. It is therefore important for a manufacturer considering the purchase of an EDM or ECM machine to ascertain high machine utilization before purchasing such a machine.

Of the electrically assisted processes, wire EDM machines provide the most rapid removal rate since material need be removed only around the periphery of the cavity to be generated. This is also true with conventional milling if it is carried out with a small tool around the periphery of a shape to be cut out.

If a tool manufacturer intends to machine hard steels and carbides, the only realistic choice is to purchase either an EDM or and ECM machine. In cases where the part to be machined has a number of cavities separated by thin walls, such as a die for microcircuit components, conventional machining is not possible as machining must be performed without any forces applied between the tool and the workpiece. In this case an EDM or ECM machine is indicated.

Market Behavior and Trends for EDM Machines

Sales of EDM machines by U.S. and foreign manufacturers in the United States are shown in Table 6-9. The figures shown are for combined sales of conventional and traveling wire EDM machines. It appears that the number of machines sold has been relatively constant from 1968 through 1975. This is somewhat misleading, however, since the U.S. economy has not been constant; in fact, business in those years was slow.

Table 6-9. EDM Machines Sold in the U.S. (Conventional and Wire EDM)[29]

	U.S. Production		Imports	
Year	Number of Units	Cost in ($ millions)	Number of Units	Cost in ($ millions)
1968	1,126	20.9	NA	NA
1969	1,149	16.3	NA	NA
1970	922	12.8	NA	NA
1971	913	10.3	NA	NA
1972	860	9.5	195	1.5
1973	1,011	15.7	146	2.2
1974	884	17.4	230	4.6
1975	NA	NA	218	6.0

Hence, constant sales actually indicate a rise in popularity. Figure 6-15 depicts the average unit cost plotted for each year for both U.S. and foreign-made machines. Unit cost was high in 1968, and decreased monotonically until 1972. In 1968 the machines were relatively new; as manufactured cost reductions were instituted, the price decreased. Around 1972, computer control and wire EDM became popular, resulting in price increases. High inflation rates from 1973-1975 also added to the price increases.

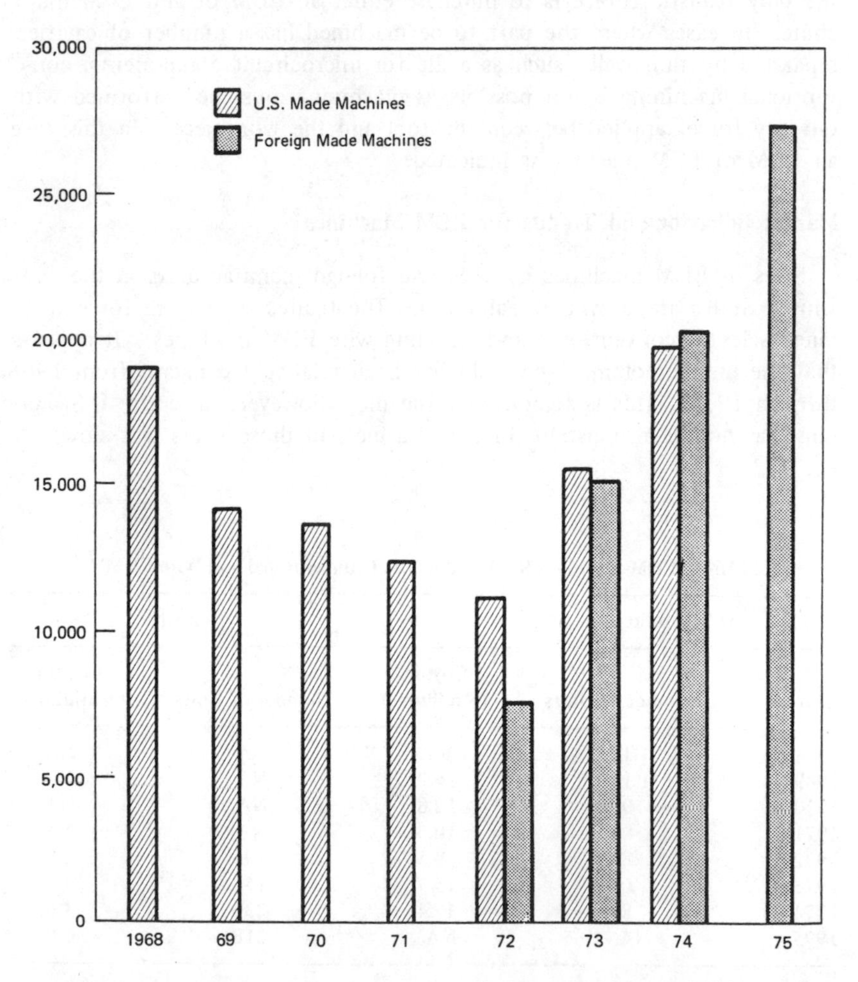

Figure 6-15. Changes in unit cost of U.S. and foreign-made EDM machines.[2]

According to the industry, it is absolutely essential for tool manufacturing firms to acquire EDM machines if they intend to remain competitive. EDM machines are indispensable for cutting the hardened tool steels, and with increased demand for tools to manufacture the parts needed for miniature circuits, capabilities for producing intricate shapes in hard steels have become mandatory. Hence, it is expected that the demand for EDM machines should be high, and the market is nowhere near saturation.

Machine cost is expected to increase as demand for more sophisticated, computer-controlled machinery increases, even though the cost of the actual components of the machine may decrease. As a striking example of decrease in component costs, the computer used by one manufacturer's machine cost $28,000 at the end of 1975. Around July of 1976 the same computer was expected to sell for $10,000, but at the same time the machine costs went up slightly because of improved accuracy and circuit reliability.

It is believed that demand for conventional EDM machines will remain high and more or less follow the GNP. For traveling wire EDM, on the other hand, the demand should increase rapidly in the next few years as toolmakers acquire these machines and as computer prices decrease. In the area of large EDM machines with electrodes to 25 tons for the auto industry, the demand should increase slightly as economic conditions improve. To date, EDM machine manufacturers have promoted their machines mainly for tool construction. As labor costs go up, demand for computer-controlled EDM machines to replace conventional machines shop units as well as machines for production use should increase.

For imported machines, U.S. demand is for two types (see Table 6-10): (1) the very low cost machines that are no longer available; and (2) the highly sophisticated units. The unit cost trend shown in Figure 6-15 also supports this. It should be noted that a discrepancy exists between Tables 6-9 and 6-10. The number of units sold is 218 in Table 6-9 and 216 in Table 6-10. Those figures were obtained from the Department of Commerce and are preliminary.

Market Behavior and Trends for ECM Machines

Sales of ECM machines by domestic manufacturers in the United States are shown in Table 6-11. The figures shown were obtained from the U.S. Department of Commerce; data for 1972 or later were not available, for it was felt that the number of machines sold was too small to warrant a special listing.

Table 6-10. Breakdown of Imports by Country of Origin of EMD Machines for 1975 (Conventional and Wire EDM)[30]

Country	Number of Units	Cost ($ thousands)	Average Cost/Unit
Switzerland	161	4670	29
Japan	37	370	10
France	1	5.	5
Spain	2	10	5
United Kingdom	3	22	7.3
West Germany	12	895	74.6
Totals	216	5972	

Table 6-11. ECM Machines Sold in the U.S. (U.S. Production)[29]

Year[a]	Number	Total Cost ($millions)
1968	88	88
1969	92	5.2
1970	117	3.8
1971	85	2.8

[a]No data available after 1971. Data for ECM machines was combined with data for other machine tools.

According to the industry, the future for ECM is not very optimistic. Major ECM manufacturers in the noncommunist countries have dropped to three: (1) Chemform in the United States; (2) Bosch in West Germany; and (3) Hitachi in Japan.

Apparently, ECM manufacturers are not selling machines to entice buyers to buy more machines. They will only sell a machine if they can make a 25% profit. However, although an ECM machine may sell for as low as $100,000, an average machine will sell upwards of $300,000, which puts them out of reach of many firms.

The main advantage of the ECM machine over other machines is high rate of metal removal of hard metals–*i.e.*, rough cutting of hard metals. The ECM operation must be followed in most instances by another polishing type of operation. Although the ECM machine may be used to do this operation also, it is too costly. It must, therefore, be supplemented with an EDM machine. The EDM machine can do the job of the ECM machine with a lower initial cost and more accurately, but more slowly. In an economy of slow markets, the EDM is the more economical choice. The need for ECD machines is created by conventional machining of hard materials; as EDM replaces these conventional machines, the ECD market should become smaller.

Until recently, the ECM industry has depended on sales to the aerospace industry. With the slowdown of the latter, the ECM industry finds itself without a client. The ECM industry may and will attempt to enter the petroleum industry where the demand for drilling bits is high and could possibly be made at a lower price, using ECM.

In conclusion, the U.S. ECM industry must find new markets if it intends to survive. The high initial cost of the ECM machine, coupled with the problems associated with its operation (such as corrosion, toxic fumes) are definite barriers to its expansion. The ECG and ECD markets, with their limited range of applications are not expected to expand greatly.

In Japan, it appears that EDM machines of domestic manufacture are inferior in accuracy and versatility to their U.S. and Swiss counterparts. Wire EDM is being manufactured for sale locally by Hitachi, under license by AGIE Industrial Electronics Ltd. of Switzerland. The Japanese ECM market, on the other hand, is well developed with an estimated 300 units in service and an estimated current demand of 50-60 units per year. While there is no aerospace industry in Japan, Hitachi, the major ECM manufacturer, has diversified its market to include the auto industry, tool and die makers, and the aluminum extrusion industry.

REFERENCES

1. Lazarenko, B.R. and N.J. Lazarenko. "Electro Erosion von Metallen," *Gosenergoisdat,* Moscow (1944).
2. "ECM, ECD, ECG Simplified," Chemform, 1410 S.W. 8th Street, Pompano Beach, Florida 33061.
3. "Fundamentals of EDM," Colt Industries, Elox Division, Griffith Street, Davidson, North Carolina 28036.
4. Krishman, A. "Arc-Free Relaxation Generator for Spark Erosion Machining," *J. Inst. Eng. (India), Div. Mech.* 52 PTME6(11) (July 72).
5. King, F.L. "EDM Goes Adaptive," ASTME technical paper, MR68-119, Dearborn, Michigan.
6. "Application Manual," Colt Industries, Elox Division, Griffith Street, Davidson, North Carolina.
7. "Elox EDM Machine Tools," Colt Industries, Elox Division, Griffith Street, Davidson, North Carolina.

8. "Elox TCV Power Supplies," Colt Industries, Elox Division, Griffith Street, Davidson, North Carolina.
9. "Andrew Model 123 CNC Wire EDM," Andrew Engineering 5520 County Road 18, Hopkins, Minnesota 55343.
10. Description and Specifications–Andrew Model 123, Andrew Engineering, 5520 Country Road 18, Hopkins, Minnesota.
11. Kurr, R. "Rationalization in the Making of Cutting Tool," Publication 294, AGIE Industrial Electronics Ltd., Losone-Locarno, Switzerland.
12. "Shapes Cut by Electric Discharge," AGIE Industrial Electronics, Ltd., Losone-Locarno, Switzerland.
13. "A Guide to the Best in EDM," AGIE Industrial Electronics, Ltd., Losone-Locarno, Switzerland.
14. "Algorithm EMT 20," AGIE Industrial Electronics, Ltd., Losone-Locarno, Switzerland.
15. "AGIEPAC," AGIE Industrial Electronics, Ltd., Losone-Locarno, Switzerland.
16. "Plainfield Said to be 1st Tool Shop Buying New Agietron Wire-Feed EDM," *Am. Metal Market/Metal Work. News* (January 13, 1975).
17. "Eltee Pulsitron EDM," Eltee Pulsitron, 26 Fairfield Pl., West Caldwell, New Jersey 07006.
18. "Electrojet Electrical Discharge Machines," Publication #EM-315-1, Cincinnati Milacron, Cincinnati, Ohio 45209.
19. "Cincinnati Little Scotsman EDM Machine, Publication #EM-134,6, Cincinnati Milancron, Cincinnati, Ohio.
20. "Cintrojet T-IV Power Supply," Publication #EM-308-4, Cincinnati Milacron, Cincinnati, Ohio.
21. "Chemform Special ECM, ECD and EG Machines for All Industries," Chemform, 1410 S.W. 8th Street, Pompano Beach, Florida 33061.
22. "ECM Modular Head," Chemform, 1410 S.W. 8th Street, Pompano Beach, Florida 33061.
23. "Industrial ECM, L2500, L3500, L5000," Chemform, 1410 S.W. 8th Street, Pompano Beach, Florida 33061.
24. "Electrolytic Grinding Principles and Practices," Everite Machine Products Co., 2005 E. Huntingdon Street, Philadelphia, Pennsylvania 19125.
25. "Everite Electrolytic Production Plunge Grinder," Everite, 2005 W. Huntington St., Philadelphia, Pennsylvania.
26. "ECD-A-250, A500, A-1000," Chemform, 1410 S.W. 8th Street, Pompano Beach, Florida 33061.
27. "ECD-500, 1000, 2000," Chemform, 1410 S.W. 8th Street, Pompano Beach, Florida 33061.
28. Lee, W. and G. Feick. "Techniques and Mechanisms of Chemical, Electrochemical, and Electrical Machining of Ceramic Materials," NBS Special Publication #348, National Bureau of Standards, Washington, D.C.
29. National Machine Tool Builders Association, Westpark Dr., McLean, Virginia. Private communication, March, 1976.
30. U.S. Department of Commerce, Washington, D.C. Private communication, March, 1976.

7

EMERGING TECHNIQUES IN ELECTROFORMING OF METALS*

INTRODUCTION

The status and development over the past 20 years of two technically important methods of electrically induced reshaping of metal are discussed in terms of their principal applications and current market. They are the electromagnetic pulse and the underwater spark discharge methods, known respectively as electromagnetic forming (EMF) and electrohydraulic forming (EHF). Both techniques are an outgrowth of the use of exploding wires which began in the 1950s as a method of generating high-energy pulses of very short duration. In both EMF and EHF, deformation of the metal into the desired shape takes place as kinetic energy imparted by the energy pulse drives the metal against a fixed mating part or die. The two techniques have complementary advantages in that EMF is suited to high electrical conductivity, more plastic metals while EHF works best on highly refractive metals of low electrical conductivity. Automatic assembly is much easier with EMF and this has opened a viable, though not growing, market for its service in mass-produced goods such as automobile parts. On the other hand, EHF has been limited to the small aerospace market for exotic manufacturing steps which can be accomplished in no other way.

Some of the most promising techniques recently developed in metal forming have been the electromagnetic and electrohydraulic methods of applying high-energy pulses to metal parts to form them into shapes precisely defined by either a die or a mating part. The energetic level of the pulse and to some extent its propagation geometry can be well controlled through the parameters of the driving electrical circuit. In electromagnetic forming, a strong magnetic field is excited in a small space

*By: Norman Lord

between a work coil and a conductive metal workpiece. The workpiece is then deformed against a die as this field exerts a pressure to expand its domain. In electrohydraulic forming, a spark discharge takes place underwater creating a shock wave which propagates to a workpiece of any ductile material mounted nearby. This piece is then deformed into a die to reproduce its shape.

These two techniques have complementary advantages. Electromagnetic forming involves simpler workpiece mounting methods so that it can easily fit into a mass production process with very high repetition rates. However, it is usually limited to high conductivity metals. Electrohydraulic forming involves simpler workpiece mounting methods so that it can a vacuum on the other, is not suited to high repetition rate arrangements. It is, however, suited to low-conductivity metals with very high refractory properties. Both techniques provide excellent reproductions of die shapes so that their products can be used in higher precision assemblies which cannot use conventional stamped products. They are techniques which were developed for high-technology industrial processes where the extra expense of the equipment and special care in its operation can be absorbed in the final product. The economic importance of these techniques has not grown past the initial developmental stages when a small industrial niche for their use was carved out during the 1960s. Some promise of further commercial progress may be drawn from the fact that although, in each case, the inventors of the technique have left the marketing organizations, each organization has been able to continue under new management.

Both electromagnetic and electrohydraulic forming were developed as outgrowths of the experiments on exploding wires started in the 1950s. With such experiments they shared a widespread technical interest in the 1960s centered on the apparent capability for controlled handling of very high-energy, short-lived energy pulses.

Forming a metal part to a new shape requires that it be stressed past the elastic limit of the material so that the new shape is established as a permanent deformation. The high impulse energy of a forcing or stamping process has been the accepted means of reshaping metal in ordinary mass production programs. This involved repetitive contact between the work material and the forge or stamper. Electromagnetic forming would not require such contact and hence the forming process would be much more reproducible among a large number of parts.

The electric circuit for electromagnetic forming is primarily an energy storage bank of high voltage capacitors, a dc voltage supply and switches to allow first a build-up of stored energy in the capacitor, and then a high current discharge through the forming coil. The workpiece being shaped is either a tube coaxial with a cylindrical forming coil or a flat plate adjacent to a spiral forming coil. The electromagnetic pulse is short enough so that high kinetic energy is imparted to the material of the workpiece which then takes up most of it for deformation beyond the elastic limit. Some typical parameters of the electromagnetic forming process are shown in Table 7-1.

There are two important auxiliary items usually associated with the process equipment. To concentrate the electromagnetic field at specific areas of the workpiece and also provide a rigid mount to withstand the reactive electromagnetic pressure, the forming coil is usually married

Table 7-1. Characteristic Parameters of Electromagnetic Forming[1]

Electric Energy Storage Bank		
Energy per Discharge	6.0 kJ	
Voltage	8.3 kV	
Capacity	180 μF	
System Characteristics		
Pulse Frequency	7.6	18.8 kHz
System Inductance	2.2	2.5 μH
Maximum Forming Coil Current	100 kA	
Maximum Physical Stress on Workpiece		
Pressure (psi)	2830	4830
Velocity (ft/sec)		
Displacement (in.)	0.140	0.150
Material of Workpiece: Aluminum, Copper, Mild Steel		

to a field shaper made of beryllium copper alloy. When the workpiece must be precisely shaped, a rigidly mounted die defining the shape is used. The workpiece is then deformed at high velocity against the die.

The principal market for electroforming has been the automotive industry where it has been used, for example, on the following parts and operations:

- assembly of air conditioner components;
- high-pressure hoses;
- dust covers for shock absorbers;
- heat exchangers;
- universal joint yokes; and
- wheel, cam and gear shaft attachments.

In addition, there is a fair market for electroforming in electrical equipment for high-voltage fuses and heavy-duty electrical connection. Aircraft manufacturing has also found use for the process. The equipment, which can be sized in terms of capacitor bank modules of 6,000 J each generally ranges from 12-84 kJ capacity and $25,000-$100,000 cost. It is currently being actively marketed by Maxwell Laboratories of San Diego.

At General Dynamics/Fort Worth EHF was developed along with EMF in the 1950s. Another company developing the process was Rohr Industries, which eventually marketed equipment for it under the name Soniform. To provide electrical energy to the underwater arc discharge, Soniform equipment uses the same circuit as used in electromagnetic forming. However, the required energy and forming pressures are much higher and the arc discharge is a single direct current pulse instead of a short-lived oscillation. Table 7-2 shows the characteristic parameters of the system.

Table 7-2. Characteristic Parameters of Electrohydraulic Forming[2]

Energy per Discharge	62.5 kJ
Peak Voltage	18.5 kV
Peak Current of Discharge	1250 kA
Capacity	380 μF
Average Power Supply	25 A 440 V

Typical workpiece material and required maximum pressure exerted by shock wave

Material	Pressure (Psi)
Aluminum (6061-0)	10,000
Aluminum (6061-T6)	35,600
Steel (1010)	30,000
Stainless Steel (304)	42,000

Because of the need to use a water medium to propagate the shock wave against the workpiece, electrohydraulic forming is difficult to adapt to a high repetition rate manufacturing process. However, it does have a complementary advantage over EMF in that the electric conductivity properties of the workpiece material have no influence on the process dynamics. As a result, the more refractory metals, which are of low conductivity, such as hastalloy, inconel, or titanium, are suitable for EHF. The manufacturing processes that form its market are therefore centered in the aerospace industries which have much more stringent mechanical requirements to meet. EHF is an inherently expensive process and as a result has not been able to develop a market in industries that require low-cost mass production methods such as automobile manufacturing.

The original group at Rohr, which developed Soniform, left in 1970 to form an independent company Soniform, Inc. of La Mesa, California. They still market the equipment as their prime specialty. A second firm, Electrohydraulics of Fort Worth, Texas, also markets similar equipment as one of many lines.

ELECTROMAGNETIC FORMING (EMF)

This process forms metal by the direct application of an intense, transient magnetic field. Pressure is applied to the workpiece without mechanical contact by a high magnetic field that is built up in between, and a forming coil which has received a strong pulse of electric current. The major application of EMF is the single-step assembly of tubular parts, usually being tightly attached to other tubular parts; it is used to a lesser extent for the shaping of tubular parts and the shallow forming of flat stock. Metals with high electrical conductivity are formed directly; poorly conductive metals are formed with the aid of highly conductive "drivers."

The basic process is covered by U.S. Patent 2,976,907 issued to G. W. Harvey and D. F. Brower in 1958. Additional patents cover various aspects of the process and equipment.

EMF was developed primarily in the General Atomics Laboratory of General Dynamics in the late 1950s as a technique closely allied to EHF.[3] After Harvey and Brower obtained their patent, the work was organized in the Magneform Department and the equipment was marketed as Magneform. The energy unit, which could be more or less common to both techniques, had been modified in several ways to improve operational safety and reliability. Initial models supplied 6000 J and these were used as modules that could be aggregated for greater power in later models.

Maxwell Laboratories (9244 Balboa Avenue, San Diego, California 92123) now owns and markets the electromagnetic forming service and equipment for it under the name Magneform.

As an exotic physical process of apparent industrial application EMF attracted a great deal of fundamental research in the 1960s. In particular, compared to other forms of high-energy metal forming, it appeared to offer a broader range of options for automated production.[4] However, the range and scope of its operations could not be extended beyond those developed in the beginning at General Atomic and very little new work has been reported during the last few years.

Physical Principles

The most essential elements of the EMF process are shown in Figure 7-1. The circuit (d) provides a pulse of electric energy to the work coil, shown in (a) (b) (c) for the three distinct alternative geometries, by first charging the condenser with switch 1 closed and switch 2 open, then discharging the condenser through the work coil with switch 1 open and switch 2 closed. The pulse will decay exponentially with a half-life, T, and induce eddy currents in the workpiece. For small T, usually a few microseconds, these currents will penetrate only to a skin depth

$$D = \left[\frac{\rho T}{\pi \mu}\right]^{1/2}$$

where D = Penetration
T = Pulse Duration
μ = Magnetic Permeability of Workpiece
ρ = Resistivity of Workpiece

For D much less than the workpiece thickness, the magnetic field is confined to the space between the coil and workpiece, creating a transient mutual repulsion. If the coil is clamped with sufficient rigidity, the workpiece is deformed by the magnetic field pressure. Pressed to a die it will take the die's shape. Enclosing a mating part, a tubular workpiece will crimp to fit tightly around it. Typical parameters involved in the process for energies below 6 kJ are shown in Table 7-3. At the maximum values of discharged energy, free forming of the workpiece takes place until impact against an unyielding cylindrical steel frame surrounding the workpiece. The impact occurs 50-100 μsec after the initial pulse.

The eddy-current fields of the workpiece cancel all magnetic fields except those in the region between the forming coil and workpiece. The magnetic field in the region interacting with the workpiece eddy current exerts a net pressure on the workpiece away from the forming coil. As the surface of the workpiece moves inward under the influence of this pressure, energy is absorbed from the magnetic field to do the work of forming.[1] In spite of the eddy current resistance, the magnetic field eventually permeates the workpiece and the forming force drops to zero. To apply most of the available energy to forming, and the smallest possible fraction to resistance heating of the workpiece by its eddy-currents, the forming pulse is kept short to minimize the skin depth, D. In most forming applications, pulses have a duration between 10 and 100 μsec.

An early theoretical treatment of these considerations was provided by Baines *et al.*, who based their analysis and verification experiments on

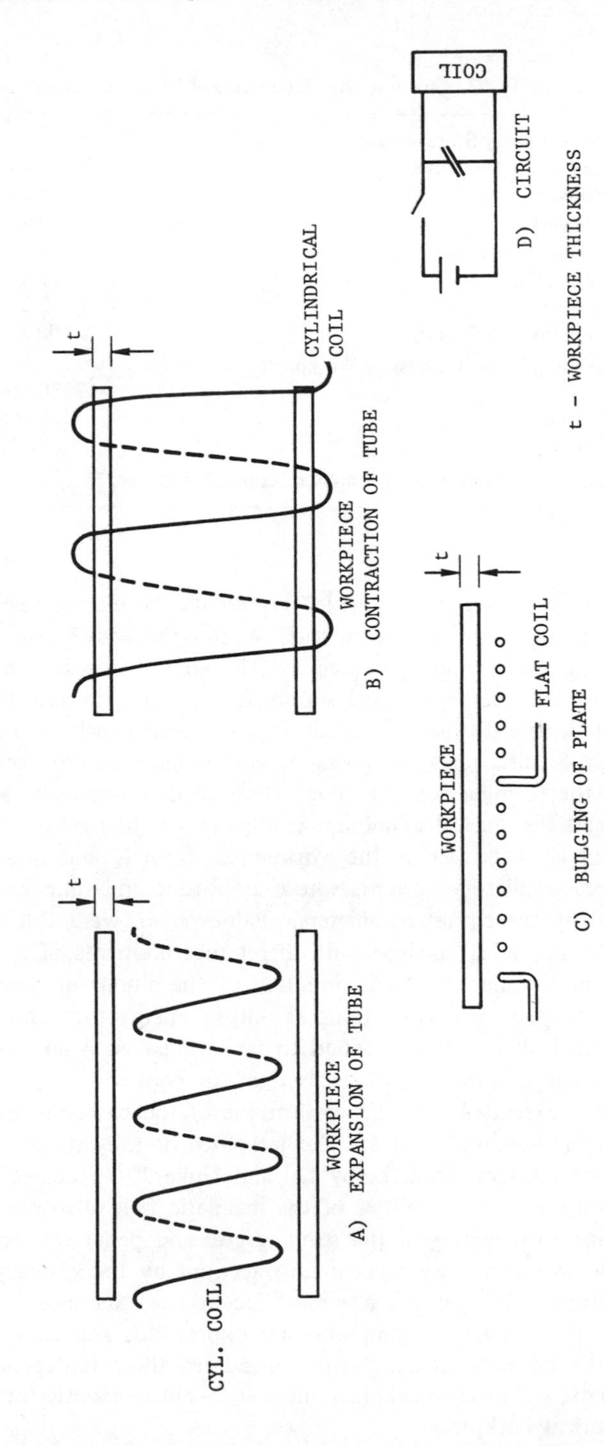

Figure 7-1. Electromagnetic metal forming: physical principles.[1]

Table 7-3. Characteristic Parameters of Electromagnetic Forming[1]

Electric Energy Storage Bank	
Energy per Discharge	6.0 kJ
Voltage	8.3 kV
Capacity	180 μF
System Characteristics	
Pulse Frequency	7.6 → 13.8 kHz
System Inductance	2.2 → 2.5 μH
Forming Coil Current	100 kA
Maximum Physical Stress on Workpiece	
Pressure	2820 → 4830 psi
Velocity	266 → 539 ft/sec
Displacement	0.140 → 0.150 in.
Materials of Workpiece: Aluminum, copper, mild steel	

the circuit of Figure 7-2, which represents the electromagnetic relation between the forming coil current, i_1, and the eddy current, i_2, in a cylindrical aluminum workpiece.[5] The energy storage capacitors, c_1, are charged by the rectified supply from a variable auto-transformer T_1, a step-up transformer T_2, diode D and limiting resistance R_L. Closing the switch, S, discharges C_1, giving an instantaneous short-lived pulse of current i_1 in the forming coil. Through the mutual inductance, M, between this coil and the coaxial aluminum workpiece cylinder within it, a secondary current i_2 is induced in the workpiece. Both i_1 and i_2 are short-lived damped oscillatory currents whose amplitude and time variation are determined by the circuit parameters. Baines *et al.* were able to verify the simple theoretical analysis with direct measurements of i_1 and observations of the mechanical deformation of the aluminum workpiece.[5] Process efficiency was calculated using as output energy the total plastic deformation work done. It was found to range between 4 and 14% and to be higher for aluminum workpieces than for copper.

More extended formal investigations of the basic process were undertaken in subsequent years. The influence of the rate of workpiece deformation was accounted for by Lal and Hillier,[6,7] Fluerasu[8,9] and Epechurin.[10] The realities of the magnetic field distribution, which leaks through the winding of the forming coil and penetrates to the inside space of the workpiece, were taken into account by Barbarovich,[11] Dehoff[12] and Bron and Segal.[13] The most recent work has been done by Al-Hassani[14] using very comprehensive expressions and computer analysis of effects, and Belyi *et al.*,[15] who considered the interdependence of electromagnetic and mechanical phenomena to calculate exactly the deformation of cylindrical workpieces.

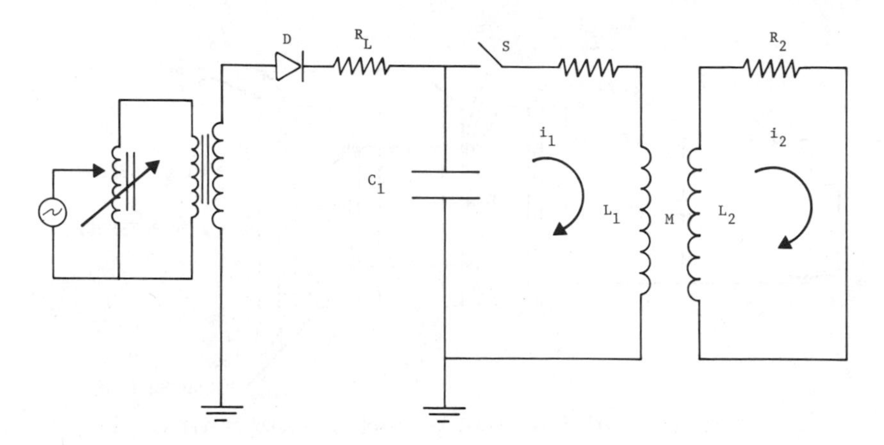

Figure 7-2. Circuit diagram of magnetic forming process with no field shaper.[5]

Practical Characteristics

Due to the short duration of the magnetic impulse, the pressure must be high enough to impart sufficient kinetic energy to the workpiece during the pulse to do the desired forming. Any resistance to the motion of the workpiece during the impulse reduces the amount of useful forming energy transferred. The peak pressure should therefore be several times that necessary to exceed the static yield strength of the workpiece and to overcome any other constraints for the duration of the impulse.

For efficient use of stored energy, coils are designed to minimize stray inductance and to avoid flux concentrations. Field patterns are ordinarily controlled with field shapers, which are massive current-carrying conductors. They are not necessarily directly connected to the primary forming coil but may be inductively coupled to it, as illustrated in Figure 7-3.

In either case, current is induced in the workpiece via mutual inductance and is a momentary oscillating current as analyzed and verified by Baines *et al.*[5] Frequency of the oscillation is roughly inversely proportional to the square root of the product of the storage bank capacitance and the forming coil inductance. The amplitude of the pulse decreases proportionally to the sum of two energy losses: resistive loss in the primary circuit and the deformation work reflected through the secondary circuit back into the primary circuit. Pressure on the workpiece produced by the coil or field shaper is proportional to the square of this current and is thus entirely positive, approximating a damped sine waveform.

Most of the forming energy is provided by the first wave; succeeding waves transmit less energy to the workpiece because of their lower energy

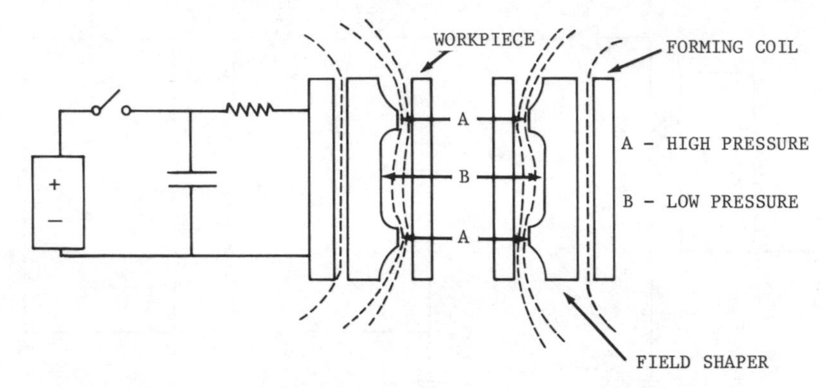

(a) LOCALIZED FIELD CONCENTRATION BY FIELD SHAPER

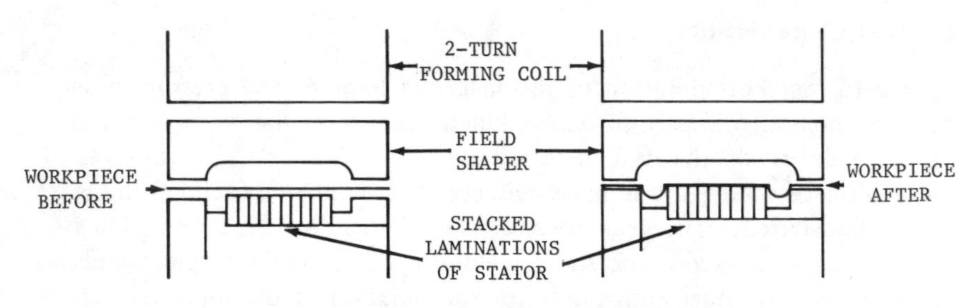

(b) ELECTROMAGNETIC FORMING OF A STEEL HOUSING AROUND STATOR LAMINATIONS, USING A FIELD SHAPER

Figure 7-3. Use of field shaper as intermediate coupling between forming coil and workpiece.

content and because of the progressively widening gap between coil and workpiece as it deforms away from the coil. Whatever energy of the electrical discharge is not transferred to the workpiece first as kinetic energy, then deformation energy, appears as resistance heating and represents the loss.

The critical importance of the field shaper is shown in Figure 7-3b. Brower[1] used this design to form in place the housing for the stator assembly of an electric motor. In a single EMF operation, the laminations were bound rigidly in place without the use of rivets or bolts. The energy requirement was 40 kJ to generate the high pressure that formed the concentric grooves aligned to the inner surface of the laminations. Because such a high pressure would have also reformed the laminations, the field shaper was used to concentrate the pressure at only the workpiece regions where it was required. To counter field leakage through the standard forming coil, a special two-turn coil, providing a higher frequency current pulse, was used. The stator laminations assembly was produced at a rate of 240/hr using manual loading and unloading.

Similar assemblies were mass produced at a rate of 600/hr on the same machine equipped with semiautomatic feed. In 1969, the machine cost was $80,000 and labor costs, direct and overhead, were about $10/hr. If the capital costs were amortized over one million operations, costs per assembly were $0.096 ($0.08 for capital and $0.016 for labor).

Handling the electromagnetic field was a central concern of much of development work at that time. Schenk[16] carried out detailed investigations on test aluminum tubular samples to arrive at methods to optimize field concentrator designs when several annual zones of the workpiece are to be compressed simultaneously. Similar investigations were carried out by Al-Hassani *et al.*[17] and Gilbert and Lawrence.[18] Davis *et al.*[19] studied prototypical beryllium coil assemblies on a theoretical basis and, using an analog computer, arrived at precise values for the pressure distribution between a coil and flat plate. Brower[1] optimized electroforming machine design for aluminum, copper, brass and steel tubular workpiece.

The constraints upon the workpiece limit the opportunities for electroforming. In addition to being a good electrical conductor (resistivity less than 15 μohm/cm), the workpiece must provide a continuous electrical path. Hence, any slots or other openings in the piece that interfere with current flow across its surface will reduce and distort the forming pressure. Workpieces other than uniform tubular shaper or flat blanks usually require specially designed field shapers that can be opened or expanded, and the time required for loading and unloading is greatly increased.

To summarize the requirements for favorable electroforming results on contemplated work, the following features of the process must be considered:

- Pressure is applied directly to the workpiece through the magnetic field medium without mechanical contact.
- Most of forming takes place after pressure impulse has ended. Metal is rapidly accelerated, gaining a large amount of kinetic energy by moving only a short distance during impulse. This energy subsequently does the actual work of forming.
- Metals must have high conductivity so their walls can be deformed; otherwise very high power is required.
- Ratio of masses of pieces involved in the forming process must be considered along with relative yield strengths and elastic properties.
- Magnetic field exerts a hydrostatic pressure which is relatively independent of the spacing between workpiece and forming coil. In

swaging and expanding no torque is applied to workpiece as in spinning and rolling.

- Peak pressure applied by field is limited by strength of forming coil material to much lower values than those in shearing, punching and upsetting.
- Magnetic fields cannot be easily "shaped" to fit all contours. It is impossible to apply high pressure in arbitrarily chosen areas while applying a low pressure in immediately adjacent areas.
- Process is not limited in repetition rate by mechanical inertia of moving parts. Timing of magnetic impulse can be synchronized with microsecond precision. Machines can function at thousands of operations per minute. Strength of magnetic impulse can also be precisely controlled.

EMF Equipment and Operational Parameters

The equipment for EMF consists primarily of the energy source with simple associated circuits. The assembly of work coil and workpiece mounting must be individually designed and constructed for each specific job.

The instantaneous power needed in the forming coil may be as high as 1000 MW and the most satisfactory way of achieving this level is to use a store and discharge principle. Capacitors, as indicated in Figure 7-1d, have been the most satisfactory means of energy storage. Oil-filled paper dielectric capacitors can store energy at densities up to 1900 J/ft^3. At this level, they will average millions of charge-discharge cycles before failure. Charging rates must be sufficient to meet the needs of the desired repetition rate. Production rates used have been 250-1000/hr. A rate of 10,000 is considered feasible. Standard commercial EMF machines have capacitor-bank energy sources of 6-84 kJ. Some characteristics of typical commercial EMF energy sources are shown in Table 7-4.

Table 7-4. Characteristics of Typical Commercial EMF Energy Sources[1]

	Maximum Energy Output (kJ)		
	6	12	60
Current Pulse			
Peak Current (kA)	100	200	1000
Peak Voltage (kV)	8.3	8.3	8
Repitition Rate (cycle/hr)			
At Maximum Energy Setting	600	600	600
At Low Energy Setting	1800	1800	1800

Work coils and field shapers which incorporate the workpiece mounting have three primary factors to be considered:

1. Size
2. Electrical characteristics
3. Strength

The coils must accept repetitive discharge of large electrical energy pulses lasting 10-100 μsec and generate field-forming pressures up to 50,000 psi. Under this stress they must suffer only elastic deformation so that, restored after its cessation to their original demensions, they will present to each workpiece in sequence the same electromagnetic field environment.

An energy discharge can be characterized by its peak pressure, which is that of the first wavefront, and by the duration of wavefront. Time between successive waves changes slightly as the workpiece deforms (changing its own inductance and the mutual inductance of its coupling to the forming coil) and as heating changes the resistance of the electrical circuit. These effects are small although there has been academic interest in their study aimed at fundamental understanding of the EMF process.

Peak pressure is approximately related to other process variables in three ways:

1. directly proportional to the energy of the electrical impulse from the capacitor bank;
2. inversely proportional to the resistivity of the workpiece and forming coil; and
3. inversely proportional to the total of the volume of the workpiece and the field shaper penetrated by the electromagnetic field (skin effect) and the volume between the coil and workpiece surfaces.

As specific example to illustrate typical conditions, Figure 7-4 shows the forming pressure, workpiece velocity and workpiece displacement as measured during the expansion forming of identical aluminum alloy tubes under three different combinations of conditions. The electrical parameters for these conditions, denoted A, B and C, are presented in Table 7-5.

For condition A, maximum pressure was reached in 23 μsec and maximum velocity in 44 sec so that most of the deformation or displacement occurred after the peak pressure wave. Under condition B, two narrow pressure waves occurred but the deformation result was similar. In condition C, free forming took place until impact of the workpiece against a die. With higher input voltage, the capacitors were charged to nearly twice the levels of A and B. Hence, there was a higher maximum pressure and workpiece velocity and therefore more stored kinetic energy at peak velocity.

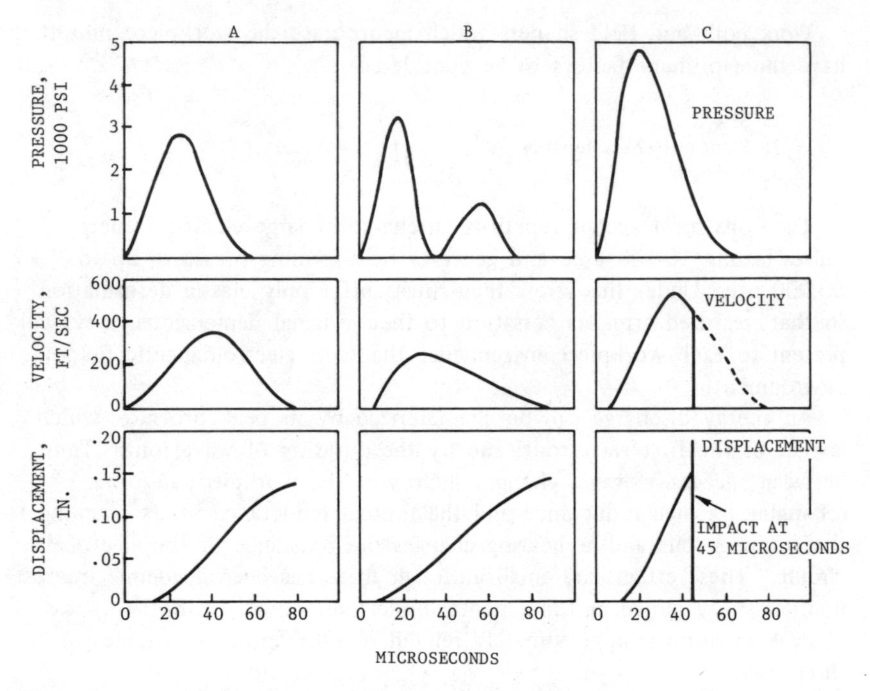

Figure 7-4. Electroforming aluminum alloy 6061-0 tubes 2 in. O.D., 049 in. wall, 2.125 in. long.[1]

Table 7-5. Circuit Parameters for Conditions in Figure 7-4[1]

	Condition		
	A	B	C
Energy-Storage Bank			
Energy (kJ)	2.86	2.86	5.50
Voltage (kV)	5.65	9.6	7.80
Capacity (μF)	180	60	180
System Characteristics			
Pulse Frequency (kHz)	7.6	13.8	7.8
System Inductance (μH)	2.45	2.22	2.45
Maximum Values			
Pressure (psi)	2820	3200	4830
Velocity (ft/sec)	289	266	539
Displacement (in.)	0.1 0	0.142	0.150
Values at Peak Velocity			
Displacement (in.)	.077	0.49	0.108
Stored Kinetic Energy (ft/lb)	81.1	69.4	283
Stored Energy Density (ft/lb/in.^{2})	5.9	5.1	20.6

Jansen of Maxwell Laboratories, the current owner of Magneform, studied the influence of frequency and geometry on the efficiency of tubular workpiece compression.[20] He obtained theoretical relationships among radial displacement, velocity, applied magnetic pressure and operational efficiency. Similar studies were undertaken by Lawrence[21] and most recently by Maeda and Higuchi,[22] who took into account the influence of relations among the various inductances of the EMF circuit.

The diameter of the workpiece section being formed determines the working surface of the field shaper or coil if a shaper is not used. A clearance of 0.050 in. is ordinarily needed for insulation. Coil materials must have an electrical resistivity less than 30 x 10^{-6} μohm/cm. The total impedance of the coil $(R + \omega L + \frac{1}{\omega C})$ must protect the capacitor bank from too rapid a discharge. Mechanical strength for coil is usually obtained by a multilayer design incorporating in the support structure sufficient mass and rigidity to withstand repeated forming impacts. The high rigidity minimizes the deflection under load. Constructing the coil presents a basically different problem depending on the three different forming processes shown in Figure 7-1: expansion, contraction and bulging. Some work coil assemblies are shown in Figure 7-5.

An average forming force as high as 50,000 psi is produced on the surface of the workpiece in a pressure pulse lasting 10-100 μsec, and the reaction to this forming force must be sustained by the coil. It must sustain repeated thrusts of forming energy and be as massive and rigid as possible to minimize its deflection under load. In practice the coil ordinarily receives additional support from a metallic "coil body" (usually of beryllium copper) and insulating structure as shown in Figure 7-5. To be capable of exerting high force, coils must be designed to avoid unnecessary field (pressure) concentrations. Careful tailoring of the field distribution over the coil conductors is an essential part of the design of a high-pressure coil.

Prototypical Example of Electromagnetically Formed Parts

Some examples that illustrate a large measure of the adaptability of the EMF process are shown in Figure 7-6. In the expansion case (Figure 7-6a), the pressure of the field in the thin annular space between coil and workpiece cylinder forms the cylinder against a cylindrical die which has drilled holes and is slightly shorter than the workpiece. As a result, the workpiece takes the inside shape of the die and in addition has a complete flange and two holes precisely pierced to match those of the die. The forming equipment had a capacity of 6 kJ and cost about $18,000. Using manual loading and unloading, the system was operated at a

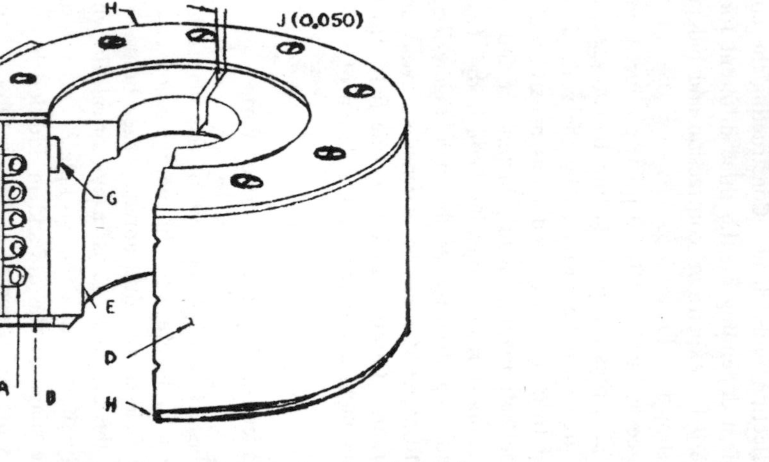

General-Purpose Compression Forming Coil
With a Removable Field Shaper

A–Primary conductor. B–Beryllium copper coil body. C–Fiberglass reinforcement. D–Steel housing and magnetic shield for coil. E–Beryllium copper field shaper. F–Fiberglass insulation. G–Fiberglass reinforcement. H–Fiberglass supporting member. J–Slot 0.050 in. wide in field shaper.

Heavy-Duty, Wafer-Type
Compression Forming Coil

A–Primary conductor. B–Beryllium copper field shaper. C–Water passage. D–Fiberglass insulation. E–Steel backup plates. F–Press bolt. G–Shaper press bolt. H–Shaper insulation.

Figure 7-5. Example of work coil assemblies.[1]

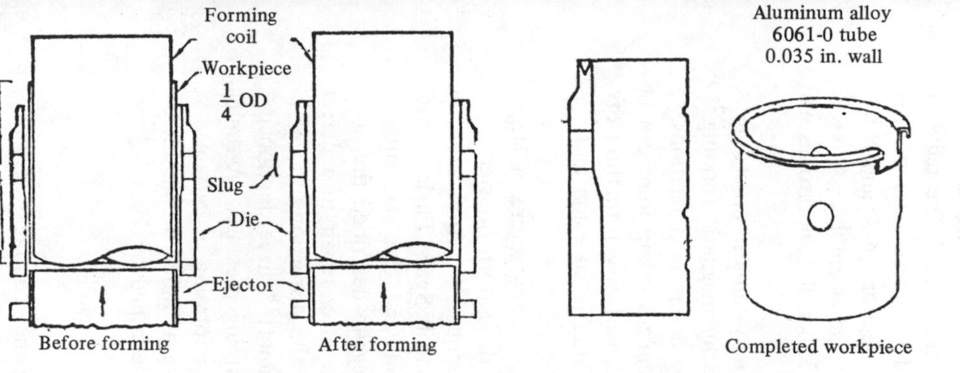

Forming and piercing a tubular part in one operation by EMF

Expansion coil forms, flanges and pierces a length of tubing against a single-piece die. Forming equipment had capacity of 6 kilojoules and cost ~ $18,000. Production rate 240 per hour using manual loading and unloading.

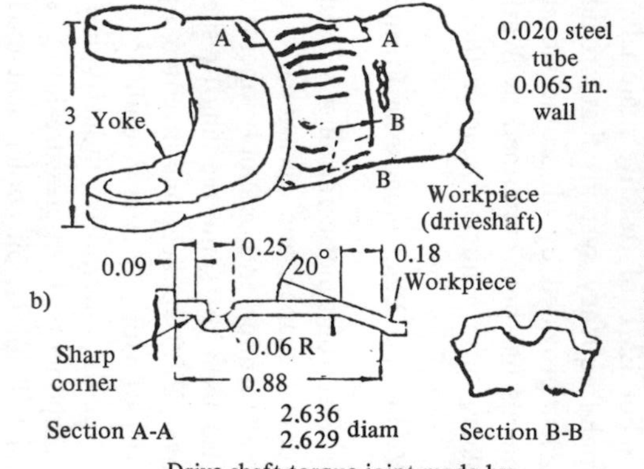

Drive-shaft torque joint made by electromagnetic forming

Torque joint formed to assemble drive shaft for a passenger car to universal joint yoke. Tests showed joint to be as strong as the drive shaft. Substituting EMF for welding reduced noise and vibration in car.

In magnetic forming, the reduced end of the tube was wrapped tightly around the rectangular protrusions in round portion of yoke as shown in Sections A-A and B-B, providing both longitudinal and axial strength in the assembly.

Figure 7-6. Examples of electromagnetically formed parts.[1]

production rate of 240 per hour. In the construction case shown in Figure 7-6 we have an important automobile industry application. A torque joint is formed by crimping the shaft onto a cylindrical extension of the universal joint yoke. Absolute resistance to longitudinal and rotational shippage was obtained by special ridges on the yoke. The reduced end of the shaft tube was wrapped tightly around the rectangular protrusions on the round portion of yoke, as shown in Sections A-A and B-B to provide both longitudinal and axial strength in the assembly. A few examples of the recent procedures developed by Maxwell Laboratories and cited by Zittel are shown in Figure 7-7.[23]

Buehler and von Finckenstein have developed a systematic procedure for the production of sheathings and crimp joints by magnetic forming.[24,25] With the aid of calculation results and empirically determined relations, the tool and machine parameters required for joining a crimp joint can be estimated rapidly and simply. They arrive at the necessary formation of the surface profile on the inner part and the obtainable adhesion and resistance to longitude loosening forces on the joint.

Forming and sizing problems encountered in the manufacture of large hardware required in the space program, such as the Saturn booster rockets, have been solved by electroforming techniques.[26] Application of an electromagnetic hammer has been used at Marshall Space Flight Center for the removal of distortions from large bulkheads and gore segments (sections of missile envelope). Blanc and Dekerlegand described the distortions of these bulkheads due to weld shrinkage after repair of a defect and the specific EMF techniques used to correct the distortion.[27]

For EMF of resistive materials, James and Philpott[28] have developed high-frequency systems that provide sufficient forming pressure. A similar assist provided by ultrasonic vibration to facilitate forming of certain metals has been described by Watkins.[29] Swaging a finned aluminum sleeve onto a nuclear reactor fuel rod has been described by Harrell.[30]

Market for Electromagnetic Forming in Industry

Electromagnetic forming is used chiefly to expand, compress or form tubular shapes, and occasionally to form flat sheet, often combining several forming and assembly operations into a single step. The method is also used for some piercing or shearing operations. Brower enumerates over 2000 applications invented by Magneform by 1968 when it was still part of Gulf General Atomic.[31]

In the automotive industry EMF is used to assemble air conditioner components, high-pressure hoses, shock-absorber dust covers, rubber boots on ball joints, oil cooler heat exchangers and accessory motor packages. Universal joint yokes, drive linkages, wheels, cams, gears and various other

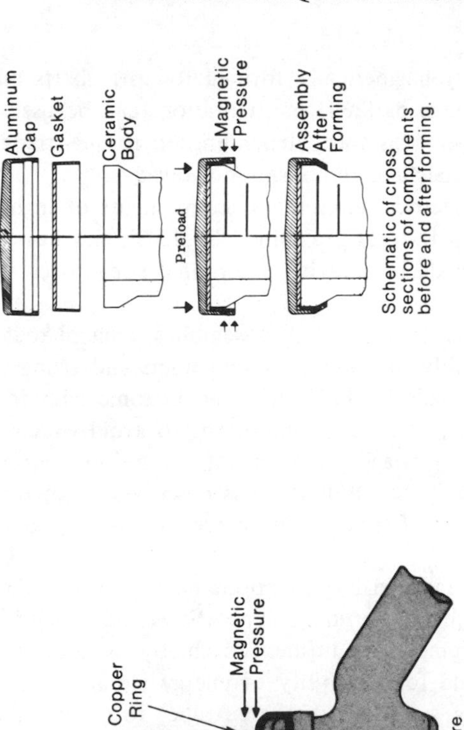
Before Forming
After Forming
Magnetic Pressure
15°
Preload
Gasket
Drawn Steel Cover
Preload
Aluminum Casting
Electromagnetic forming of automotive fuel pump cover.
Fuel pump lip design for crimping cover with conventional crimping machines.
Copper Ring
Magnetic Pressure
After Forming
Before Forming
Protective rubber boot on automotive idler arm assembly.
Aluminum Cap
Gasket
Ceramic Body
Preload
Magnetic Pressure
Assembly After Forming
Schematic of cross sections of components before and after forming.
Magneform machine crimps aluminum caps on ceramic insulators on high voltage surge diverter assembly line. Above: Cap and insulator details.

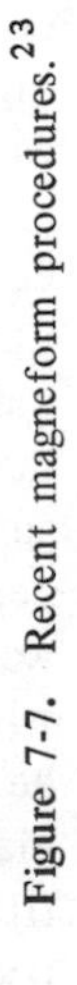

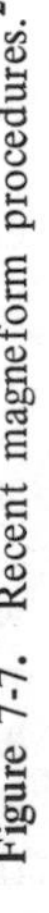

Figure 7-7. Recent magneform procedures.[23]

linkages or fittings are electromagnetically formed to drive shafts on torque tubes. Splines, pockets or knurl configurations can be used in the fittings to provide torque resistance, depending on torque-strength requirements, type of materials and dimensions involved.

In the manufacture of electrical equipment, components of high-voltage fuses, insulators and lighting fixtures are joined by EMF, and heavy-duty electrical connections are made by swaging a terminal sleeve over a conductor cable.

Aircraft applications include torque-shaft assemblies, control rods and linkages, forming and assembly of cooling system ducts and sizing of tubing. Axially loaded joints made by EMF are used in some aircraft applications and in actuator rods, where it is important to avoid excess weight. These assemblies are made by swaging the end of a tube into circumferential grooves in the second tube. Fuller[32] has described magnetic forming at De Havilland Aircraft of Canada using energy levels high enough to form thick steel plates.

A wide range of diverse industrial applications have been described as early as 1969 for Europe (*Metal Forming*, June 1969) and Japan.[33] Capital costs for EMF equipment are influenced chiefly by the requirements for repetition rate and for durability of energy storage capacitors, discharge ignitrons (mercury vapor arc discharge tube) to control the surge, and forming coils. The 1969 cost of a standard (600 operations per hour) energy-storage bank for production use in the 12-84 kJ range was about $25,000-100,000. Energy-storage banks for experimental use, where durability, speed, safety and control features are less critical, can be built for about $500-1000/kJ of capacity. Wildi,[34,35] manager of Magneform Engineering, estimated at the time that some 200-250 industrial Magneform machines had been installed in the U.S. and other countries between 1962 and 1970. This would imply an average annual market of about 30 machines or sales revenue from $1-3 million.

There is apparently still a viable market for the EMF process and equipment among some leading firms. The only U.S. firm marketing the equipment is still Maxwell Laboratories. Their list of customers includes:

- General Motors Corporation
- Oldberg Manufacturing Company
- Ford Motor Company
- IBM Corporation
- Chrysler Corporation
- Eastman Kodak Company
- Nissan
- Toyota
- Lockheed
- Westinghouse Electric Company

- Grumman Aerospace Corporation
- General Electric Company
- Amphenol
- Alcoa

Magneform products are usually simple in shape and fairly small. However, some are fairly complex and require extensive work prior to EMF, which primarily supplies the final precise shape. Some of these shapes, shown in Figure 7-8, are likely to be made on a job-lot basis by Maxwell Laboratories as a machining service.

ELECTROHYDRAULIC FORMING (EHF)

The use of controlled explosions to provide very high energy in shaping metal parts received considerable attention beginning in the late 1950s and is still being actively pursued.[36,37] However, the process dynamics are hard to control and practically impossible to incorporate into an automatic procedure so that its use has always been limited to special critical items that would warrant the high level of expenditure and special attention. The alternative of electrical discharge under water appeared to offer good control along with the availability of comparable force levels applied to the metal being formed.

At General Dynamics/Fort Worth (GD/FW) electrohydraulic forming was being developed as a companion process to electromagnetic forming in the late 1950s. A 5000-joule spark discharge unit had been constructed as a device to detonate explosives for explosive forming. Evaluating energy release by the unit in air and water both with and without wire bridging the arc terminals, the power of the high-energy spark was observed to be capable of significant metal forming. Higher voltages were tried but they introduced problems for small diameter parts and eventually GD/FW settled on a 4000-volt unit using bridge wires.

Several other companies were developing the process independently at this time, Rohr Industries among them where Gilbert C. Cadwell did very early work (1961) and electrohydraulic equipment was patented and marketed under the name Soniform. He developed the technique for Rohr's primary interest in the manufacture of aerospace parts.[38] A major advantage was its utility for refractory metals which characteristically had low conductivity. The electrohydraulic process was also developed with similar equipment at universities and at the U.S. Army Redstone Arsenal.[39,40]

Only Rohr Industries actively marketed the equipment. Eventually the company sold the patent rights back to Cadwell who left them in 1970 to form the Soniform corporation, which still actively markets the equipment and process as its main specialty. A second firm, Electrohydraulics

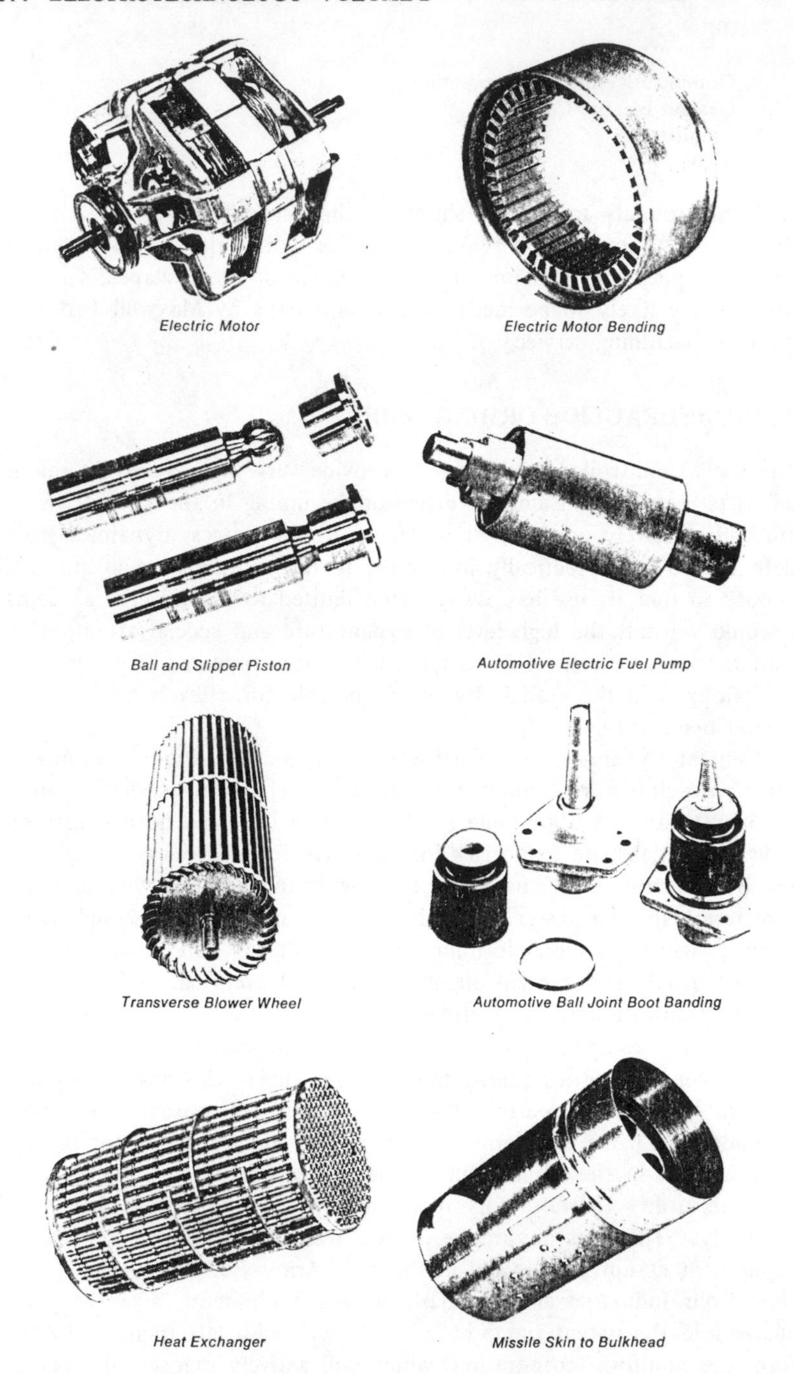

Figure 7-8. Current Magneform products (from Maxwell Laboratories Catalog).

of Texas, markets the process along with many other allied services such as metal spinning.

Physical Principles

In utilizing the behavior of sheet-metal targets subjected to explosive generated shock waves research workers have used a sheet metal diaphragm gauge as a highly sensitive experimental measuring technique.[37] The gauge consists of a circular sheet-metal diaphragm with edges clamped tightly and placed over a circular cavity in a rigidly clamped die. The shock causes the flat diaphragm to deform into a shape somewhere between a cone and a section of a sphere with maximum deflection being used to determine the severity of the shock wave. Transferral of this process to a water medium for shock wave propagation and electric arc to generate the shock wave provided the essential elements of electrohydraulic forming.[2]

The dynamic elements of EHF are illustrated in Figure 7-9 which presents an idealized configuration of electric circuit and mechanical parts. The direct energy source for the process is a high-voltage capacitor bank. The capacitors are charged over a relatively long interval by a powerful dc source and then discharged across the gap to produce a short-lived high-peak energy underwater arc discharge. If the external circuit connected to the gap is properly designed, the discharge interval will be sufficiently short. Water or other fluid is then rapidly ionized and heated to extreme temperatures by bombarding electrons from the arc. Due to the inertia of the water mass, the heated gas bubble cannot expand at the rate of temperature increase and the pressure builds rapidly to create a shock wave projected toward the workpiece.

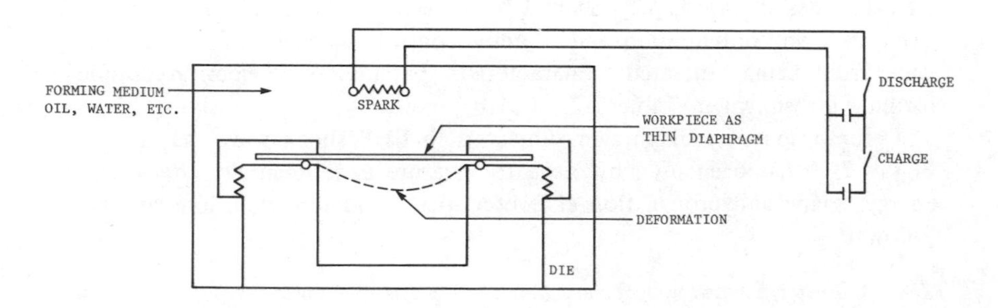

Figure 7-9. Electrohydraulic metal forming: physical principles.[2]

The shock wave accelerates the unclamped regions of the workpiece downward with sufficient force to exceed its elastic limit and permanently deform it. After the shock wave is reflected, the deformed workpiece regions continue their motion due to acquired kinetic energy. If high kinetic energy is available for the final impact against a clamped die, then very fine detail of the die's surface will be faithfully reproduced on the workpiece.

When the arc is generated, with the external electrical circuit correctly designed, a high current will flow across the gap within a very short time. Water in the current path is rapidly ionized and heated to extreme temperature. Due to the inertia of the water mass surrounding the arc channel the heated gas or plasma in the channel cannot expand as rapidly as the temperature increases and, as a result, initial pressure for a shock wave is built up corresponding roughly to the perfect gas relation

$$p = \frac{w\,RT}{V}$$

where p = pressure of gas, psi
w = weight of gas, lb
R = gas constant in lb-in./lb-degree Rankine
T = absolute temperature of gas, Rankine
V = volume of gas in arc channel, cubic inches

Release of the pressure takes place through propagation of the shock wave spreading radially outward through the water.

The very high temperature, T, promised theoretically that high Carnot efficiencies might be achieved in forming metal exposed to the shock wave. This would overcome one of the drawbacks of electromagnetic forming, that of relatively low empirically observed efficiency in energy use. The electrical requirements to assure the required high current arc had been developed in exploding wire research and circuit resistance and inductance were kept at a minimum.[41] Frequently, a thin wire or ribbon, placed across the arc gap, is shaped to guide the establishment of an arc along its own configuration and thereby control the shape of the shock wave front being generated. Characteristic parameters of electrohydraulic forming are shown in Table 7-2.

Deformation of a diaphragm subjected to EHF shock waves as in Figure 7-10 has been used by Kegg to measure experimentally shock wave energy, shape and propagation characteristics.[2] Some conclusions are as follows:

1. Distance between spark and water surface is unimportant beyond a minimal spark depth.
2. No effect was found in varying the initiating wire material.

3. Forming efficiency *increases* as the initiating wire diameter *decreases.*
4. Forming efficiency increases as the spark gap increases.

Franke has added a piezoelectric measurement of the shock wave to directly relate the energy in the shock wave as propagated to the energy deforming the workpiece.[42] This adds a substantial element of control to the operation. Guenther *et al.* described the use of foils to impart high acceleraton and deformation to mylar and lucite.[43] Concepts for generalizing theoretical treatment of EHF were developed by Oyave & Masaki, based on laws of similarity.[44] They verified their formulation to stored energy releases up to 19 kJ. Preumer, in a comprehensive review of explosive forming, has described methods of adapting it to realistic production problems.[45]

Kegg[2] was able to develop, on the most basic considerations, the following relation for Δ, the deflection of the diaphragm

$$\Delta = K\frac{r}{T}\,\frac{(1/2\ CV^2)^m}{R^n}\left[\frac{g}{\sigma\rho}\right]^{1/2}$$

where K, m, n = empirical constants to be determined
r = radius of die cavity, in.
T = workpiece thickness, in.
C = capacitance of energy-storage bank, farads
V = charge voltage, v
g = gravitational constant, $in./sec^2$
R = distance of arc from workpiece, in.
σ = workpiece flow strength in tension, psi
ρ = density of work material, $lb/in.^3$

Using aluminum, cold-rolled steel and stainless steel the following values for the constants were found:

K = 0.00295
m = 0.363
n = 0.882

The experimental results are summarized in Figure 7-10.

Practical Assembly

An assembly which arranges appropriately for producing the spark, workpiece and constraints is shown in Figure 7-11.[46] The workpiece and spark gap are mounted so that the distance between the clamped ends is rigidly constrained by thick-walled brass or steel vessels. The spark gap terminals are adequately insulated from each other, and the metal outside envelope is used as a ground terminal. The shock wave is shaped, in part, by a conical section surrounding the axis of the spark so that there is

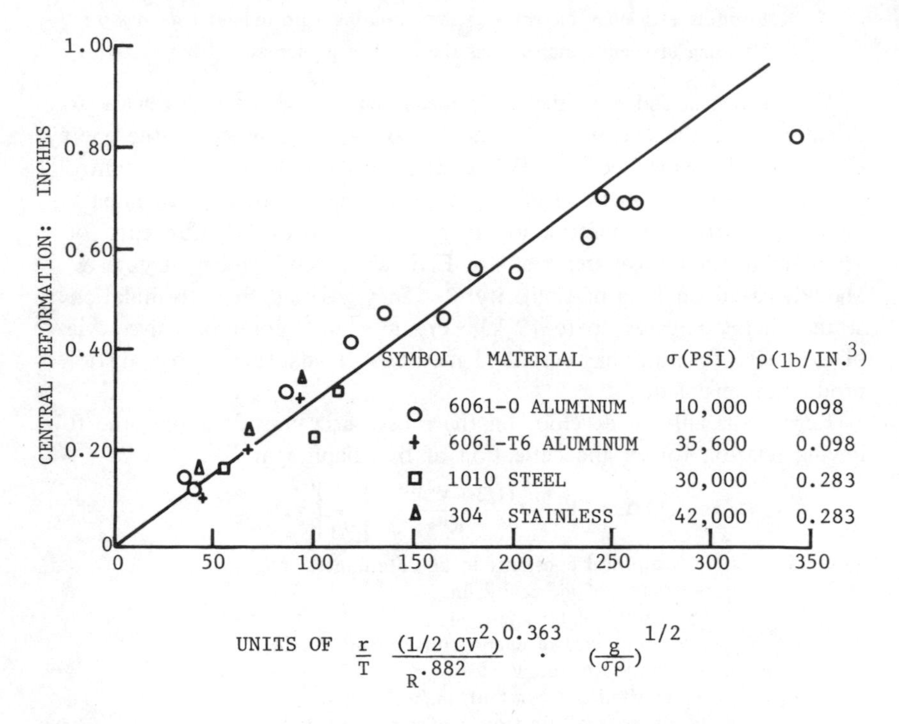

Figure 7-10. Deformation of metal diaphragms by electrohydraulic shock waves.

some focusing, and the wavefront travels as approximately pure longitudinal along an enclosing cylinder to subject the workpiece to a pressure that is instantaneously uniform across its plane. There is provision either to bleed air from or to evacuate the space between the die and the workpiece.

Control and focusing the shock wavefront by properly shaped walls in the water chamber is an important addition to the rudimentary apparatus of Figure 7-8, which became a highly developed technique under Cadwell's leadership for the Rohr Soniform process.[38,47] Another addition to the technique with somewhat similar efficiency improvement is the use of a double spark as developed by Inoue and Nishiyama.[48] Some new methods of adapting the technique for production were developed by Harrison and Grainger and Hollingum.[49,50] However, as argued by Cadwell, the most economical metal-forming role that can be filled by EHF is forming exotic, hard materials to difficult contours and close tolerances.[47]

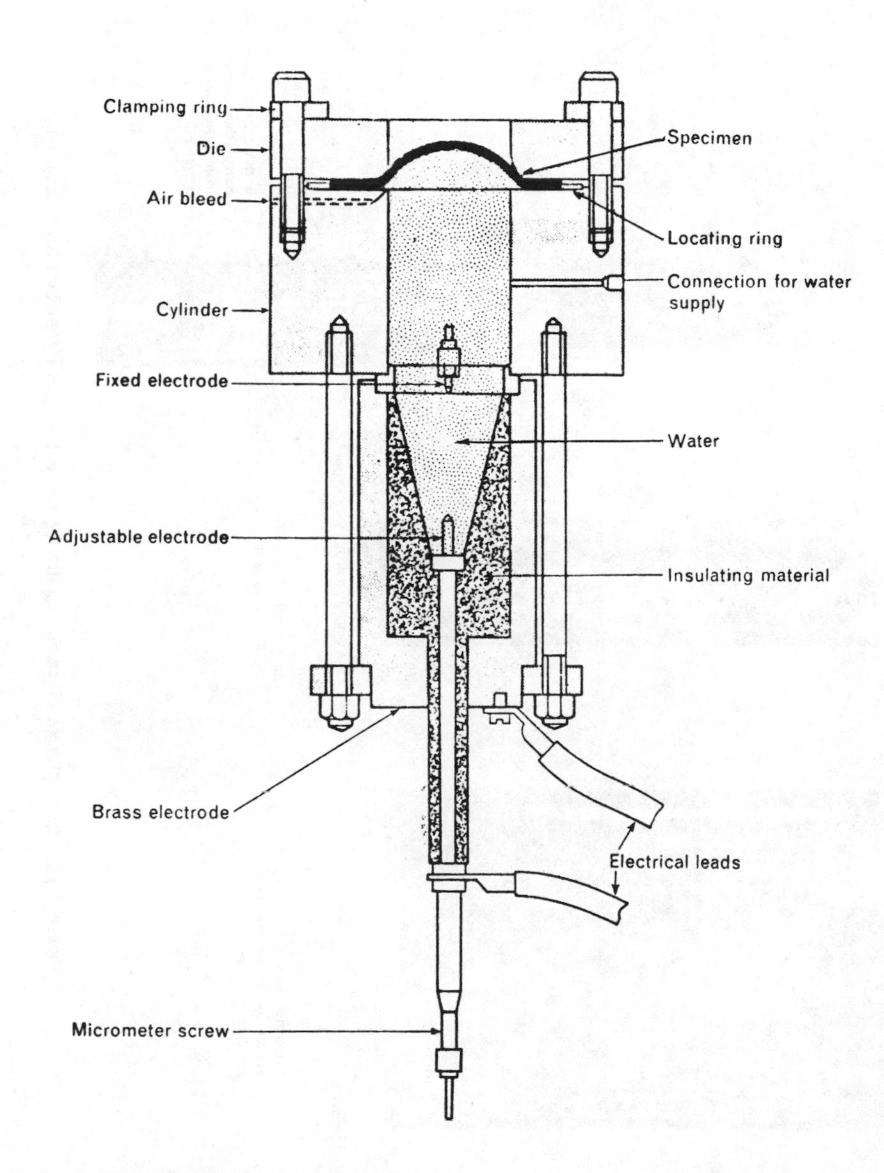

Figure 7-11. Practical assembly for discharge-type electrohydraulic forming of a plate.[46]

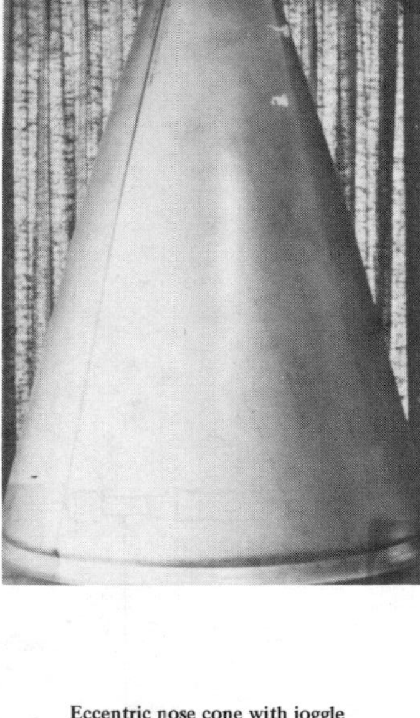

Eccentric nose cone with joggle at each end formed into molded stainless steel cone.

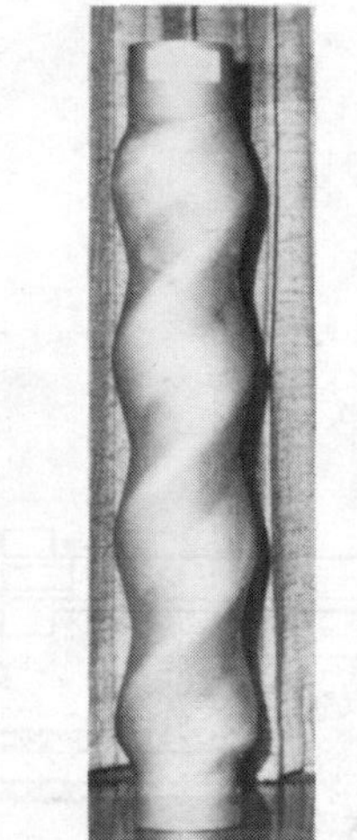

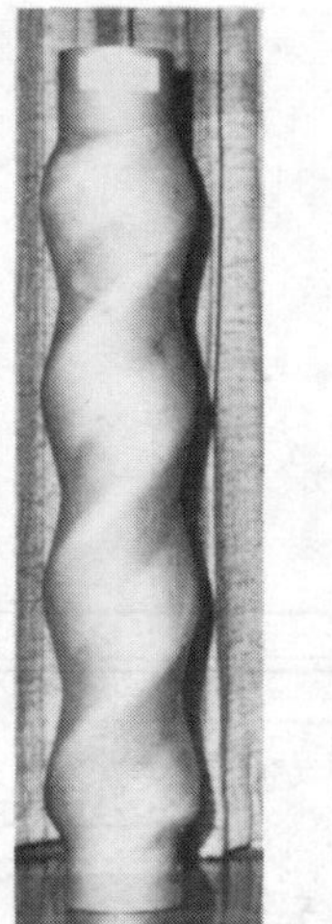

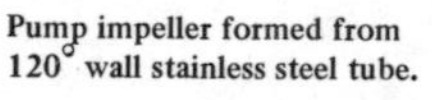

Pump impeller formed from 120° wall stainless steel tube.

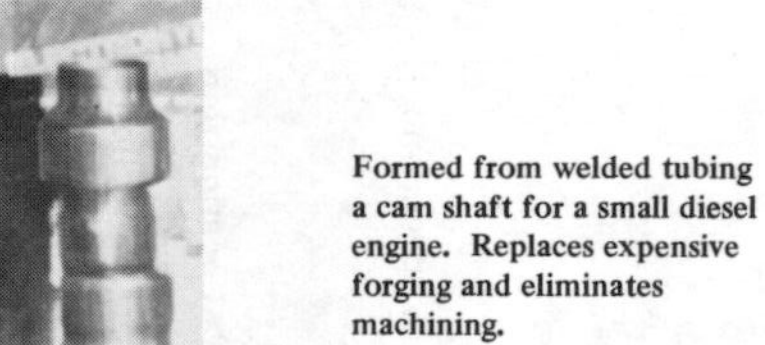

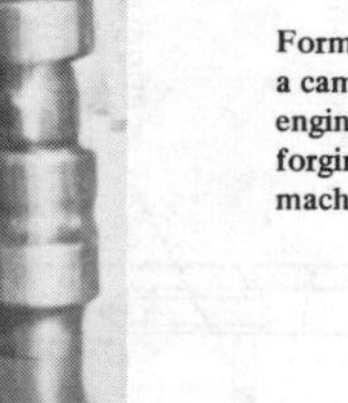

Formed from welded tubing a cam shaft for a small diesel engine. Replaces expensive forging and eliminates machining.

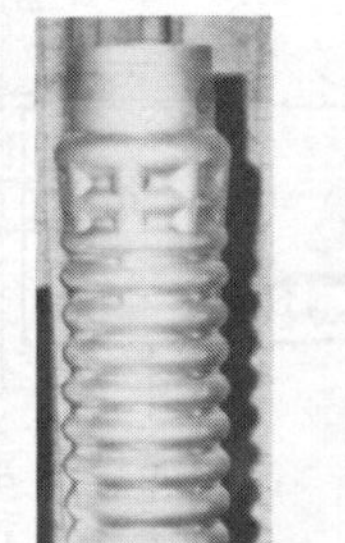

Both vertical and horizontal convolutions formed into stainless steel tube in one operation.

Figure 7-12. Examples of electrohydraulically formed shapes (from Soniform catalog).

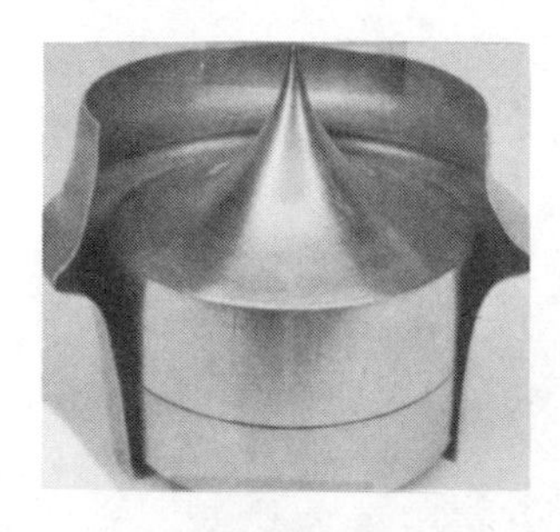

A pressure wave bouncing off of a reflector will free form a tubular workpiece as shown.

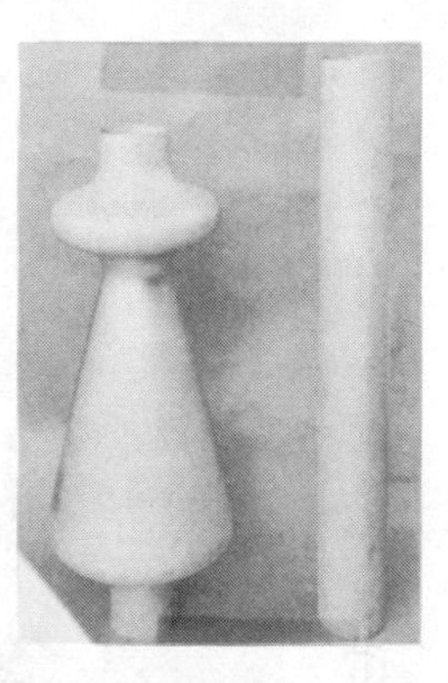

A 4½″ welded stainless steel tube (right) is formed into an experimental rocket nozzle with a 13¾″ apron (left).

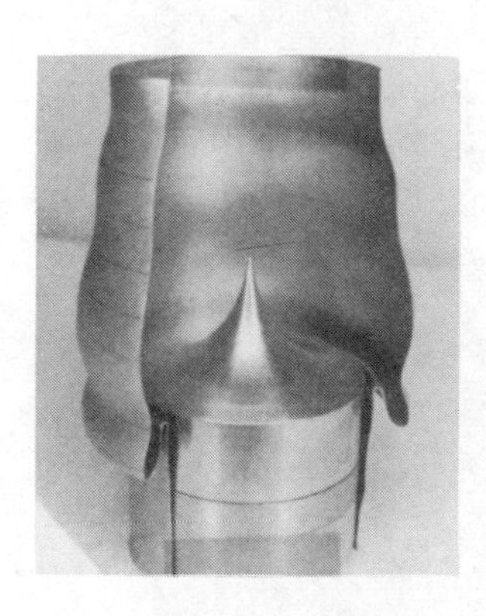

A primary pressure wave first expands the tube. Then a reflected secondary wave turns the tube back on itself. A complex free-formed shape of this type can be duplicated within 0.06 inches.

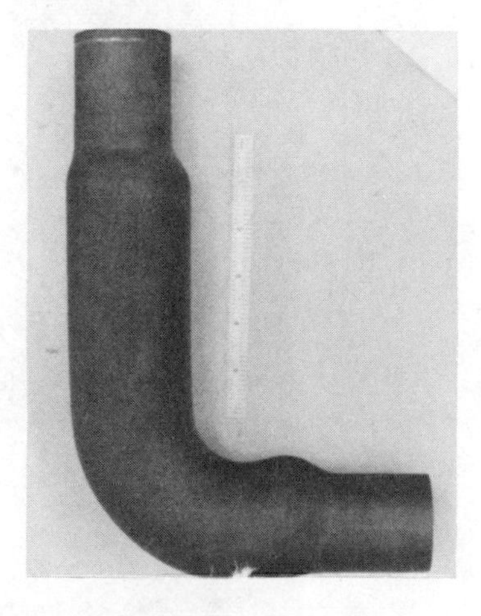

A 2-inch welded Hastelloy tube is preformed with a 4-inch bend radius. It is then placed in the SoniForm die and expanded to 2¾″ diameter with a 105° bend on a 2″ bend radius.

Figure 7-13. Examples of electrohydraulically formed shapes (from Soniform catalog).

A needle-like reflector is used for accurately forming a 2-inch tube over its entire 32-inch length.

Figure 7-14. Example of electrohydraulically formed shapes (from Soniform catalog).

Prototypical Examples of Electrohydraulically Formed Shapes

By using a combination of a wire-guided arc shaper and carefully machined sonic reflector shapes, the Soniform EHF process has shown that it is possible to meet the requirements of a wide variety of desired workpiece shapes. To illustrate, some typical formed pieces are shown in Figures 7-12 to 7-14.

Esterby described production EHF machinery, capable of handling tubes up to 12 in. diameter and 2 ft long, which operates with energy pulses up to 100 kJ.[51] Pruemer and Wolf applied the technique to enable sheet metal to be deep drawn, shaped and embossed in a single operation.[52] Watkins demonstrated that auxiliary ultrasonic vibration was of substantial assistance in electrohydraulic forming of certain metals.[29] In this technique, however, ultrasonic wavefronts are superimposed on the shock wavefront, and dimension effects must be carefully controlled.

Market for the EHF Process in Industry

By 1969 Cadwell (with Rohr industries, heading the Soniform group) had marketed his EHF techniques to form more than 300 different precision parts for the aerospace industry.[47,53] Examples of the difficult metals worked are Inconel 718, Titanium and Hastelloy. The Hughes Aircraft division of Hughes Tool has used EHF on precipitation-hardened 15-7 molybdenum.[54] There was some interest by the automobile industry in EHF parts; however, their mild steels with relatively lower material integrity are much harder to form by sonic wave than the tougher but more uniform aerospace metals. In addition, auto parts manufacturing and assembly operate under a much tighter cost constraint. This prevented significant growth into the auto market for EHF, which is much more expensive than EMF for each part processed.

The EHF process has stayed mainly in its original role of forming those metallic parts which could be formed in no other logical way. Soniform has remained a small firm, and in 1971 the estimated annual market was for no more than $500,000. The only other actively marketing firm, Electrohydraulics of Texas, offers a very similar process with machines ranging in size from 5-100 kJ at voltages between 9 and 20 kV. However, they offer a host of other services, and it is unlikely they have been able to develop an appreciable extension of the market.

There is some hint of a new market in a particular technical aspect of EHF. Soniform was very successful in cold-expanding without cracking commercially pure titanium tubes because EHF displaces metal rather than stretching it and, as a result, imparts a wrought structure similar to the effect of forging. EHF parts, like forgings, have distinct flow lines in the material, as can be seen in some of the examples in Figures 7-11 and 7-12. Jensen and Jorgensen have recently made use of this property in forming a housing out of sheet metal for AC motors replacing what would normally be a cast housing required for its rigidity.[55]

COMPARISON OF EHF, EMF AND OTHER METAL-FORMING TECHNIQUES

A comparison of EMF and EHF characteristics in their most appropriate operations is summarized in Table 7-6. Tubes and flat plates can be deformed in both EMF and EHF. However, in EHF, tubes can only be expanded; a contraction is possible but the EHF shock wave and workpiece geometry would be difficult to arrange for axially symmetric results. For EHF the workpiece must be elaborately mounted, generally with a vacuum on the side away from the fluid medium. Electrohydraulic forming is most valuable for imparting fine shape details of an enclosing die. For this work it is best to have the vacuum which would be awkward to incorporate into the electromagnetic forming process. Because of the difficulty of mounting the workpiece with fluid on one side and vacuum on the other, EHF is probably limited to the slow repetition rate which can be achieved with manual assembly. The peak electrical requirements of EHF as shown below are higher than those of EMF but can vary more widely depending on the materials used. A comparison is shown in Table 7-7.

Since EHF is not limited to high-conductivity metals, many resistive highly refractory metals such as titanium and inconel are treated without difficulty. Moreover, the milder metals pose some difficulty to EHF so that there is considerable complementarity with respect to EMF.

Of other electroforming techniques, electrodeposition from chemical solution is satisfactory for quality on work where it is applicable. However, some metal alloys are not suitable for the process and there is always the long time required to carry out electrolytic deposition of appreciable thickness. The classic stamping process cannot achieve the same precision of die shape reproduction, although for economy and repetition rate, it cannot be surpassed by either EHF or EMF.

If the energy requirements are low enough, they can be supplied in a hydraulic arrangement by a compressed air-activated piston. Higher energy requirements can be met with chemical explosives although with far less control. In summary, EHF and EMF are useful for highly precise work on a variety of metals. The customers for EHF are generally aerospace companies which have more stringent requirements and expect higher costs.

EMF customers are usually the auto companies and electric equipment industries.

Table 7-6. A Comparison of EMF and EHF Operational Characteristics

Operational Issue	EMF Characteristic	EHF Characteristic
Workpiece Material	High electric conductivity, mild strength metals	High-strength metals and alloys, Conductivity-independent
Workpiece Size	Relatively small with largest dimension under 1 foot	Wide variation in size, capable of forming pieces up to several feet
Assembly Operations	Very fast capability	Limited to manual set-up
Electric Power Requirements	High-voltage pulse, usually 6 kJ of energy	High-voltage pulse, usually 60 kJ of energy
Special Operational Advantages	Ideal for axially symmetric deformation of tubes	Specially wrought metal effects which impart extra strength to finished product

Table 7-7. Comparison of Electromagnetic and Electrohydraulic Forming Requirements (from Soniform Catalog)[1,2]

	Electromagnetic	Electrohydraulic
Electric Circuit Requirements		
Energy per Discharge (kJ)	6.0	62.5
Peak Voltage (kV)	8.3	18.5
Capacity (mF)	180	380
Peak Discharge Current (kA)	100	1,250
Maximum Pressure Exerted on Workpiece Material (Psi)		
Aluminum, Copper, Mild Steel	5,000	
Aluminum (6061-0)		10,000
Aluminum (6061-T6)		35,600
Steel (1010)		30,000
Stainless Steel (304)		42,000

REFERENCES

1. Brower, D. F. "Electromagnetic Forming," in *Metals Handbook* Vol. 4, *Forming* (Metals Park, Ohio: American Society for Metals, 1969), pp. 256-264.
2. Kegg, R. L. "A Study of Energy Requirements for Electric Discharge Metal Forming," *J. Eng. Ind.* 86(13):127-133 (May 1964).
3. Smith, K. F. "Metal Forming by Electrical Energy Release–Trends and Applications," SAE Pub. 650190 (1965).
4. Seyler, L. "Electromagnetic Metal Forming Offers Options for Automating," *Automation* 17(5):69-72 (May 1970).
5. Baines, K., J. L. Duncan and W. Johnson. "Electromagnetic Metal Forming," *Proc. Inst. Mech. Eng.* 180 Pt 1(4):93-110 (1965-6).
6. Lal, G. K. and M. J. Hillier. "Electrodynamics of Electromagnetic Forming," *Internat. J. Mech. Sci.* 10:491-500 (1968).
7. Lal, G. K. and M. J. Hillier. "Expansion of a Thin Free Tube in Electromagnetic Forming," *Int. J. Prod. Res.* 8(1):59-64 (1970).
8. Fluerasu, C. "The Use of Transient Parameters in the Study of Electromagnetic Forming," *Rev. Roum. Sci. Tech Ser Electrotech Energ.* 14(4):565-585 (1969).
9. Fluerasu, C. "Electromagnetic Forming of a Tubular Conductor," *Rev. Roum. Sci. Tech Ser Electrotech Energ.* 15(3):457-488 (1970).
10. Epechurin, V. P. and V. Shalygin. "Measuring Deformation Rate of Metal Blanks Under Magnetic Pulse Pressure-Working," *Meas. Tech.* 14(7):1016-1018 (July 1971).
11. Barbarovich, Yu. K. "Calculation of Forces in Magnetic Forming," *Electrotekh.* (7):45-49 (July 1971).
12. Dehoff, A. "Determination of Magnetic Pressure During Electromagnetic Forming," *Electric* 25(7):275-277 (July 1971).
13. Bron, O. B. and A. M. Segal. "Multiwinding Inductors of Various Shapes for the Magnetic Forming of Components," *Electrotekh.* (3):22-25 (March 1971).
14. Al-Hassani, S. T. S., J. L. Duncan and W. Johnson. "On the Parameters of the Magnetic Forming Process," *J. Mech. Sci.* 16(1):1-9 (February 1974).
15. Belyi, J. V. and F. Gorkiv. "Electromechanical Processes During Magnetic Impulse Forming of Metals," *Khoorost, Electromekh.* (4):442-447 (April 1974).
16. Schenk, H. "Investigations of Simple and Composite Field Concentrations for Magnetic Forming of Tubes," *Baender Bleche Rohre* 10(4):226-230 (April 1969).
17. Al-Hassani, S. T. S., J. L. Duncan and W. Johnson. "Techniques for Designing Electromagnetic Forming Coils," *Proc., 2nd Internat. Conf. of Center for High Energy Forming*, June 23-27, 1969.
18. Gilbert, L. E. and W. N. Lawrence. "Design of Electromagnetic Swaging Coils," *Proc. 2nd Internat Conf. of Center for High Energy Forming,* June 23-27, 1969.
19. Davis, M.E. *et al.* "Investigation of Developing Effects in a Magnetic Forming Beryllium Coil Assembly," IEEE-Region III Convention 7th Proc. (1968).

20. Jansen, H. "Influence of Frequency and Geometry on the Efficiency of the Electromagnetic Compression of a Metal Tube," *Proc. 2nd Internat. Conf. of Center for High Energy Forming*, June 23-27, 1969.
21. Lawrence, W. N. "Scale Modeling Calculations for Electromagnetic Forming," *Proc. 2nd Internat. Conf. of Center for High Energy Forming,* June 23-27, 1969.
22. Maeda, T. and T. Higuchi. "Study on Electromagnetic Press," *J. Fac. Eng. Sec. B.* 32(4):743-757 (December 1974). "Magnetic Forming," *Metal Forming* 36(6):163-166 (June 1969).
23. Zittel, A. "Getting the Most Out of Electromagnetic Metal Forming," *Assembly Eng.* (September 1976).
24. Buehler, H. and E. von Finckenstein. "Contribution to the Production of Sheathings and Crimp Joints of Tubular Parts by Magnetic Forming," *Baender Bleche Rohre* 10(7):413-418 (July 1969).
25. Buehler, H. and E. von Finckenstein. "Dimensioning of Crimp Joints for Joining by Magnetic Forming," *Werkstatt Betr.* 104(1):45-51 (January 1971).
26. "Electromagnetic Metal Working," *Tool Prod.* 34(5):60-64 (August 1968).
27. Blanc, M. and D. Dekerlegand. "Electromagnetic Forming Techniques Used to Correct Contour Distortions on Saturn S-1c Bulkheads," *Proc. 2nd Internat. Conf. of Center for High Energy Forming,* June 23-27, 1969.
28. James, T. E. and J. Phillpott. "High Frequency Systems for Magnetic Forming of Resistive Materials," *Proc. IEE (Lond) Conf. on Electrical Methods of Machinery Forming and Coating,* March 17-19, 1970.
29. Watkins, M.T. "Physics Contribution to New Technology," *Cont. Physics* 9(5):447-474 (September 1968).
30. Harrell, J. D. "Magnetic Pulse Forming Moves into Production Applications," *West. Mach. Steel World* 57(9):27-31 (September 1966).
31. Brower, D. F. "Industry Develops Niche for Magnetic Pulse Forming," *Metal Prog.* 93(4):95-98 (April 1968).
32. Fuller, R. F. "Magnetism Tamed for Forming," *Can. Metal/Mach. Prod.* 30(4):61-63 (April 1967).
33. Aoki, Y. and M. Kosaka. "Application of Electrohydraulic Metal Forming Equipment to Various Industries and its Development for Mass Production," *Proc. 2nd Internat. Conf. of Center for High Energy Forming,* June 23-27, 1969.
34. Wildi, P. "Up-to-Date Look at Magnetic Forming," *Metalwork. Prod.* 113(9):64-66 (February 1969).
35. Wildi, P. "Some New Developments in Magnetic Pulse Forming," Paper MF 70-152, SME Collect. Papers, Vol. 70, Book 7 (April 1970).
36. Chace, W. G. and H. K. Moore. "Exploding Wires," *Proc. 3rd Conf. on Exploding Wire Phenomena* (New York: Plenum Press, 1964).
37. Ghosh, S. K. "On the Measurement of Dynamic Strain and Forces in the Explosive Forming of Aluminum Domes," *Aluminum* 52(9): 569-571 (September 1976).
38. Cadwell, G. C. "Electrohydraulic Metal-Working in Standard Technique," ASTME Creative Manufacturing Seminars, Tech. Paper MF 68-539 (February 1968).

39. Al-Hassani, S. T. S., J. L. Duncan and W. Johnson. "Magnetohydraulic Forming of Tube. Experiment and Theory," *Internat. J. Mech. Sci.* 12(5):371-392 (May 1970).
40. Avery, R. M. "Equipment Used in Conjunction with Electrohydraulic and Magnetohydraulic Forming," ASTME Creative Manufacturing Seminars Tech Paper MF-68-538 (February 1968).
41. Chace, W. G. and H. K. Moore. "Exploding Wires," *Proc. of 1st Conf. on Exploding Wire Phenomena* (New York: Plenum Press, 1960).
42. Franke, H. J. "Controlled Metal Forming by Spark Discharge Under Water," *Metall.* 27(2):134-137 (February 1973).
43. Guenther, A. H. *et al.* "Acceleration of Thin Plates by Exploding Foil Techniques," in *Proc. 2nd Conf. on Exploding Wire Phenomena*, W. G. Chace and H. K. Moore, Eds. (New York: Plenum Press, 1962).
44. Oyave, M. and S. Masaki. "Development of Laws of Similarity for Electrospark Forming," *Japan Soc. Mech. E-Bull.* 11(45):545-553 (June 1968).
45. Pruemer, R. "Explosive Forming," *TZ Prakt Metallbearb.* 68(12): 437-440 (December 1974).
46. Van Oss, J. F. *Chemical Technology*, New York (1975).
47. Cadwell, G. C. "Electrohydraulic Forming of Components for Industrial Use," SME Collected Papers, Vol. 70 Book 7, Paper MF-70-151 (April 1970).
48. Inoue, T. and U. Nishiyama. "Double Spark Method of Electrohydraulic Forming," *Proc. 2nd Internat. Conf. of Center for High Energy Forming,* June 23-27, 1969.
49. Harrison, D. and A. T. Grainger. "Development and Application of Electrohydraulic Forming as a Production Process," *Proc. IEE (Lond) Conf. on Electrical Methods of Machinery Forming and Coating,* March 17-19, 1970.
50. Hollingum, J. "Electrical Methods Move Into a Wider World," *Metwork Prod.* 114(4):51-54 (April 8, 1970).
51. Esterby, J. W. "Electrohydraulic Forming of Metals," *Metals Mat.* 2(1):15-18 (January 1968).
52. Pruemer, R. and Wolff. "Some Applications of Electrohydraulic Forming," *Baender Bleche Rohre* 15(1):19-22 (January 1974).
53. Cadwell, G. C. "Electric Metalworking," *Proc. 2nd Internat. Conf. of Center for High Energy Forming,* June 23-27, 1969.
54. "Sonic Energy Makes Impossible Shaper Profits for New Company," *West. Mach.* 10 (December 1970-January 1971).
55. Jensen, P. K., J. Jorgensen and L. Alting. "Hydraulic Forming," *Sheet Metal Ind.* 53(7):55-56;59-61 (July 1976).

8
TRENDS IN TITANIUM MANUFACTURING*

INTRODUCTION

The primary titanium metal industry throughout the world uses either of two metallothermic reduction processes for quantity production of virgin metal: reduction of purified titanium tetrachloride by magnesium in the Kroll process and by sodium in the Hunter process. The metal from either process is produced in a porous sponge form. There are nine operational plants–three in the U.S., two in Japan, one in England, and three in the U.S.S.R. Primary process is reduction of titanium tetrachloride by metallic magnesium (80% of world capacity) or by metallic sodium (balance).

The market for titanium in the U.S. was sponsored by and is still primarily dependent on military aircraft requirements, although its use in the chemical process industry is growing continually. U.S. consumption of domestic titanium sponge in 1976 was about 55% of annual capacity of 24,000 short tons with 80% being used by aerospace and 20% by industrial markets. All major plants have captive production of primary reactants for chlorination and reduction by electrolysis of the by-product $MgCl_2$ or NaCl. Energy consumption for production of titanium ingot is about 67,000 kWh(E) and 80 x 10^6 Btu per metric ton.

Titanium has also been produced on a limited production scale by thermal decomposition of titanium tetraiodide and on a laboratory scale by electrolytic processes using fused salts as the electrolytes. Two fused salt electrolysis programs continue to be in pilot plant development in the U.S. In addition, basic research in electrolytic behavior of titanium is also being investigated. Table 8-1 presents a review of the research, development and commercialization efforts made for the recovery of titanium.

Current R&D activity in titanium process technology is being pursued mainly at two levels. The problems or deficiencies of the established

*By: Marcel Barbier, B. Bovarnick

Table 8-1. Review of Efforts Toward Titanium Recovery

Source Material	Process (es)	Developers	Outcome	Remarks
$TiCl_4$ + Mg	Kroll	U.S. Bureau of Mines	Industrial	
$TiCl_4$ + Na	Hunter	U.S. Bureau of Mines	Industrial	
Ti-Iodide	Van Arkel, deBoer	Foote Mineral Co.	Industrial	For very high purity metal
TiO_2	Reduction by K, Ca, Al, C, H	Metal Hydrides, Inc.	Abandoned	Contamination of Ti by Ca, Al. Low yields of hydrogen reduction in plasma reactors
Oxynitride, Oxycarbide, Halides	Reduction, electrolysis, disproportionation (of subchlorides)	Many, including U.S. Army	Abandoned	
$TiCl_4$	Electrolysis in fused salts	Timet, Dow-Howmet	Pilot-plant stage, continuing	
$TiCl_4$	Metallothermic using alkali, alkali earth metals, or Al	Several	Unsuccessful	High reactant costs, low efficiency, low purity
Ti-Fluorides	Electrolysis in fused salts	Horizons	Working but not developed to plant stage	

process technology of metallothermic reduction are well recognized. Continued engineering advancements are being made to improve the productivity of existing facilities, to increase the yield of high-quality product, and to decrease energy consumption and process costs. Thus, cost-effectiveness of the existing process is enhanced and any potential alternate process faces greater difficulty in being accepted.

The most promising alternative process is electrolysis of titanium tetrachloride in diaphragm cells with fused halide electrolytes. This process is only at the small pilot plant cell level of development. Potential benefits could be realized from fused salt electrolysis to overcome many of the problems attributed to the established process technology. Indicated energy consumption for fused salt electrolysis is about 18,000 kWh(E) per metric ton less than by conventional technology. Projected costs of titanium production by fused salt electrolysis for fixed capital and variable operating expenses are believed to be only marginally competitive with conventional technology. The findings of past R&D on fused salt electrolysis are being developed in pilot-scale units.

Further independent R&D on physical chemistry of electrochemical reactions is also being studied for both low- and high-temperature electrolytes. Table 8-2 summarizes development work in titanium metallurgy and Table 8-3 gives a breakdown of electrical energy consumption for the Kroll process and for fused salt electrolysis.

The status of technology in the foreseeable future for the production of titanium metal appears to be a continuation of the established methods–magnesium in the reduction of titanium tetrachloride by the Kroll process, which is predominant, and by sodium in the Hunter process, although the latter could be somewhat limited by the availability of sodium and environmental constraints. In addition, considerable effort is being directed to the electrolysis of titanium tetrachloride using molten salt electrolytes in diaphragm cells to determine whether they have potential for future installation of increased capacity. The need for increased capacity will depend on military commitments for a new round of advanced-performance aircraft, although chemical process equipment and utility heat exchanger applications are taking an increased share of the supply.

This chapter presents an assessment of recent trends in titanium manufacturing in the United States, a review of the research and development on titanium, which led to its current manufacturing technology, and the review of on-going work which might lead to improvements or changes in process technology. Its objectives are:

1. to identify the contribution of past R&D on processes for winning titanium metal;
2. to judge the opportunity for new process technology;

Table 8-2. Developments in Titanium Metallurgy

Process	Kroll Process		
Developer	Oregon Metallurgical Corp.		
Innovative Items	Horizontal six met reactor. Molten magnesium charging, withdrawal of magnesium chloride, recycle of Mg and Cl, distillation of Ti. Sponge, inert gas sweep of exhaust vapors		
Process	Fused Salt Electrolysis of Chlorides		
Developer	1) National Lead Co. and Titanium Metals Corp.	2) Dow-Howmett	3) U.S. Bureau of Mines
Innovative Items	Small cells operation and design, semicommercial cells 80 kg/day, alkali salts, diaphragm cell, inert atmosphere	Pilot scale cell	Cell for high purity Ti
Process	Fused Salt Electrolysis of Fluorides		
Developer	Hoch (preliminary work)		
Innovative Items	Yttrium oxide-calcium oxide for containment. High-temperature operation.		

Table 8-3. Electrical Energy Consumption for $TiCl_4$ Reduction (kWh/metric ton)

Process	Kroll (Magnesium)	Electrolysis
Dry Room Sponge Removal	12,600	12,600
Sponge Distillation	7,200	7,200
Magnesium Electrolysis	18,000	–
Electrolysis	–	12,000
Consumable Materials		
Chlorine	5,800 (1.5 ton/ton)	3,900 (1 ton/ton)
Magnesium	10,000 (0.1 ton/ton)	–
Carbon	– (1 ton/ton)	– (1 ton/ton)
Total	53,600	35,700
Gain vs. Kroll	–	17,100

3. to determine the status of on-going process R&D; and
4. to indicate the apparent prospects for commercialization of the most advanced processes.

The various methods for titanium production will be analyzed briefly and the reasons and circumstances responsible for their success or failure on the technical or commercial plane indicated. Also, areas of present development and factors on which their further exploitation depend, will be outlined.

RESEARCH AND DEVELOPMENT EFFORTS UP TO THE KROLL PROCESS

Reasons for dependence of titanium technology on R&D

The titanium industry, more than any other modern engineering structural metal, has depended primarily upon progress of R&D for winning of higher-purity metal from its precursors and its ores. This primary dependence on R&D arises as a result of mechanical strength properties of titanium, which are very sensitive to residual impurities. Almost pure metal is required to attain sufficient ductility for engineering applications.

Historical Path of R&D

Titanium was discovered by Gregor in 1791 and was confirmed a few year later by Klaproth. Throughout the nineteenth century, successive efforts by several investigators failed to obtain the metal in sufficient purity to be ductile. The first successful result is attributed to Hunter at General Electric, in 1910, who followed the earlier path taken by Deville in his work about 1850 and later by Nilson and Petersson in 1887. Hunter reacted relatively pure titanium tetrachloride with metallic sodium in an air-tight bomb; however, Hunter's original titanium was only 99.5% pure so that it was still brittle at room temperature, although ductile at elevated temperatures. This approach has constituted one of the main directions for R&D efforts and has become one of the important production processes in the industry.[1-3]

In subsequent research, van Arkel and de Boer in 1925 learned that thermal decomposition of titanium iodide on a hot filatment results in deposition of very high-purity metal, which was ductile at room temperature. Furthermore, the liberated iodine could be recombined with crude titanium feed stock at lower temperatures to form the iodide. Thus, a cyclic iodide deposition process could be operated in an evacuated closed chamber until the hot surface could no longer be maintained at temperature or until the feed stock was exhausted. This concept was developed to a limited commercial-scale process during the 1930s in Europe and during the 1940s in the United States. It is still the preferred route for making very high-purity metal.[4]

The most important direction taken in titanium R&D was the work of Kroll beginning in the 1920s, which culminated in his discovery of magnesium reduction of titanium tetrachloride in the 1930s. He achieved successful reduction to metal in 1937, in Germany, and subsequently brought this discovery to the United States. Industrial interest in titanium R&D in the United States began in a serious way with the work of the U.S. Bureau of Mines in the late 1930s, and was accelerated considerably when Kroll joined the agency in 1940 after his escape from Germany. The investigation continued, and U.S. Patent 2205854 was granted to Kroll in 1940. The development of his process to a larger scale was pursued at the Bureau of Mines during World War II.[5]

The publication of the Bureau of Mines report, "Metallic Titanium and its Alloys" in 1946 demonstrated the first pilot production of ductile metal, which had a projected cost compatible with its prospective applications. The first Conference on Titanium was organized in 1948 by the Office of Naval Research on the results of the government-sponsored metallurgical development program carried out in conjunction with the Bureau of Mines.[6] The program results established that titanium had outstanding properties for potential engineering applications:

- exceptional resistance to salt water and chloride corrosion;
- excellent strength/weight ratio;
- good retention of mechanical properties at elevated temperatures up to 500°C; and
- insensitivity to embrittlement on heating to temperatures of 700°C.

The availability of metal with these properties attracted attention for many applications. Naval engineers could foresee its use in construction of shipborne components and equipment. Aircraft and engine designers recognized the potential merit of these properties for high-performance military aircraft. Additionally, the potential for titanium to substitute for stainless steels offered significant interest to strategic planners, who are always concerned about the country's dependence on imported chromium for corrosion-resistant alloys.

Applications of Titanium in Technology

The success of the Kroll process led to the expectation that titanium would become available in substantial quantities. The government agencies then identified a number of prospective applications which titanium could benefit and which, in turn, could provide a substantial demand for it to support its large-scale development. The Navy was interested in it for condenser tubing, shafting, ships' propellers, pump rods, hull plates, pontoons and ships' superstructure fabrications. The Air Force looked at it for structural members of airframes, parts for high-temperature rocket, ram-jet, and turbojet engines and jet turbine blades. With these initial prospects, commercial participation began in 1948 with DuPont's installation of the first commercial sponge plant. In addition, an intensive R&D program was sponsored by the government with many laboratories, to improve or overcome several apparent limitations of the Kroll process.

Gradual Development of the Kroll Process

From initial batches of grams, the Bureau of Mines gradually scaled up the process to a semicommercial capacity of 120 kg/day in 1952. By this time, commercial plants, beginning at DuPont in 1948, had been put into production, so that the Bureau of Mines Kroll pilot plant would be shut down in 1954. The Kroll process at that time became the most important route for production of titanium and has continued its dominance. Although several companies began production of sponge in the 1950s, market changes forced most of them to shut down; today, only three sponge plants are operational in the U.S. As seen in Table 8-4, there are only six plants in the free market countries and three in the U.S.S.R., although the latter is believed to be building a fourth plant.[7]

PROBLEMS OF KROLL PROCESS

The metallurgical quality of the early sponge reduction product, usually characterized by the Brinnel hardness of a melted button of the sponge, was in the range of 150-180 Brinnel or higher, which was at or in excess of the upper limit of metallurgical acceptability for subsequent metalworking, transformation to mill products and fabricated shapes for engineered parts. Among the causes of the high hardness are the extremely stable unreduced compounds arising from the high reactivity of elemental titanium with oxygen, nitrogen, chlorine, or carbon. The stability of these compounds had been extensively demonstrated by the long history of unsuccessful research efforts to obtain the metal in the pure state.[2]

The Kroll process results in a porous spongy metal mass. The direct consequence of the entrapment of magnesium and magnesium chloride in the interstices of the titanium sponge. Elaborate treatments are required to remove them from the sponge. In addition, trace residuals of oxygen

Table 8-4. Titanium Sponge Plants[8]

	Reductant	Estimated Capacity (metric tons/yr)
United States		
Timet	Mg	15,000
RMI	Na	7,000
Ormet	Mg	2,400
Total		22,400
England		
ICI	Na	3,600
Japan		
Toho	Mg	5,000
Osaka	Mg	5,000
New Metals[a]	Na	2,000
Total		12,000
U.S.S.R.		
3	Mg	30,000

[a]Na-based sponge is clouded by possible ban on leaded gasoline and environmental restriction relative to disposal of NaCl. New Metals closed the plant shortly after start-up due to ban on leaded gas.

and chlorine are also present, as well as some carbon, nitrogen and hydrogen. Another handicap of the Kroll sponge was its relatively high cost due to its batch scheduling, consumption of large quantities of magnesium and chlorine as process consumables, raw materials and direct consumption of energy production or recycling these process reactants. These and other undesirable characteristics, which have stimulated R&D on alternative processes, can be summarized as follows:

- high consumption of energy, almost exclusively electricity;
- high consumption of high-cost process reactants;
- high fixed capital costs, especially to recover and recycle process reactants;
- batch processing;
- limited size of process batches;
- slow production rate of batch processing;
- high direct operating costs;
- extensive cleaning treatment and sophisticated consolidation processes of process product sponge to obtain solid forms with sound metallurgical quality for engineering applications; and
- low yields and high scrap rates of requisite downstream consolidation processes.

RESEARCH ON ALTERNATIVES TO KROLL PROCESS

The research programs undertaken to overcome or avoid these undesirable features followed both conventional chemical and process engineering approaches as well as taking some very exotic directions. The process routes pursued by workers in the field have extended from concepts for reduction of titanium dioxide itself by other metals such as potassium, calcium, or aluminum and by carbon and hydrogen. None has been effective, resulting in impure metal with either several percent of residual oxygen or contamination by the reductant or both. Some efforts have led to consideration and investigation of other intermediate compounds other than the tetrachloride, such as oxynitride, oxycarbide, or lower valence halide compounds which might then be treated for subsequent reduction to the metal. These attempts employed a wide range of chemical treatments ranging from reduction reactions of the oxide, reduction of the chloride, electrolysis using a wide variety of electrolytes for dissociating chlorides, fluorides, or oxides, and disproportionation of the lower chlorides. While each approach has offered some promise and shown some success, none has yet been able to produce ductile metal at a cost competitive with the Kroll magnesium reduction or the similar Hunter sodium reduction of the tetrachloride (see below for details on the Hunter process).

The U.S. government's change in emphasis from high-performance aircraft to missiles in 1957 resulted in withdrawal of its extensive support for process R&D on titanium. This, in turn, was followed by the termination of most laboratory programs on process research. Among the various alternative processes investigated is the electrolysis of titanium tetrachloride with fused eutectic salt electrolytes. The investigation of electrolyte behavior of titanium continues to be the subject of several R&D programs. Extensive research was expended on this method from the 1940s until the 1960s. Although most of these programs have been terminated, follow-up R&D and pilot-plant efforts have been continued by Titanium Metals Corporation of America (TMCA) or TIMET and by a joint venture between Dow Chemical and Howmet. For several years, both programs have announced imminent economic success but neither has yet been committed to the construction of a production-scale plant. In addition, other R&D programs on electrolysis are also being pursued at the laboratory stage on both chloride and fluoride systems.

Processes Based on Titanium Tetrachloride

Sodium Reduction (Hunter Process)

The first direction taken in the development of alternative processes to the Kroll process was to return to the sodium reduction of $TiCl_4$, the first successful route developed by Hunter in 1910. Among the advantages presumed to be offered by the Hunter process was that sodium was thought to be lower in cost than magnesium.[2] The sodium was expected to have greater efficiency as the reductant. The reduction temperature in the sodium route was known to be lower than for magnesium so that this reactant offered the benefit of easier handling in the molten state. Hence, engineering and operating a sodium reduction process was expected to be more readily accomplished.

From the research it has been learned that the sodium reduction proceeds by the reduction of the titanium tetrachloride to the dichloride in the first step, resulting in a molten mixture of the two metal salts. The second step is the final reduction of the dichloride to metallic titanium, which is formed intimately mixed with the solidified sodium chloride. The sodium salt reaction product is then leached from the titanium. After gaining this knowledge, the U.S. Bureau of Mines was able to employ close control of process step temperatures and obtain large titanium crystals of very low hardness, demonstrating that metal of very high purity was produced. Separate research programs paralleling those at the Bureau of Mines resulted in two proprietary adaptations of the Hunter sodium reduction process and led to separate plants being put into operation. The first was accomplished by Imperial Chemical Industries, Ltd. (ICI) in Englang and is run by Imperial Metals Industries, Ltd., which is partially owned by and closely affiliated with ICI. The second was in the U.S. and is the plant of Reactive Metals, Inc. (RMI) that was originally owned jointly by P.R. Mallory, who developed the process for the plant, and Sharon Steel. Both companies have since withdrawn, and RMI has become a joint venture of National Distillers & Chemical Corporation and U.S. Steel Corporation.[3,9]

Hydrogen Reduction

Efforts were also applied to the hydrogen reduction of titanium tetrachloride. Analysis of the thermodynamics of this reaction readily showed that extremely high temperatures on the order of 3500°K were required for the reaction to proceed to completion.[10] To attain the requisite temperatures, the reaction must be carried out in the vicinity of the hot zone of an electric arc. A number of experimental workers pursued this direction and have claimed success beginning with Wintraub in 1912 (U.S. Patent No. 1046043), but their work has been difficult to reproduce and verify. Therefore, their claims have since been disregarded. An extension of this approach has been the utilization of plasma reactors instead of the electric arc. While higher instantaneous energies can be attained with a plasma reactor, the dual problem of very low conversion efficiency, coupled with the reverse reaction of the products on exiting out from the hot plasma zone results in very low process yield and extremely high process costs. These conditions have led to this process also being rejected as a viable route for production of titanium.

Fused Salt Electrolysis

The process approach, which has probably seen the most research effort and which some believe offers the greatest promise, has been the electrolysis of titanium tetrachloride in a fused salt electrolyte to metallic titanium. In the history of U.S. titanium research, there have been several important research programs on electrolysis of titanium as shown in the listing of patents granted (see Appendix). The important programs could be categorized into two groups according to their dependence on chloride or fluoride chemistry. Of these, the only one that worked with fluoride chemistry was Horizons Titanium Corporation. Horizons was able to bring its R&D program up to, but did not extend it to, the pilot-plant stage.[11-13] This work was completed during the 1950s and was the first to be described in the literature. Although it has long since been abandoned, its patents and publications are frequently referenced in the literature. On the other hand, several groups working with chloride chemistry were able to attain the pilot-plant stage. Of these, two groups known to have programs still active on fused salt electrolysis of titanium tetrachloride, are TIMET and Dow-Howmet.[14] The electrolysis of fused chloride salts is the principal research direction of future potential production processes to compete with or displace the Kroll or Hunter methods.[15,16]

Thermal Disproportionation of Subchlorides

Another direction investigated for conversion of tetrachloride to metal is thermal disproportionation to the subchlorides as intermediate products which, in turn, can be shown to disproportionate to the metal. This process concept has been described and patented by Ferraro of the U.S. Army (U.S. Patent No. 2874040, 1959). R&D on discrete steps of the process scheme has been investigated individually and has shown that they all can be accomplished. However, it is questionable that it has been carried through as an integrated process beginning with the tetrachloride and actually finishing with metallurgical quality titanium, as the overall process is so elaborate and complex. Research on this process approach has also been abandoned.

Other Chloride Processes

Other reduction process routes have also been considered from the titanium tetrachloride as feed stock. These are essentially metallothermic using alkali or alkali earth metals as reductants. Aluminum has also been considered as a potential reductant. The alkali and alkali earth elements generally incur the penalty of high cost of the consumable reactant or lower reaction efficiency or both compared to either magnesium or sodium. In addition, they have not been able to produce metal of sufficient purity to have ductility. As might be expected, none has demonstrated suitable results to justify continued development.[2,17]

It should be noted that in the period after World War II, the anticipated military and commercial potential for titanium stimulated many research efforts under government sponsorship to compete with or to avoid infringing on the patents of the Kroll process. In this milieu, many projects were undertaken that were technologically naïve and did not have the benefit of mature theoretical analysis. They were pursued, however, because the availability of high-purity titanium chloride in substantial quantities from the titanium dioxide pigment industry provided an excellent-quality intermediate feed stock that conceptually only required one more step to be converted to the metal. Many purported research efforts in this category fall more into the classification of seeking invention or discovery, rather than systematic scientific investigation.

Other Process R&D

Thermal Decomposition of Titanium Iodide

The process of thermal decomposition of titanium tetraiodide was discovered by van Arkel and de Boer, as has been discussed in a preceding section. The principal R&D efforts in the U.S. for development of this process to large-scale operation were pursued by Foote Mineral Company. When the availability of sponge titanium with good metallurgical quality was limited, in the early 1950s, a supply of high-grade metal was provided by iodide titanium crystal rod. This premium-grade material played an important part in supplying specimens for laboratory investigations of physical metallurgy and mechanical properties. However, the cost of the iodide rod was about 40-50 times that of sponge. The high cost of iodide titanium arose from the cost of elemental iodine, the low productivity of the process, and the sensitivity of the metallic materials of construction of the process reactor to attack by the iodine. Thus, as soon as lower-cost sponge became available, its economics caused the displacement of iodide titanium. By the mid-1950s, this process was shut down and has not been operated again in the United States.[3,4]

Reduction of Titanium Oxide

It can be shown that calcium, calcium carbide, magnesium aluminum and carbon have thermodynamic potential for reduction of titanium dioxide to elemental titanium. As these elements are readily available, they have all been considered as reductants for titanium. In general terms, each would result in titanium with a significant residual oxygen content at the reaction equilibrium point, since none except the carbon has a volatile oxide product. To drive the reduction reaction to completion would require a substantial excess of the reactant element. This would result in the entrapment of this reactant metal and the by-product oxide in the product titanium, except for the carbon, which would react to become titanium carbide. In particular, the calcium reduction was investigated by Metal Hydrides, Inc. It was found that the excess of calcium required to reduce the oxygen to the lowest possible levels resulted in contamination of the titanium. In the case of aluminum, it was found that an aluminum-titanium intermetallic compound is formed. Thus, the results of direct elemental reduction of titanium dioxide were unsuccessful, and this approach was terminated.[2,3]

Production of Titanium Metal from Other Compounds

The possibility of producing pure titanium metal from any of a number of intermediate compounds has been examined by T.A. Henrie.[17] He has shown that in addition to the compounds and reactions discussed in the preceding sections, conversion of titanium oxide from ore can follow several possible routes (Figure 8-1). He has classified the sequential steps after raw material as crude intermediate, pure intermediate and final product. He has recognized that the possible routes that have not been actually performed (routes other than the ones shown on the left of the chart) must be treated as only potential processes because of the lack of adequate information. His review implies that there is little justification to investigate any of these more speculative processes.[17]

CURRENT STATUS OF RESEARCH AND DEVELOPMENT

New Developments of the Kroll Process

From the initial successful production of metallic titanium sponge in batches of grams, the Kroll process has benefited from continued R&D to improve its productivity. Current standard operating practice results in batch sizes of one to two metric tons. The as-produced crude metal has a sponge structure that is readily made with residual impurities of:

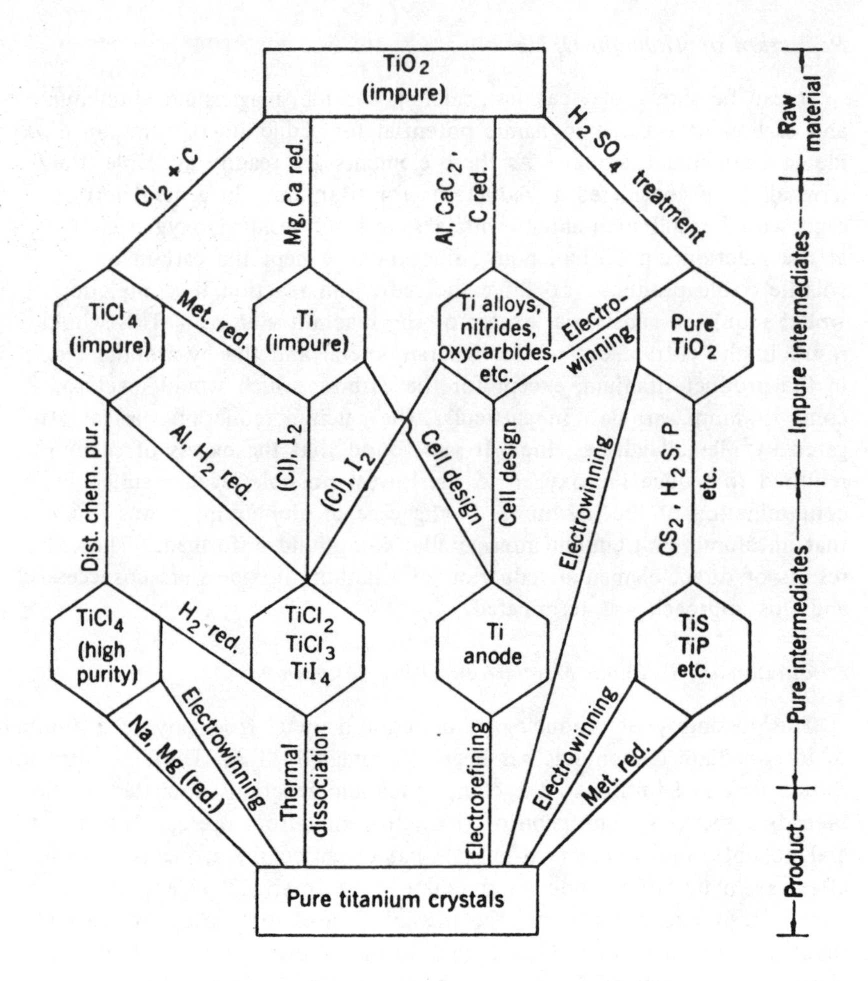

Figure 8-1. Possible routes for titanium processing.[17]

- oxygen, less than 1000 ppm;
- nitrogen, less than 100 ppm;
- hydrogen, less than 300 ppm after leaching out and less than 100 ppm after distilling out the entrapped unreacted magnesium and by-product salt;
- iron, consistently less than 1000 ppm and as low as 100 ppm; and
- chlorine, generally less than 1000 ppm.

The resulting metallurgical quality of the sponge, as indicated by the Brinnel hardness of less than 100, is quite good, although some grades require that oxygen be added back into the melt to attain requisite mechanical properties in the ingot. Still further refinement of sponge chemistry could be realized should it be found necessary.[8]

A more recent improvement in productivity of the Kroll process was the Oregon Metallurgical Corporation development of increased reactor batch capacity from one to two metric tons to about six metric tons by the use of an enlarged horizontal reactor. The larger-sized batch and reactor also included the development of:

- charging molten magnesium to the reactor;
- withdrawal of molten magnesium chloride from the magnesium electrolysis cells;
- increased utilization and recovery of magnesium and chlorine by high-temperature distillation of the sponge; and
- faster and deeper distillation of the sponge by using an inert gas sweep to exhaust the distillate vapors instead of a vacuum pumping system.

The full benefits of these improvements are now being established since the plant of Oregon Metallurgical Corporation, which had been shut down since 1972 after several improvements, was put back on-line at the end of 1976.

Assessment of energy consumption for production of titanium ingot via the Kroll process, reduction of titanium tetrachloride by magnesium, requires analysis of the energy at each step in the process flow chart. The energy is consumed as electrical and thermal, depending on the operation to be performed. The major steps in the process flow chart are:

1. mining and beneficiation of the ore to concentrate;
2. chlorination of the ore concentrate and purification of the crude chloride;
3. magnesium reduction of the chloride by the Kroll process;
4. dry room removal of the sponge from the reactor;
5. crushing and purification of the sponge; and
6. electrolysis of the magnesium chloride product from the sponge reaction step to obtain elemental chlorine for recycle back to the chlorinator, and magnesium for recycle back to the Kroll reactor.[18]

In addition, energy equivalents of consumable materials must also be accounted for; these step energy demands are shown in Table 8-5.

Limitations of the Hunter Process

Experience with the titanium sponge produced by the Hunter process reduction of the tetrachloride by sodium has shown the Hunter process product to be the equivalent of the Kroll product.[8] Despite the absence

Table 8-5. Energy Consumption in Production of Titanium Sponge by Kroll Process Magnesium Reduction[10]

Process Step	Energy: Electrical (kWh/metric ton)	Energy: Thermal (10^6 Btu/metric ton)
Ore Concentration		
Extraction		5
Beneficiation	1,250	
Chlorination		
Primary Reaction	1,150	
Purification	2,400	
Kroll Reduction		
Reactor Heating		40
Dry Room Sponge Removal	12,600	
Sponge Distillation	7,200	
Magnesium Electrolysis	18,000	
Consumable Materials		
Chlorine: 1.5 tons/ton	5,800	
Magnesium: 0.1 tons/ton	10,000	
Carbon: 1.0 tons/ton		30
Ingot Melting	6,000	
Miscellaneous Plant Energy	3,000	—
Totals	67,400	80

of plant process data, indications are that the cost of sodium-reduced sponge is about equivalent to that of magnesium. However, Hunter plants are dependent on an external source of metallic sodium for make-up of process losses. Provision must be made for disposal of the excess sodium chloride by-product from an existing plant or any new plant. Thus, installation of a sodium-reduction Hunter process plant requires an association with a nearby source producing sodium either in excess of its own needs or as a by-product of other chemical process operations in sufficient quantity to provide at least two tons of sodium for each ton of titanium sponge capacity anticipated for the Hunter process plant, without captive electrolysis of the sodium chloride.

An example of the impact of limiting the supply of sodium was the New Metals plant in Japan, which was based on sodium reduction.[8] The supply of sodium for New Metals came as an adjunct of the production of lead compounds for leaded gasoline. The ban on leaded gasoline shortly after the plant went into production caused the loss of its sodium supply and forced the plant to shut down.

An additional concern is the environmental impact upon disposal of the sodium chloride reaction product from the sodium reduction. Environmental control regulations have imposed severe restrictions on the disposal of potentially harmful chemical waste products. Hunter process plants would be subjected to such controls on the sodium chloride waste. No new source would be allowed to dispose of sodium chloride, and the existing plant in the U.S. (Reactive Metals) must satisfy the environmental control authorities for approval of its disposal procedures since each unit mass of titanium results in five times that quantity of sodium chloride.

Progress in Fused Salt Electrolysis

Among the various R&D programs on processes for winning titanium that have been investigated since the commercialization of the Kroll and Hunter processes, the most promising is the fused salt electrolysis of titanium tetrachloride.[14] There are several advantages in using titanium tetrachloride as the feed material for the electrolysis. The use of the chloride avoids the problem of eliminating residual concentrations of other compounds such as oxygen from oxide or carbon from carbide, both of which have been suggested and investigated as candidate feed materials for electrolytic reduction. High-purity material is readily available as an interim product of the important chloride process route for the manufacture of titanium dioxide pigment. If an electrolytic process plant were to come on-stream, the opportunity is present for its raw material to be purchased without having to be fully integrated back to the extraction of titanium from the ores.

In the United States, R&D on electrolysis of titanium tetrachloride was undertaken during the 1940s by a limited group of industrial laboratories. These laboratories included National Lead Co., New Jersey Zinc Co., Dow Chemical Co., and E.I. DuPont de Nemours and Company, among the more important.[3] It should be noted that all four companies were actively engaged in electrolytic processes for other metals, and all were directly or indirectly involved in the pigment industry. In addition, the U.S. Bureau of Mines had also undertaken its own program with the initial objective of electrorefining low-grade sponge and scrap and subsequently extended the program to the electrowinning of titanium metal. Except for the Bureau of Mines research, these programs have been treated with the utmost confidentiality as proprietary commercial activities.[11,19-25] As a result, the open published literature is very restricted, and technical articles have generally not been released for publication until several years after protective patents were obtained. In some cases, technical information was never released for publication.[26]

Examination of the available information indicated that in general terms, the research led to similar directions for the several groups that investigated the problem. The results of the research of three laboratories have been reported. The first to publish its experiences was the National Lead Co., beginning with the experimental investigation of small cells reported first by Alpert and his co-workers[27] and followed by the subsequent work by Opie and co-workers on features of cell operation and design.[13,28] In particular, Opie and Moles found that by confining the electrolytic reduction of $TiCl_4$ to the interior of a perforated basket-cathode, the electrolyte between the anode and the cathode can be kept free of reduced chlorides of titanium;[13] this eliminates undesirable reactions of reduced titanium chlorides with brickwork or oxygen in the cell atmosphere, so that semiwalls are not needed. A picture of their basket-cathode is shown in Figure 8-2, and a drawing of the cell where it is used is presented in Figure 8-3. Table 8-6 gives the basket-cathode cell operating characteristics.

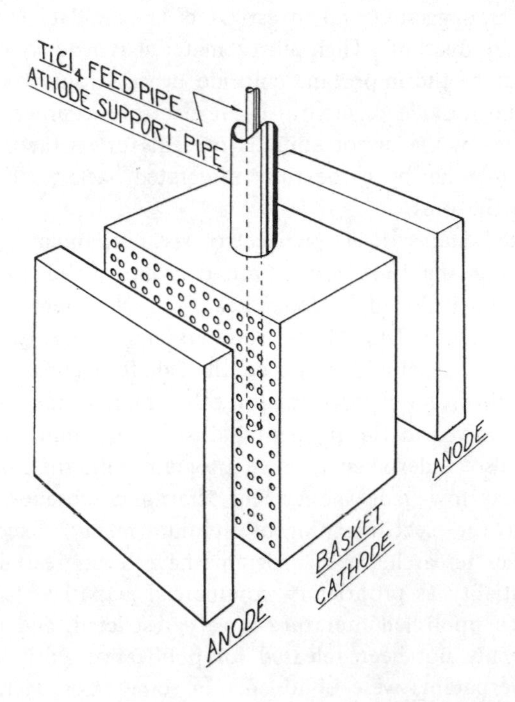

Figure 8-2. National Lead's basket-cathode.[13]

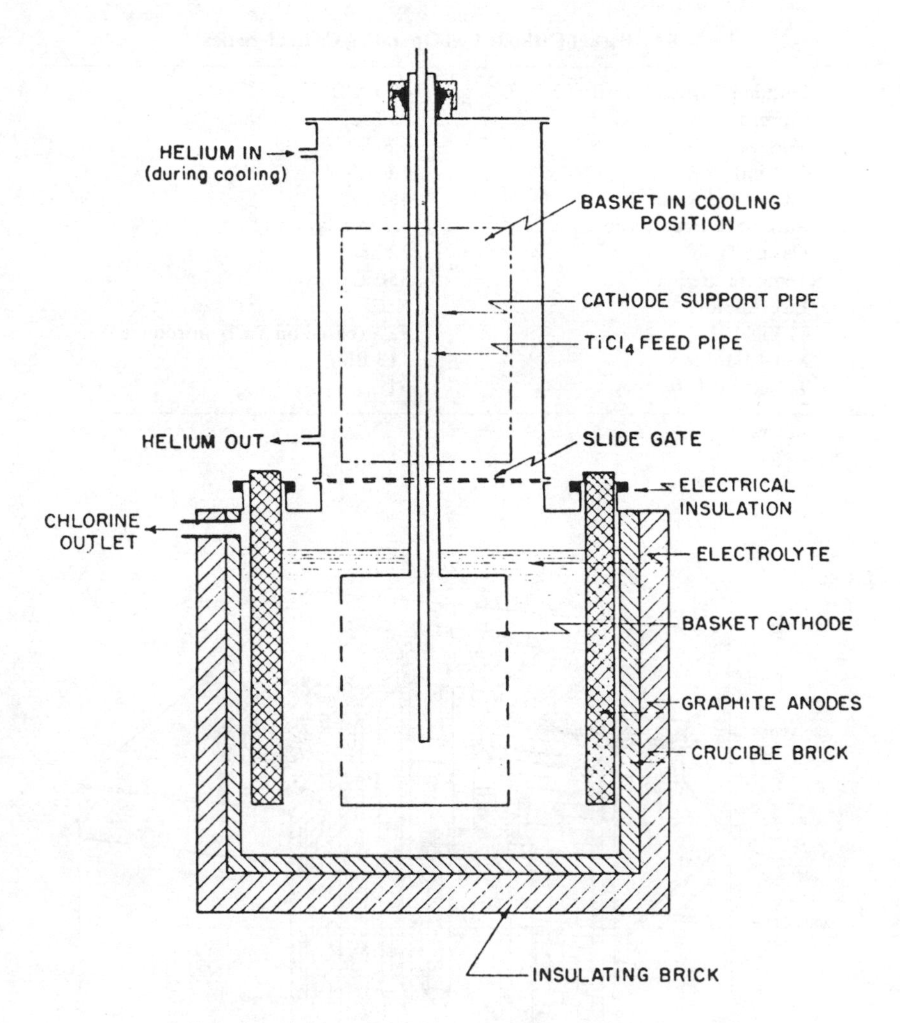

Figure 8-3. Schematic drawing of National Lead's basket-cathode laboratory cell.[13]

A general view of an experimental National Lead Co. pilot plant basket-cathode cell assembly design is presented in Figure 8-4.[14] Most recently, a summary of the extended development of this work by the Titanium Metals Corporation to the semicommercial size electrolysis with capability of 400 kg/day from five cells was presented at the 1973 meeting on Production of Titanium Metal held by the National Materials Advisory Board.[14]

Table 8-6. Basket-Cathode Cell Operating Characteristics[12]

Cathode Current Density	400 A/ft^2
Current	750 A
Voltage	6.8-7.2 V
Cell emf	2.6-2.9 V
$TiCl_4$ Feed Rate	485 ml/hr
Current to Feed Ratio	6.1 faradays/mol
Plating Time	72 hr
Temperature	850°C
Electrolyte	NaCl
Ti Yield	93% (based on $TiCl_4$ introduced)
Metal Hardness	113 Bhn
Ti Current Efficiency	61%

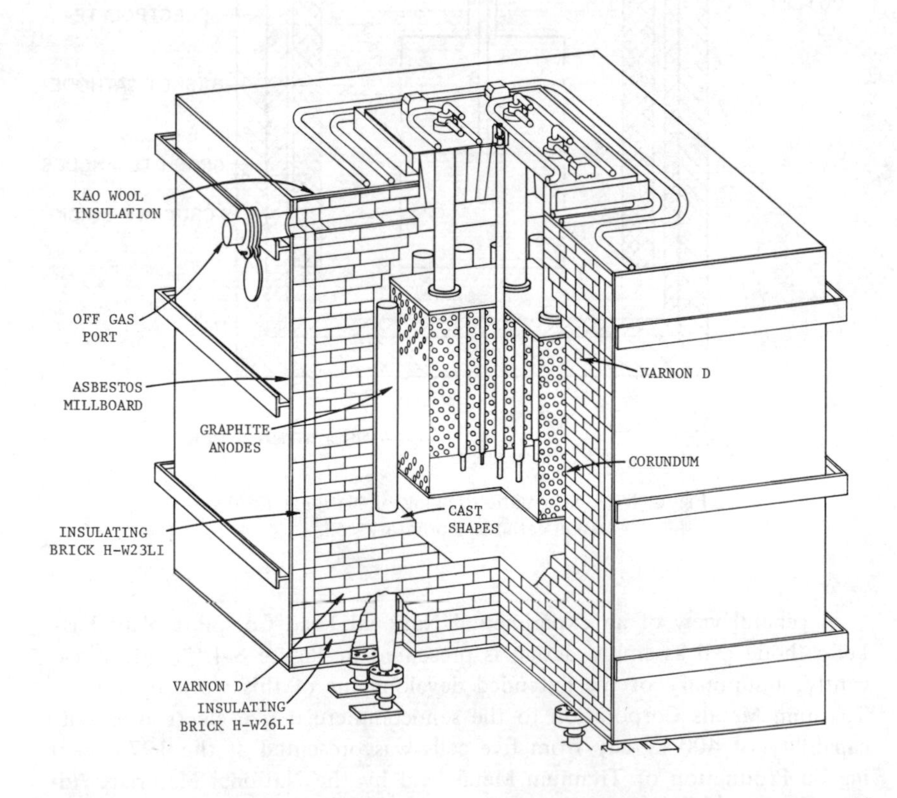

Figure 8-4. Experimental Titanium Metals Corporation pilot-plant cell assembly design.[14]

The second to publish was the New Jersey Zinc Co., on the results obtained by Rand and Reimert on research with small-scale laboratory cells.[29,30] Their cell, which uses an alumina diaphragm, was made of metal,[29] and the salts were the LiCl-KCl-NaCl eutectic plus several percent Ti ion, operating at 550°C with continuous $TiCl_4$ feed (Figure 8-5).

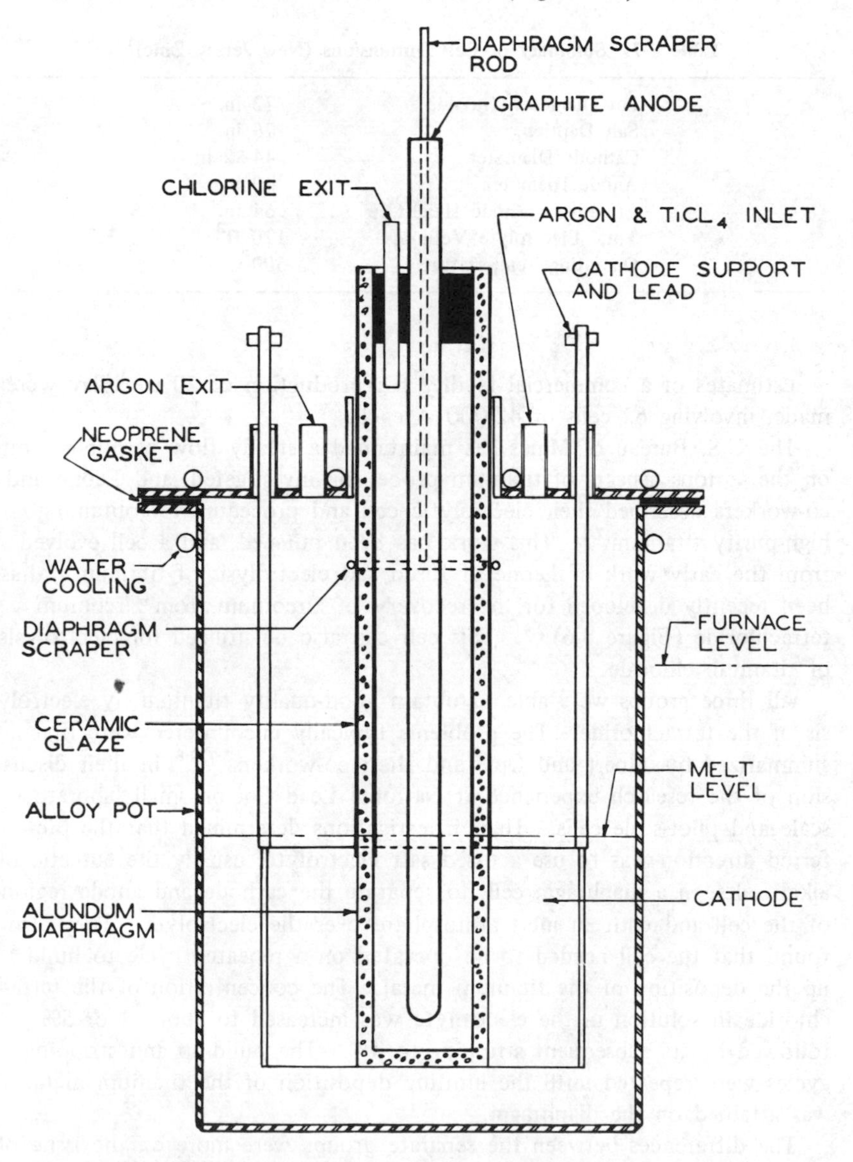

Figure 8-5. Schematic cross section of the New Jersey Zinc laboratory cell.[29]

The principles learned became the basis for design and operation of a subsequent pilot-scale cell, which in turn was described in 1968.[12,26] The large cell developed for pilot plant purposes by the New Jersey Zinc Co. had dimensions outlined in Table 8-7.

Table 8-7. Summary of Cell Dimensions (New Jersey Zinc)[11]

Pot Diameter–Outside	72 in.
Salt Depth	76 in.
Cathode Diameter	44-52 in.
Anode Diameter	40 in.
Effective Cathode Height	64 in.
Total Electrolyte Volume	120 ft^3
Operating Temperature	500°C

Estimates of a commercial facility for production of 20 tons/day were made, involving 62 cells of 42,000 A each.

The U.S. Bureau of Mines has maintained a steady flow of publications on the various aspects of titanium processing investigated, and Leone and co-workers described their electrolytic cell and procedure for obtaining high-purity titanium.[11] This work has been pursued, and a cell evolved from the early work of Leone of fused salt electrolysis of titanium[11] has been recently developed for the recovery of zirconium from zirconium tetrachloride (Figure 8-6).[24] This cell can also be utilized for electrolysis of titanium chloride.

All three groups were able to obtain good-quality titanium by electrolysis of the tetrachloride. The problems typically encountered were first summarized by Alpert and Opie and their co-workers[27,28] in their discussion of the research experience at National Lead Co. on small laboratory-scale and pilot-scale cells. Their investigations determined that the preferred direction was to use a fused salt electrolyte, usually the eutectic of alkali salts, in a diaphragm cell, to separate the cathode and anode regions of the cell and with an inert atmosphere over the electrolyte. They also found that the cell needed to be operated on a repeated cycle to build up the deposition of the titanium metal. The concentration of the tetrachloride in solution in the electrolyte was increased to about 4 or 5% followed by its subsequent stripping to 1%. The build-up and stripping cycles were repeated until the limiting deposition of the titanium metal was attained on the diaphragm.

The differences between the separate groups were more on the issue of engineering design of specific features of the cells and their operating

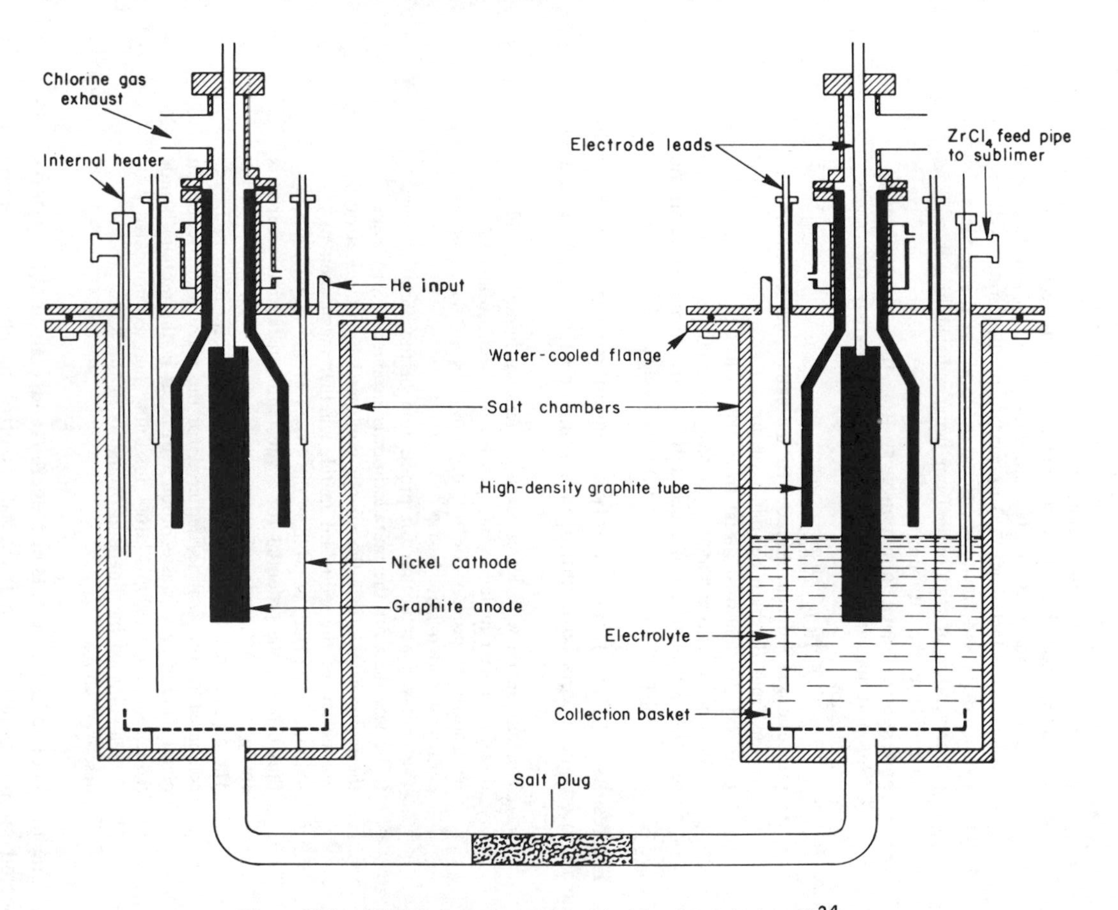

Figure 8-6. USBM zirconium tetrachloride electrolysis cell.[24]

procedures, rather than on their general principles. The problems typically encountered were summarized by Opie and Svanstrom who listed them as:[28]

- getting the titanium into solution in the bath;
- controlling oxidation and reduction reactions associated with the multiple valencies of titanium, which result in low electrical efficiency and including: (a) prevention of dichloride oxidizing to trichloride and trichloride to tetrachloride at the anode and (b) prevention of trichloride from being reduced to only dichloride at the cathode;
- prevention of metal formation away from the cathode through disproportionation of titanium dichloride to satisfy the equilibrium:

$$3TiCl_2 \rightleftarrows 2TiCl_3 + Ti$$

- protection of reduced chlorides in the bath from reaction with oxygen from the bath container or air;
- removing the deposited metal from the bath without allowing it to become contaminated with oxygen or nitrogen; and
- selecting cell materials of construction that will not contaminate the metal product or cause cell inefficiencies.

The results obtained from electrolytic reduction to titanium tetrachloride in fused alkali salts have indicated a number of potential advantages:

1. Titanium metal with lower levels of residual impurities and therefore better metallurgical quality, has been obtained;[26] with adequate process control, this could be produced on a consistent and reliable basis;[16]
2. The energy consumption for production of the metal appears to be less than for the metallothermic production processes;[14] the major gain is that the energy consumed by the electrolytic production of the reductant metal, whether magnesium or sodium, is avoided, as well as the losses of the metal itself, thus offering the potential for a substantial reduction of process energy and costs; and
3. The capital costs of the electrolyte cell plant are decreased compared to the facilities required for the reactor and reductant metal handling and waste product disposal in the metallothermic reduction processes; this latter possibility may be mitigated by the on-going engineering improvements in the Kroll process described above.

The prospective requirements for energy and consumable materials expected in the operation of a fused salt electrolysis plant as estimated by extrapolation from the experimental papers are listed in Table 8-8. These estimates show that the electrical energy requirement could decrease by about 25%. The decrease in energy is composed of two segments. One is the difference between the energy required to operate the cell, which

Table 8-8. Prospective Energy Consumption for Titanium Production by Fused Salt Electrolysis[10]

	Energy	
Process Step	**Electrical (kWh/metric ton)**	**Thermal (10^6 Btu/metric ton)**
Ore Concentration		
Extraction		5
Beneficiation	1,250	
Chlorination		
Primary Reaction	1,150	
Purification	2,400	
Electrolysis		
Cell heating		40
Electrolysis	12,000	
Dry Room Removal	12,600	
Deposit Distillation	7,200	
Consumable Materials		
Chlorine: 1.0 tons/ton	3,900	
Carbon: 1.0 tons/ton		30
Ingot Melting	6,000	
Miscellaneous Plant Energy	3,000	—
Totals	49,500	80

has been extrapolated from the literature to be about 12 kWh/kg.[11] The other segment is the conservation of the energy equivalent of the Kroll process loss of 0.1 ton of magnesium per ton of titanium. This estimate neglects the energy equivalents for the replacement materials of construction of the cell, for which the available information is inadequate to make estimates of future cell construction.[26] Thus, while there is some prospective saving of electrical energy in titanium production by fused salt electrolysis, savings appear to be modest when all factors involved in the production of the titanium ingot are considered.

The utilization of Idaho Ilmenite ore has prompted a more general study on the titanium industries, which could be implemented in this context. Figure 8-7 presents a flow sheet for the preparation of all products and by-products recoverable from this mineral. It includes a smelting plant (including an electric furnace) to separate out the iron and recover the titanium oxide, which should then be transformed to chloride and electrolyzed with fused salts. Figure 8-8 is a more detailed presentation of the various operations needed to perform the operations outlined in the previous figure.

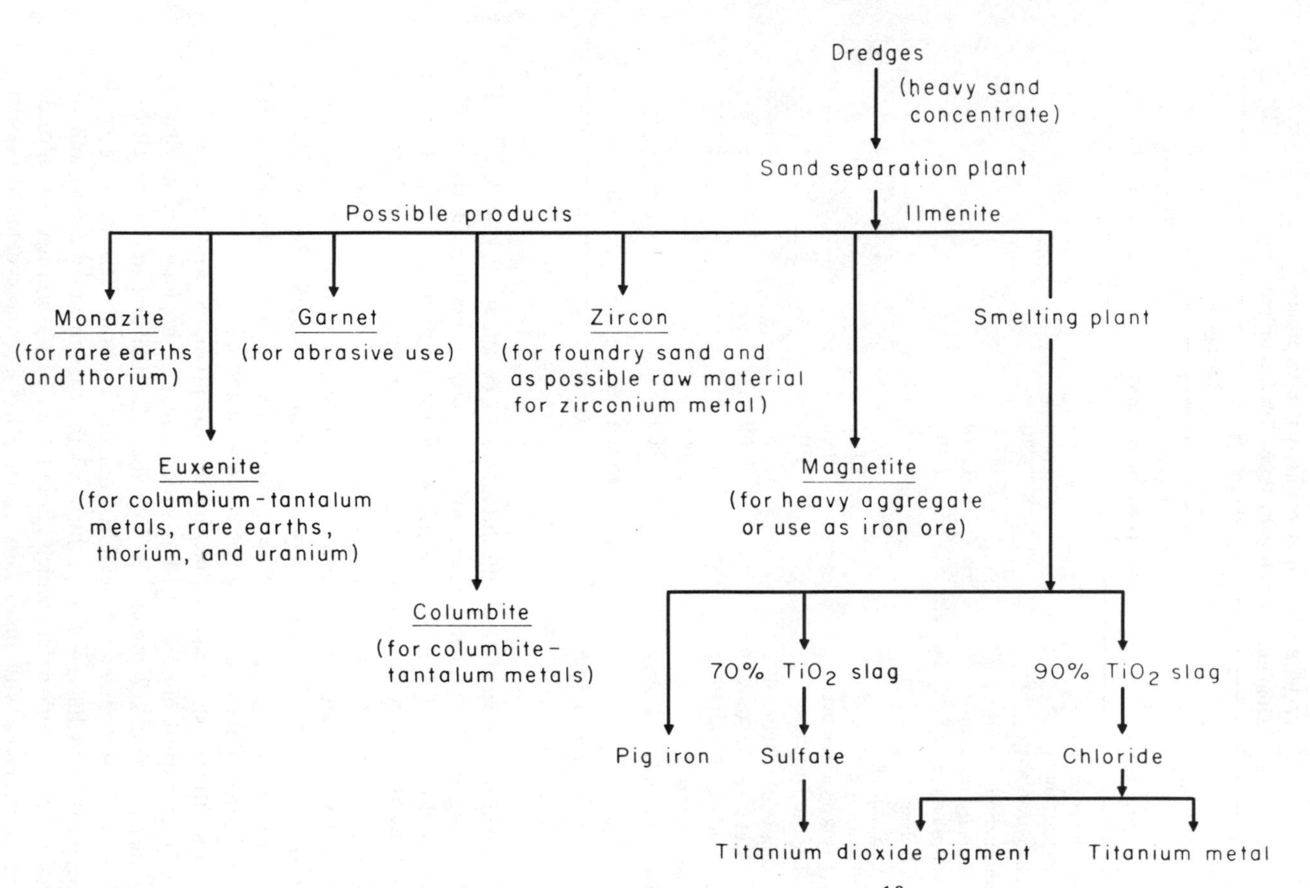

Figure 8-7. Flow sheet for utilization of Idaho ilmenite.[18]

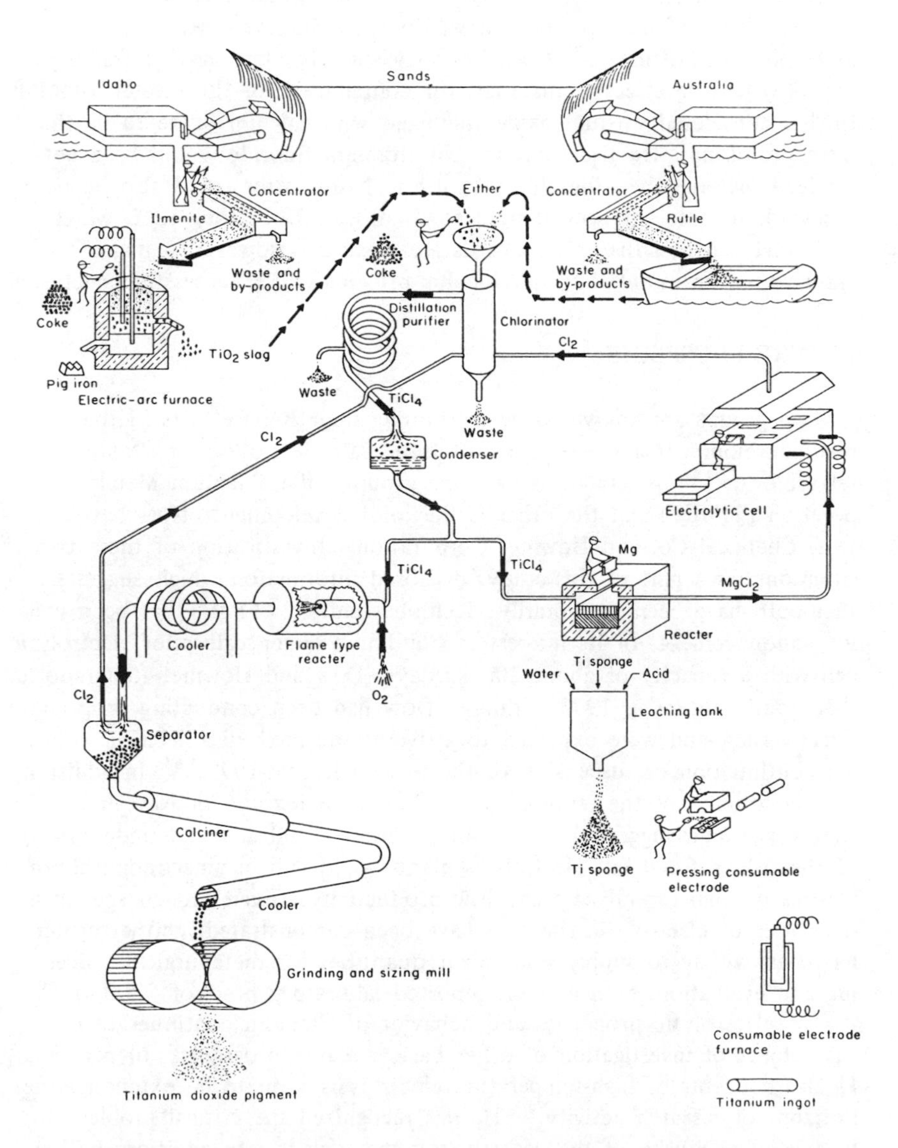

Figure 8-8. Chlorination route for the production of metal or pigment with the use of Idaho ilmenites or Australian rutile.[18]

Possibilities for High-Temperature Electrolysis

It can be shown that productivity of electrolytic cells would be substantially increased if they could be operated at high temperatures to produce molten metal. Hoch has suggested that R&D in this direction be pursued.[9,15] The major problem, which is readily acknowledged, is that the high reactivity of molten titanium would result in severe attack of the materials of construction. Hoch has suggested that rare earth refractory oxides would be effective materials for containment of the molten titanium. He has considered yttrium oxide stabilized with calcium oxide to be the most promising. He also suggests that titanium fluoride would be a better feed material than the chloride, although the latter could also be used. His work to date has been limited to small-scale laboratory R&D which has convinced him that his concepts are feasible. However, a major program would be essential to develop his principles to a successful conclusion.

FUTURE PROSPECTS

Two groups are known to be continuing their R&D efforts in the pilot plant development and evaluation of fused salt electrolysis of titanium tetrachloride.[14] As stated earlier, one group is the Titanium Metals Corporation (TIMET) and the other is the joint development effort between Dow Chemical Co. and Howmet Corporation. Investigation of these two situations, as a part of this study, disclosed information which suggests that both have been temporarily deemphasized.[10] TIMET has been making announcements of its interest in building a larger-scale pilot electrolytic cell with a capacity of about 200 kg/day. Dow and Howmet first reported their joint venture in 1973, although Dow had been conducting proprietary R&D earlier, and were expected to arrive at the next step decision point for continuation or suspension of the project in mid-1977.[16] In addition, extrapolation[26] of the reported pilot plant experience[12] of New Jersey Zinc Co. would suggest that without government subsidy and underwriting of the risk, a fused salt electrolysis plant would still be uneconomical for reasons of high capital cost and low productivity. Even though specimen quantities of electrolytic titanium have been demonstrated, neither group has been willing to supply pilot plant quantities for metallurgical processing and evaluation, even at their reported laboratory price of $200/kg.[10]

The electrolytic properties and behavior of titanium continues to be the subject of investigation of other basic research programs. In particular, Hoch's concepts of high-temperature electrolysis seem to be extending the horizons of research activity.[9] He has recognized the critical problem to be one of adequacy of the construction materials.[15] In addition, he has pointed out the potential merits of titanium fluoride rather than titanium chloride as the feed stock. The absence of an independent source of high-grade fluoride in large quantity is an additional handicap, which will impede the development and acceptance of this process concept even after it might be demonstrated to be technically advantageous.

In the United States, the titanium sponge production industry has typically operated at less than capacity. This situation was exaggerated during 1976 when production fell to the level of 50-60% capacity. The demand for sponge fluctuates dramatically with the changing requirements of the military aircraft procurement programs. Potential demand for titanium could be increased very substantially by several advanced-performance aircraft, which are at various stages of final engineering and qualification for acceptance. Until Congress appropriates the requisite funding, no action can be taken to commit the production programs. If all of these aircraft were to go into production, a substantial increase in titanium sponge production capacity would be needed. When the need arises, the opportunity will be presented for installation of electrolytic cells for high production.

REFERENCES

1. Abkowitz, S., J.J. Bruke and R.H. Hiltz, Jr. *Titanium in Industry* (New York: D. Van Nostrand Co., Inc., 1955).
2. Barksdale, J. *Titanium,* 2nd ed. (New York: The Ronald Press Co., 1966).
3. Williams, S.C. *Report on Titanium* (New York: Brundage, Story & Rose, 1965).
4. Office of Naval Research Conference on Titanium, Washington, D.C., 1948.
5. McQuillan, A.D. and M.K. McQuillan. *Titanium* (London: Butterworths Scientific Publications, 1956).
6. First International Conference on Titanium, London, 1968.
7. Third International Conference on Titanium, Moscow, 1976.
8. Peters, H.F. "Kroll Process–1973," Presentation to NMAB, 1973 Meeting on Production of Titanium Publ. NMAB-304, National Academy of Sciences–National Academy of England, Washington, D.C., 1974.
9. Hoch, M. "Critical Review–Winning and Refining," *2nd Internat. Conf. on Titanium,* Cambridge, Massachusetts, Massachusetts Institute of Technology (1972).
10. Bovarnick, B. Unpublished research, 1976.
11. Leone, O.Z., H. Knudsen and D. Couch. "High-Purity Titanium Electrowon from Titanium Tetrachloride," *J. Metals* (March 1967).
12. Myhren, A.J. *et. al.* "The New Jersey Zinc Co. Electrolytic Titanium PII Plant," *J. Metals* 38(May 1968).
13. Opie, W.R. and O.W. Moles. "Basket-Cathode Electrolyte Cell for Production of Titanium Metal," *Trans. Met. Soc. AIME* 218:696 (August 1960).

14. National Materials Advisory Board, 1974.
15. Hoch, M. University of Cincinnati, Cincinnati, Ohio. Private Communication, 1976.
16. Turner, D.H. Howmet Corporation. Private Communication, 1976.
17. Henrie, T.A. "Extractive Metallurgy of Titanium," U.S. Bureau of Mines, SI., in *AIME-TMS Conf. on High Temp. Refractory Materials,* Vol. 34, Part 1 (New York: Gordon and Breach Publishers, 1965), pp. 139-154.
18. Fulkerson F.B. and J.J. Gray. "Titanium Industries and their Relation to the Pacific Northewest," Portland, U.S. Department of the Interior (1964).
19. Fleck, D.C. *et. al.* "Reducing Titanium Tetrachloride with High-Sulfur Sodium," U.S. Bureau of Mines, RI5596 (1960).
20. Haver, F.P. *et. al.* "Development of a 10,000-Ampere Cell for Electrowinning Titanium," U.S. Bureau of Mines, RI5805 (1960).
21. Homme, V.E. *et. al.* "Crystalline Titanium by Sodium Reduction of Titanium Lower Chlorides Dissolved in Sodium Chloride," U.S. Bureau of Mines, RI6813 (1966).
22. Homme, V.E. *et. al.* "Methods for Producing Titanium Lower Chlorides," U.S. Bureau of Mines, RI6360 (1964).
23. Homme, V.E. *et. al.* "Sodium Reduction of Titanium Chloride," U.S. Bureau of Mines, RI5398 (1958).
24. Martinez, G.M. *et. al.* "Investigation of a Cell Design for Electrowinning Zirconium Metal from Zirconium Tetrachloride," U.S. Bureau of Mines, RI8125 (1976).
25. Wong, M.M. *et. al.* "Electrolytic Methods of Preparing Cell Feed for Electrorefining of Titanium," U.S. Bureau of Mines, RI6161 (1963).
26. Reimert, L.J., Palmerton, Pennsylvania, retired; previously with New Jersey Zinc Co. Private communication, 1976.
27. Alpert, M.B., F.J. Schultz and W.F. Sullivan, "Electrolytic Preparation of Titanium from Fused Salts," *J. Elechem. Soc.* 104:555 (September 1957).
28. Opie, W.R. and K.A. Svanstrom. "Electrodeposition of Titanium from Fused Chloride Baths Using $TiCl_4$ as Feed Material," *Trans. Met. Soc. AIME* 215:253(April 1959).
29. Rand, M.J. and L.J. Reimert. "Electrolytic Titanium from $TiCl_4$– I Operation of A Reliable Laboratory Cell," *J. Elechem. Soc.* 111: 429(April 1964).
30. Reimert, L.J. and M.J. Rand. "Electrolytic Titanium from $TiCl_4$– II Influence of Impurities in $TiCl_4$," *J. Elechem. Soc.* 111:434 (April 1964).

Appendix. Important Industrial Patents on Production of Titanium

Organization	Patent	Title	Author
Horizons Titanium Corp.	2707169	Preparation of Titanium Metal by Electrolysis	Morris A. Steinberg, Merle E. Sibert, Alfred A. Topinka
Horizons Titanium Corp.	2707170	Electrodeposition of Titanium	Eugene Wainer
Horizons Titanium Corp.	2723182	Method of Producing Alkali Metal Titanium Double Fluorides in Which the Titanium has a Valence of Less than Four	Merle E. Sibert, Quentin H. McKenna
Horizons Titanium Corp.	2724635	Production of an Alkali Metal Double Fluoride of Titanium	Eugene Wainer
Horizons Titanium Corp.	2731402	Production of Metallic Titanium	Alfred A. Topinka, Quentin H. McKenna, Stuart S. Carlton
Horizons Titanium Corp.	2731404	Production of Titanium Metal	Eugene Wainer
Horizons Titanium Corp.	2731406	Preparation of Electrolyte	Thomas R. Young, Robert L. Somerville
Horizons Titanium Corp.	2786809	Electrolytic Cladding	Bertram C. Raynes
Horizons Titanium Corp.	2813068	Production of Titanium by Fused Salt Electrolysis	Morris A. Steinberg, Alfred A. Topinka
Horizons Titanium Corp.	2833706	Electrolytic Method of Producing Titanium	Eugene Wainer
Horizons Titanium Corp.	2974092	Production of Titanium	Merle E. Sibert
National Lead Co.	2734003	Method for Treating Metals	Marshall B. Alpert
National Lead Co.	2734855	Electrolytic Preparation of Reduced Titanium Chloride Composition	Thomas M. Buck, Marshall B. Alpert

Appendix, Continued

Organization	Patent	Title	Author
National Lead Co.	2734856	Electrolytic Method for Refining Titanium Metal	Frank J. Schultz, Thomas M. Buck
National Lead Co.	2741588	Electrolytic Production of Titanium Metal	Marshall B. Alpert, Robert Lee Powell
National Lead Co.	2748073	Fused Salt Electrolytic Cell for the Production of Refractory Metals	Svante Mellgren
National Lead Co.	2749295	Electrolytic Production of Titanium	Kjell A. Svanstrom, William R. Opie
National Lead Co.	2801964	Cathode Assembly for Electrolytic Cells	William R. Opie, Kjell A. Svanstrom
National Lead Co.	2904479	Electrolytic Polishing of Zirconium, Hafnium and Their Alloys	Andrew T. McCord, Donald R. Spink
National Lead Co.	2904491	Apparatus for Producing Refractory Metal	Oliver W. Moles, Leonard W. Gendril, Howard R. Palmer
Titanium Metals Corp. of America	3274083	Electrolytic Production of Titanium	Linden E. Snyder
Titanium Metals Corp. of America	3282822	Electrolytic Cell for the Production of Titanium	John C. Priscu
Titanium Metals Corp. of America	3607011	Electrolytic Purification	John C. Priscu, Eldon R. Poulsen
Titanium Metals Corp. of America	3616441	Electrolytic Cell Anodes	John C. Priscu, Eldon R. Poulsen

New Jersey Zinc Co.	2780593	Production of Metallic Titanium	George E. Snow, Andrew J. McCord
New Jersey Zinc Co.	2789943	Production of Titanium	William W. Kittelberger
New Jersey Zinc Co.	2821468	Production of Titanium	Lester D. Grady
New Jersey Zinc Co.	2848397	Electrolytic Production of Metallic Titanium	Lawrence J. Reimert, Erastus A. Fatzinger
New Jersey Zinc Co.	2875033	Method of Separating Titanium Crystals	Lester D. Grady
New Jersey Zinc Co.	2892764	Production of Titanium Metals	Earl W. Andrews
New Jersey Zinc Co.	2898275	Production of Titanium	Earl W. Andrews
New Jersey Zinc Co.	2898276	Production of Titanium	George E. Snow
New Jersey Zinc Co.	2900318	Electrolyzing Device	Earl W. Andrews
New Jersey Zinc Co.	2908619	Production of Titanium	Charles E. Barnett
New Jersey Zinc Co.	2975111	Production of Titanium	Lawrence J. Reimert, Erastus A. Fatzinger
New Jersey Zinc Co.	2998373	Electrolytic Cell for Production of Titanium	George E. Snow
New Jersey Zinc Co.	3054735	Production of Titanium	Raymond L. Johnson
New Jersey Zinc Co.	3082159	Production of Titanium	Lawrence J. Reimert
E.I. duPont de Nemours & Co.	2845386	Production of Metals	Carl Marcus Olson
E.I. duPont de Nemours & Co.	284395	Electrolytic Process for Production of Titanium	Charles J. Carignan
Timax Corp.	2861030	Electrolytic Production of Multivalent Metals from Refractory Oxides	Harvey L. Slatin
Timax Corp.	2864749	Process for the Production of Titanium Metal	Harvey L. Slatin

Appendix, Continued

Organization	Patent	Title	Author
Timax Corp.	2994650	Preparation of Pure Metals from Their Compounds	Harvey L. Slatin
Timax Corp.	3137641	Electrolytic Process for the Production of Titanium Metal	Harvey L. Slatin
Dow Chemical Co.	2943033	Preparation of Lower Titanium Halides in a Molten Salt Bath	Robert D. Blue, Marshall P. Neipert
Kroll	2205854	Method for Manufacturing Titanium and Alloys Thereof	Wilhelm Kroll

9
ACETYLENE PRODUCTION BY ELECTRIC MEANS*

INTRODUCTION

Acetylene (C_2H_2) is a gaseous hydrocarbon used principally as an intermediate in the production of synthetic chemicals. All acetylene produced in the U.S. is by the carbide and partial oxidation processes; none is produced by arc processes.

In the United States over the past decade, acetylene has been substantially displaced by ethylene as a chemical reactant, since ethylene is substantially less expensive and easier to produce. However, the ethylene process consumes expensive petroleum feedstocks, which are in short supply and often imported. Therefore, it is conceivable that acetylene, produced from coal feedstocks (in abundant supplies and less expensive) and using electrical processes, might again be competitive in the chemical and plastics industry.

One electric arc process to make acetylene–the Du Pont process–has been used industrially in the United States, and two others have been developed through the pilot reactor stage. In addition, research is being conducted on an improved arc process, and research on another arc process to make calcium carbide, which is eventually hydrolyzed to acetylene, is just starting.

The Du Pont process uses hydrocarbons as feed (No. 2 fuel oil) and works with a carbon efficiency of 65%. It requires 3 kWh of electricity per pound of acetylene produced. The Du Pont plant has been shut down, however, because the chemical industry it was serving has shifted to using ethylene instead of acetylene.

The processes developed through the pilot reactor stage are the AVCO arc process and the Westinghouse plasma torch. The AVCO arc process manufactures acetylene directly from coal. It requires 3.3 pounds of coal

---

*By: Marcel Barbier, Mary Harlow, Robert Ouellette, Robert Pikul

and 4 kWh of electricity per pound of acetylene. The hydrogen is present in the coal and need not be supplied. Continuing development leading to the implementation of this process has been cancelled, due to the declining demand for acetylene.

The Westinghouse plasma torch has been operated up to 3-MW arc power with hydrocarbons as feed. By coupling three torches, arc powers as high as 10 MW could be obtained with the present state of the art. The system could further be modified with some development effort to use coal as the feedstock.

An improved arc process, the fluid convection cathode process, whereby the temperature of the reaction zone in the arc is increased, is in the early stages of physical research at Columbia University. The arc process to make calcium carbide is but a concept on which physical research is just starting.

A comparison of arc methods with the traditional processes for producing acetylene, in terms of raw materials used, energy used, and capital investment results in the following conclusions:

- The Du Pont process requires 33% less electrical energy than the carbide route but uses methane or liquid hydrocarbons in short supply;
- The AVCO process needs 10% less electrical energy than the calcium carbide route but requires over twice as much coal;
- Arc processes require a larger capital investment than the partial oxidation route, but this last process requires methane as a feedstock; and
- Research on a new high-temperature arc process appears promising but is several years away from commercial application.

Researchers involved in the Du Pont, AVCO and Westinghouse processes have stated that they have additional ideas on the industrial engineering of acetylene arc plants, which could be of value in the planning for construction of such plants. It might be appropriate for interested industrial firms to contact these process developers directly for further engineering details.

GENERAL PRINCIPLES

Acetylene (C_2H_2) is a gaseous hydrocarbon that has been widely used as an intermediate in the production of synthetic chemicals and plastics, including water-base paints, vinyl materials, dry cleaning fluids and aerosol insecticides. Acetylene is manufactured principally by three methods: from calcium carbide; by the partial oxidation process; and by arc-electric processes.[1-4]

In the first method, calcium carbide is produced by reacting calcium oxide and carbon in an electric furnace:

$$CaO + 3C \rightarrow CaC_2 + CO$$

The electrical energy consumption is on the order of 4000 kWh/metric ton of carbide.[2] The carbide is subsequently hydrolyzed to acetylene:

$$CaC_2 + 2H_2O \rightarrow C_2H_2 + Ca(OH) + 26540 \text{ kcal}$$

Twenty-six pounds of acetylene are thus produced with 64 lb of carbide. This reaction is exothermic and produces approximately 26540 kcal per mole of acetylene, which are dissipated.

The second method of acetylene production is based on the partial combustion of a hydrocarbon with oxygen:

$$2CH_4 \rightarrow C_2H_2 + 3H_2 - 90 \text{ kcal}$$

Part of the hydrocarbon feedstock is thereby burned to provide energy for the pyrolysis of the remainder:

$$CH_4 + 3/2O_2 \rightarrow CO + 2H_2O + 210.8 \text{ kcal}$$

The resulting reaction products would be as follows:

$$3CH_4 + 3/2O_2 \rightarrow C_2H_2 + 5H_2O + CO + 121 \text{ kcal}$$

Feedstocks ranging from C_1 to C_8 are utilized in the process.

The third process for obtaining acetylene employs an electric arc, which produces an acetylene triple bond molecule either by pyrolyzing a hydrocarbon, or directly from the reaction of carbon and hydrogen gas in the electric arc. Ideally, the reaction is either:

$$2CH_4 \overset{(arc)}{\rightarrow} C_2H_2 + 3H_2 \text{ (with methane as feedstock)}$$

or

$$2C + H_2 \overset{(arc)}{\rightarrow} C_2H_2 \text{ (with carbon as feedstock)}$$

In practice, however, the double-bonded ethylene (C_2H_4) is also produced, either directly or by degradation of the acetylene triple bond, subsequent to its formation in the arc. This process is quenched by cooling the walls of the reactor vessel with a water jacket, so that the gas mixture is stablized quickly.

At least seven arc process facilities have been used worldwide in the production of acetylene. Three of these are in Germany; one, utilizing the Huels I system,[5] has been used for 30 years with excellent results.

It employs vortex stabilization of the hydrogen arc, and optimum conditions are encountered when three-fifths of the hydrocarbon feed is fed through the arc and two-fifths utilized as a quench below the arc. A modified version of this system, known as the Huels II, is used in the second facility. The hydrocarbon feedstock is injected totally downstream of the arc, which is used merely to heat the hydrogen,[6] alleviating char build-up. The third facility employs the Hoechst AC system, which operates with consumable graphite electrodes and a high-current, high-intensity arc; hydrocarbon feed is again injected downstream.[7] Together, these plants produce about 200 million pounds of acetylene per year from a variety of C_1 - C_5 feedstocks.

Other facilities are in Borgesti, Rumania[1] (also a Huels-type plant, although it operates with a combustion feedstock preheat), in the USSR[8] and in Japan.[9]

The seventh facility is the Du Pont de Nemours arc acetylene plant in Montague, Michigan which operated until 1968. This facility was built to produce hydrocarbon-based acetylene as a raw material for neoprene manufacture, and was shut down when butadiene supplanted acetylene in producing neoprene. The Du Pont system has a magnetically rotated arc, substantially decreasing the electrical power requirement for acetylene manufacture.[10]

In all seven facilities, acetylene is produced by cracking hydrocarbons. Between 1966 and 1972 a considerable effort was undertaken by the AVCO Corporation of Massachusetts and sponsored by the Office of Coal Research in the United States, to develop an all-electric arc process capable of making acetylene directly from coal.[9,11] A prototype reactor was successfully operated at a level of 80-100 kW with coal feed rates of 450-700 g/min, which led to the design of a 10-MW reactor. Subsequently the engineering design for an industrial plant to produce 300 million pounds of acetylene per year was also completed.[12] Because of the decline of the use of acetylene in the chemical industry, however, further efforts in this area were postponed.

Table 9-1 gives the performance characteristics of the various reactors mentioned; each acetylene-producing facility, however, would be composed of several reactors. The concentrations are given in percent molecules (not weight). Figure 9-1 shows a schematic cut through the chambers of each type of reactor and illustrates its operation. In the following sections we will examine more closely the primary American processes.

Table 9-1. Performance of Various Arc Acetylene Reactors[11]

	Huels I	Huels II	Hoechst AC	Du Pont de Nemours	Projected AVCO Arc-Coal
Volts	7100	7100	1400	3500	8000
Amps	1200	1200	3 x 4200	3100	1250
MW	8.5	8.5	10.0	10.8	10.0
Pressure (atm)	1	1	1	0.5	0.5
Hydrocarbon (coal)[a] Feed Through Arc (lb/hr)	3100	0	0	4200	4200
Hydrocarbon (coal)[a] Feed Downstream (lb/hr)	2200	7000	4500	0	2000
Hydrogen flow (recycled) (scfm)	2150	1400	1470	3200	3000
Hydrogen Enthalpy (kW/scfm)	4.0	6.1	6.8	3.4	3.5
C_2H_2 Produced (lb/hr)	1860	1860	2200	2700	3000
Energy Consumed (kWh/lb C_2H_2)	4.6	4.6	4.6	3.0	4.0
C_2H_2 Concentration (mol %)	15.9	15.0	14.5	18	15.5
C_2H_4 Concentration (mol %)	7.1	6.4	6.5	3.0	b

[a]Coal for AVCO Process.
[b]Depends on quench used.

EXISTING PROCESSES AND APPLICATIONS IN THE UNITED STATES

Du Pont de Nemours Arc Acetylene Process

The Du Pont de Nemours process resulted from research on high-temperature reactions with a rotating electric arc. According to D. A. Babcock of Du Pont, the system utilized a water-jacketed copper tube as the anode and an internal concentric carbon rod for the cathode. The magnetic field from a surrounding direct-current field coil rotated the arc, which was struck between the two electrodes at about 7000 rps.

The arc's rotational speed was so great that the hydrocarbons, passed with hydrogen through the copper tube at a pressure of approximately

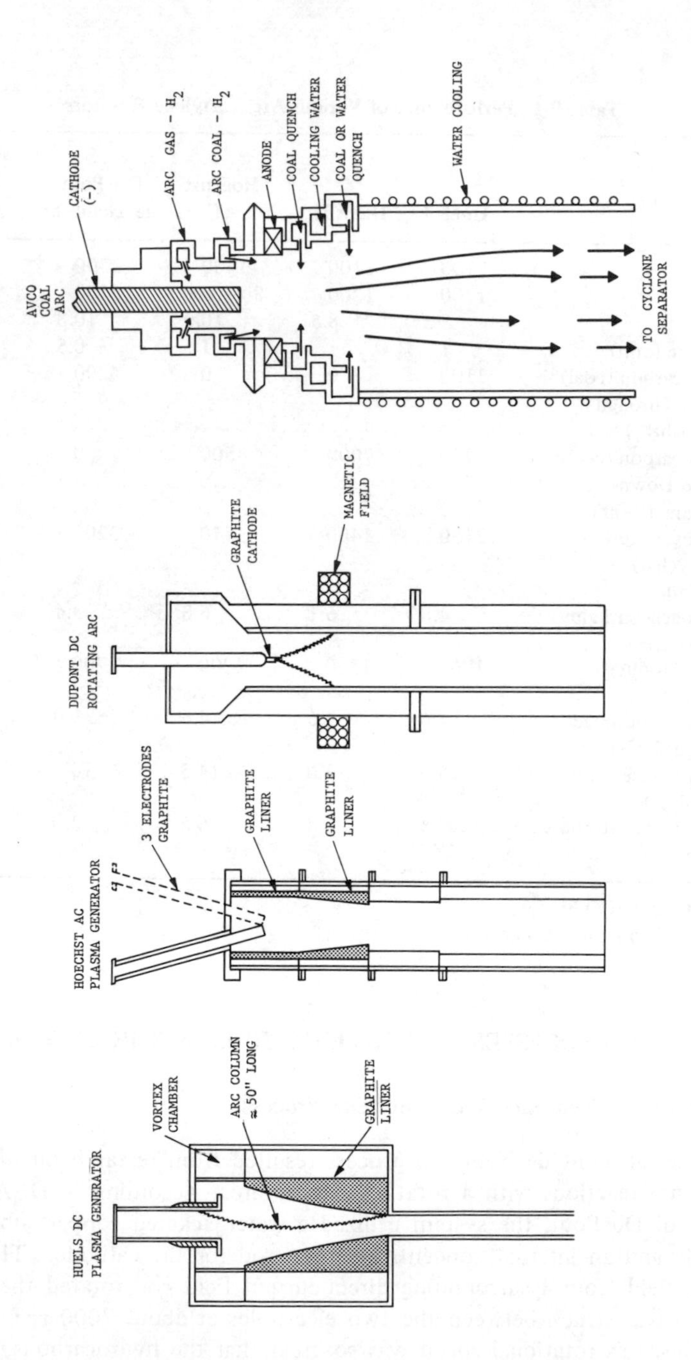

Figure 9-1. Schematics of reactors.[11]

half an atmosphere, in effect passed through a continuous plasma with minimum by-passing. Temperatures reached were around 2000°C. Quench-cooling of the reaction products with water resulted in fairly good yields of acetylene.

The final reactor design used in the Montague plant was evolved from a laboratory model and a pilot plant model erected in Louisville, Kentucky. Data on these plants are given in Table 9-2. In this table, carbon yield is the ratio of carbon present in the acetylene to carbon present in the hydrocarbon feedstock.

Table 9-2. Scale-Up of Du Pont de Nemours Reactor Condition[10]

			Montague Plant	
	Laboratory	Louisville Pilot Plant	Goals	Actual
Feed	CH_4	Natural gas	Kerosene	No. 2 oil
Acetylene (lb/hr)	7.4	100	3100	3000
Carbon Yield (%)	80	74	77	5
Preheat (°C)	None	None	500	500
Pressure (mmHg)	160	225	350	390
Anode (Diameter (in.)	1.75	3.5	11.5	11.5
Energy Consumed (kWh/lb C_2H_2)	7.0	5.3	2.5	3
Electric Power (MW)	0.052	0.528	7.8	9.3
Voltage (Volts)	162	470	2800	3000
Current (A)	320	1110	2800	3100
Magnetic Field (Gauss)	600	1450	3000	3300
Operation Time (hr/yr)	–	–	8100	7930
Unit Size (lb C_2H_2/yr)	–	–	25.3×10^6	24.3×10^6
Plant Size (lb C_2H_2/yr)			50×10^6	48.6×10^6

Process and Plant Description

In the Du Pont process, hydrocarbon liquid was fed to a vaporizer and preheater from which it passed through the arc reactor (Figures 9-2 and 9-3). Reaction products, quench-cooled by direct contact with water, were passed through a scrubber-cooler to a vacuum pump, and compressed

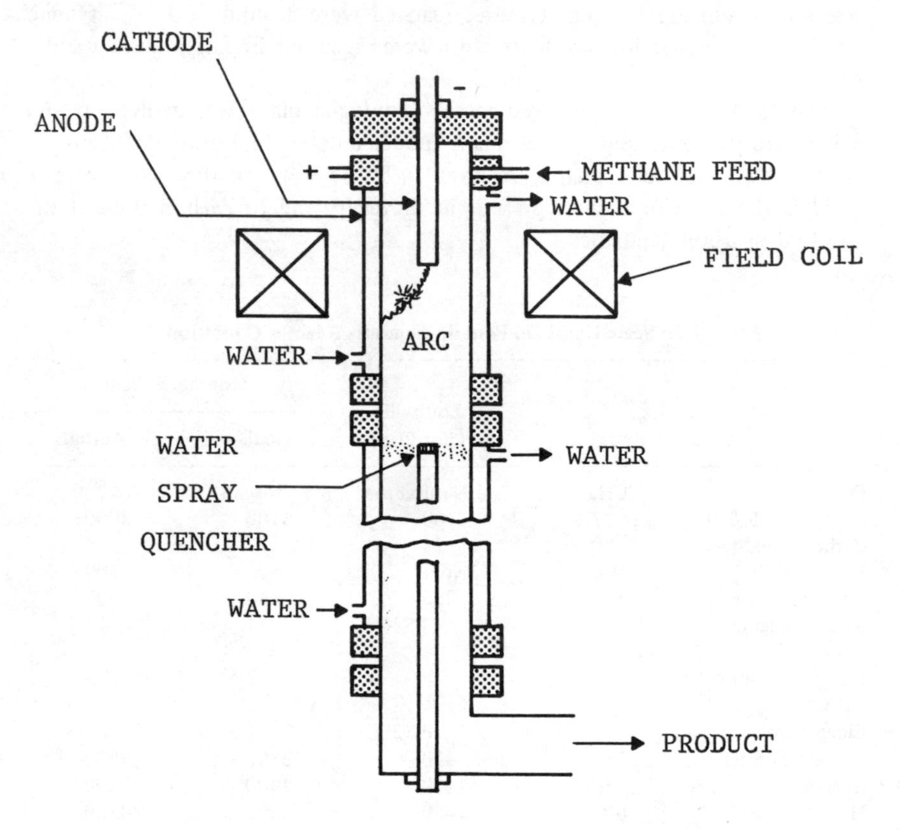

Figure 9-2. Schematic laboratory rotating arc reactor, Du Pont process.[10]

acetylene was then separated out by absorption and stripping. By-products, including carbon, high boilers, and hydrogen were burned; a substantial amount of the hydrogen was actually recycled to improve reactor performance. A flow diagram of the process is given in Figure 9-4.

Several design problems in the process surfaced. For example, it was known that carbon was consumed from the cathode tip; thus, a cathode feeder was required to maintain tip position. The cathode also had to be insulated electrically to prevent short circuiting. In addition, the anode had to be cooled externally to prevent melting, and scraped internally to remove carbon.

Other problems were identified and rectified. In the pilot plant, the arc, which could be characterized as fluctuating rapidly in voltage and current, often went out because the control response of the power supply

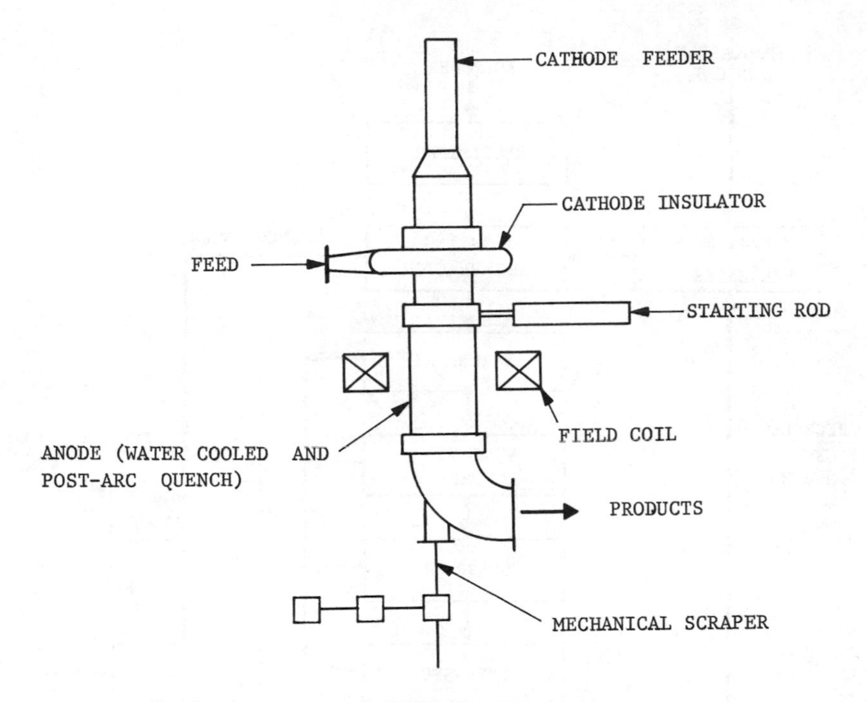

Figure 9-3. Mechanical design of plant reactor, Du Pont process.[10]

was too slow to keep up with the fluctuations. In the Montague plant, controls were used on the rectifier, which could respond to a voltage fluctuation within 8 msec, and an air-core instead of an iron core dc inductor was used to minimize hysteresis. A mercury-arc rectifier was chosen over other alternatives, although in a subsequent expansion, silicon-diode rectifiers were used successfully. To hold current relatively constant despite voltage fluctuations, the power supply transient response was also changed from a flat to a steep characteristic. The power supply circuit diagram is shown in Figure 9-5.

The laboratory reactor produced a hydrogen-rich gas containing acetylene and varying small quantities of other materials. The chemical composition of the gas is given in Table 9-3.[13] In addition, the gas contained elemental carbon in the form of dust, particles and pieces. To prevent plugging of reciprocating or centrifugal compressors with such solids, which would have resulted in excessive outage times and expensive maintenance, screw compressors were investigated. These involved two parallel screws rotating interlocked, so that when activated, one screw turned somewhat faster than the other by timing gears. These compressors had a satisfactory

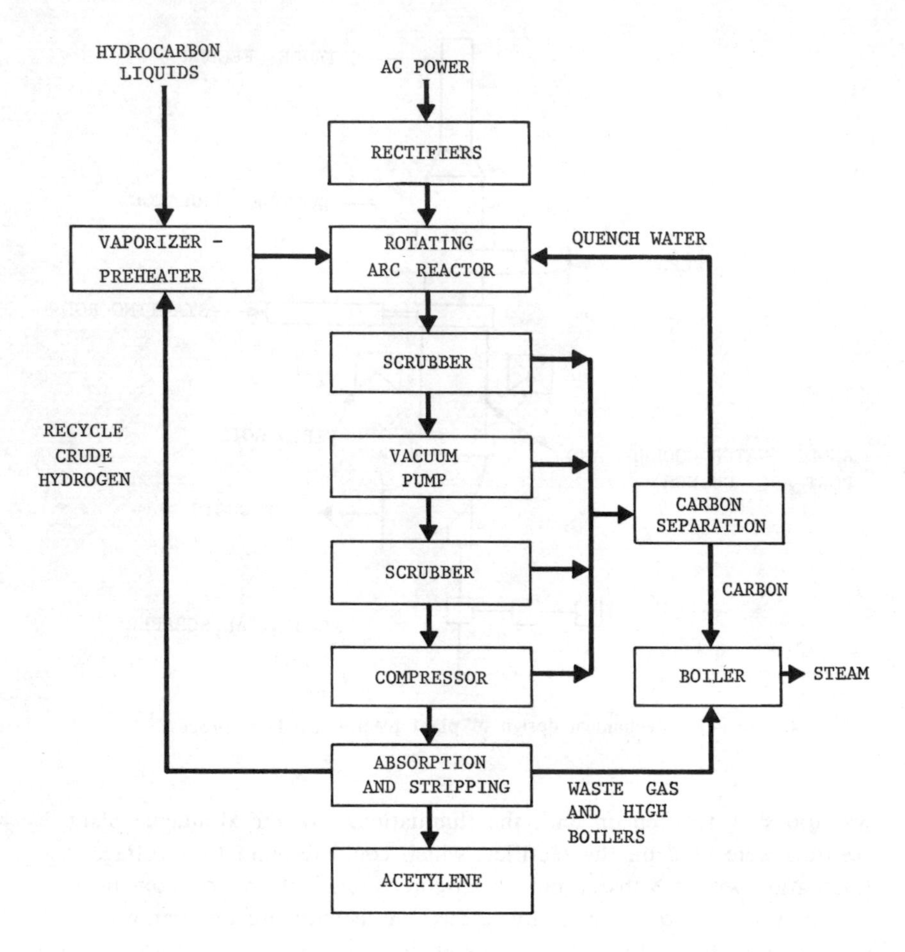

Figure 9-4. Flow diagram of the arc-acetylene process.[10]

compression ratio, and the screws had a self-wiping action which was of interest for dirty gas service.

Acetylene purification was accomplished by absorption and stripping; dimethyl formamide (DMF) was selected as the solvent. A single train of plate columns was used, consisting of a diacetylene absorber, stabilizer and stripper, followed by an acetylene absorber, stabilizer and stripper. The DMF was removed from the exit gas streams by scrubbing with water and was recovered in a refining still.

Explosion arrestors were provided by scrubbers at the inlet and discharge of the compressor train and also by the various columns in the

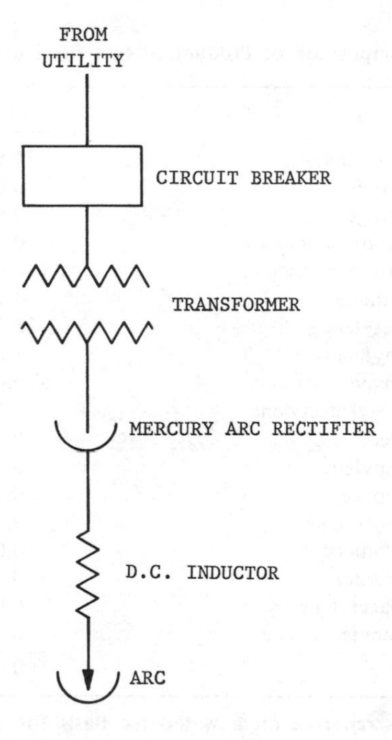

Figure 9-5. Plant reactor power supply circuit.[10]

refining train. Compressor discharge (and hence, purification pressure) was set at 150 lb/in.^{2} above atmospheric pressure–below the threshold pressure for ignition of the gas, which at this stage could contain up to 18 mol % acetylene.

Several start-up problems were encountered and resolved, including:

1. Large pieces of carbon scraped from the anode required installation of an in-line grinder to reduce carbon to a pumpable slurry;
2. Arc-overs from the cathode insulator to the feed gas plenum required redesign of the plenum;
3. Improvements in the cooling, electrical insulation and moisture resistance of the field coil were necessary; and
4. Shut-downs[10,14] due to polymer build-up in various parts of the system necessitated inhibitors and changes in operation and equipment.

Economic Considerations

Du Pont's investment for the 50 million lb/yr unit constructed at Montague in 1962 was \$8 million.[15] Adjustments for inflation, scale and

Table 9-3. Gas Composition of Product Stream for Du Pont Arc Reactor[13,a]

	Vol %
Hydrogen	74.5
Oxygen	0.06
Nitrogen	0.5
Carbon monoxide	0.5
Carbon dioxide	<0.01
Methane	4.7
Acetylene	15.2
Ethylene	3.0
Ethane	<0.01
Methyl acetylene	0.2
Allene	0.04
Propylene	0.2
Propane	<0.01
Diacetylene	0.7
Butadiene	0.03
Benzene	0.25
Triacetylene	0.02
Phenylacetylene	0.02
	100.0%

[a]The above analysis is reported on a water-free basis for a sample taken after carbon scrubbing. This scrubbing step removes 0.04 lb of elemental carbon and 0.10 lb of hydrocarbon oils per pound of acetylene when using a C_{10} feed. Carbon is reduced to 0.02 lb, and hydrocarbon oils are eliminated when using butane feed.

location would have to be considered for any new facilities. The economic data sheet on the operations of the Montague plant, which closed in 1968, is presented in Table 9-4 (the dollar value presented is for that year). According to the data, the fabrication of 1 lb of acetylene requires approximately 1.5 lb of raw material (either methane or kerosene, natural gas, oil No. 2, C-10 hydrocarbons feedstock) and approximately 3 kWh of electrical energy. (The 1.2 kWh/lb of C_2H_2 used to compress the gas was not included in this evaluation.)

The AVCO Arc-Coal Process

Process Description

A second major arc process for producing acetylene is the AVCO Coal-Arc Process which was sponsored by the Office of Coal Research, Department of the Interior in an effort to develop an arc process to convert

Table 9-4. Du Pont de Nemours Montague Plant–Economic Data[15]

	Normal Plant Operation Last 5 Months Jan. - May, 1968
Material	
Feedstock	No. 2 oil (C-10)
Yield on Carbon	65%
Electricity, dc	3.5[a] kWh/lb of C_2H_2
All other Materials [Oil for carbon disposal, DMF[b] (0.005 lb/lb), electrodes, natural gas for vaporizing fuel]	$0.96/CWT
Steam Credit (from other operations)	9 lb/lb of C_2H_2
Other Costs	
Operating Crew	5 men/shift
Maintenance and Replacement Cost (labor, employee benefits and material)	150% of operating labor and employee benefits
Other Electrical Consumption, ac (primarily for gas compression)	1.2 kWh/lb of C_2H_2
Lab, Misc. Supplies, Outside Services and Other Miscellaneous Costs	50% of operating labor and employee benefits
Cooling Water (once through)	5000 gpm
Technical Supervision	

[a]The short life estimated for the plant prevented corrective action to the preheater performance and it was not used, accounting for the difference between the R.A. Schultz paper[13] and actual performance in 1968.
[b]Dimethyl formamide.

coal to acetylene. According to an AVCO report documenting the development, several reactor designs based on different combinations of process variables (coal feed rates, gas flows, power levels, and quench arrangements) were evaluated based on three critical parameters.

1. acetylene yield (lbs of acetylene per 100 lb of coal fed to the reactor);
2. concentration of acetylene in the product stream; and
3. specific energy requirement, or SER (kWh/lb of acetylene).

The initial reactor design was for a consumable anode reactor, in which coal was used as the anode in a dc arc system. Experiments were carried out using argon as the quench medium. Although coal was converted to acetylene by this method, the values for the critical parameters were not encouraging. With the substitution of hydrogen for argon as the quench medium, however, the yield increased from 5-12%, the concentration from 3-4%, and the SERs decreased from 30 to 15 kWh/lb of acetylene.

Subsequent design changes, such as submerging the anode below the surface of the coal, resulted in continued improvement, although the limit of the submerged anode concept was reached at yields of about 16%, concentrations of 4% and SERs of 10 kWh/lb of acetylene. The char residue indicated that the reaction zone was usually one-quarter to one-half inch in depth, and any acetylene generated below the pyrolyzed surface would probably be cracked to carbon and hydrogen, upon contact with the hot char.

To avoid the entrapment and subsequent cracking of gaseous products below the pyrolyzed surface, changes were made in the electrode configuration and reactor design. In the new design, the arc-heated hydrogen reactor, coal was pyrolyzed by contacting it with hot hydrogen that had been heated during passage through an electric discharge. A plasma generator equipped with a tungsten-tipped cathode and water-cooled anode was operated using pure hydrogen without measurable erosion of the electrodes. The unit was adapted to a reactor chamber, allowing the introduction of coal directly into the plasma jet followed by immediate cooling as the jet expanded into a water-cooled chamber.

The plasma generator was a 30-kW unit, operating on hydrogen with 75% efficiency, *i.e.,* 75% of the gross input electrical energy was converted to gas enthalpy. The product stream, which was sampled by means of a probe in one of the exit lines, was analyzed by gas chromatographic techniques.

The results on the arc-heated hydrogen reactor showed significant improvement over those obtained with the consumable and submerged anode reactors. The acetylene concentration increased from 4-8%, SER values decreased from 10 to 6.8 kWh/lb, and the yields remained constant at 15-16%. That the yield did not increase over the values reported for the consumable anode reactor was traced to the fact that much of the coal feed still was not reacting with the plasma. More effective mixing of the solid coal with the plasma was therefore required to improve the acetylene yields.

To further increase the solid/plasma contact and interaction, a reactor was designed utilizing a cylindrical cathode at the center of an annular anode. This reactor could be operated at power levels as high as 120 kW with coal feed rates of 700 g/min. It was also found that the operation could be improved by operating the reactor at a reduced pressure of 4-6 in. of mercury. The results on the rotating arc reactor showed an increase in the yield which reached 18% and SERs of 6.5 kWh/lb, while maintaining an acetylene concentration of 8.0%.

The gas-coal stream leaving the anode zone still contained enough energy to convert additional coal to acetylene; a secondary coal feed was therefore incorporated into the reactor to inject coal into the hot gas stream downstream of the arc to improve the results on the rotating arc reactor. Since the gas stream was still very hot as it left the anode zone, the coal could react to form more acetylene by interacting with the hot gas while simultaneously quenching the stream with little dilution of the product.

The combined rotating and arc-heated hydrogen reactor is the same shown on the right side of Figure 9-1. In typical operation, coal was fed at the rate of 450 g/min through the arc and 200 g/min through the quench with total hydrogen flow rates of 24 standard ft^3/min, *i.e.*, at 760 mm of mercury pressure and 0°C, and power between 80 and 100 kW. The results obtained under these operating conditions were a considerable improvement for all three critical parameters: a yield increased to 35%, the concentration to 10%, and SER values were reduced to as low as 4 kWh/lb.

In a new reactor design, designated Model R4, an uncooled graphite anode was changed to a water-cooled copper anode, and the throat size was increased from 1.12-2.5 in. In addition, a provision was made for additional gas passages in the new arc head to allow flexibility in handling various reactants. The Model R4, a diagram of which is given in Figure 9-6, was later modified to accommodate a quench stream.

The minor arc gas, usually hydrogen, is fed at a positive pressure through the conduit A, to a manifold where it enters the reactor at sonic velocity through an annular ring of orifices. A second reactant, possibly methane or additional hydrogen enters similarly through pipe B. The coal, fluidized with hydrogen gas, passes through entrance port C and into the reaction zone through a series of orifices arranged concentrically around the cathode. Additional gas can be introduced through parts D and directed along the anode surface as a sheath. The magnetic coil used to rotate the arc is shown as E; the anode is designated F and is a 2.5-in. diameter copper tube, cut to size to allow quick and inexpensive replacement.

The design of the pilot plant reactor, shown in Figure 9-7, was based on the R4 and an improved version, the R6, reactors. The pilot plant reactor utilizes water-cooled electrodes, a quench section, a periodic wash cycle and the other features of the R4. The final dimensions were specified from the results of the scale-up from the R4 to the R6 reactor. Power for the reactor will be supplied at the 1 MW level with voltage control equipment to allow adjustments between 500 and 2000 V.[11] The composition of the product stream from the pilot plant is given in Table 9-5.

A summary of the results achieved or expected from the various designs of the AVCO reactors, as shown in Table 9-6, indicate the improvements

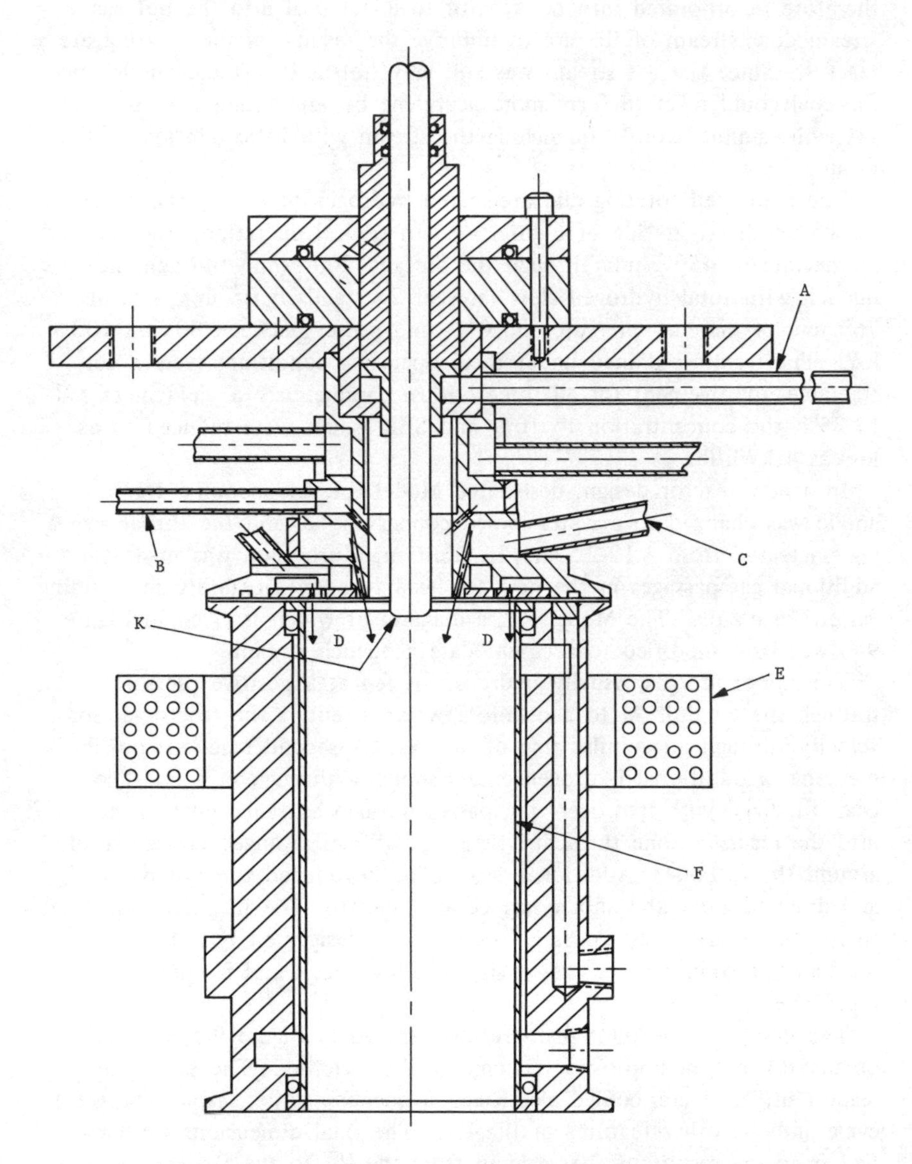

Figure 9-6. The model R4-reactor.[11]

in the critical parameters that were achieved with design changes. A generalized flow chart of the AVCO Arc-Coal Process is given in Figure 9-8.

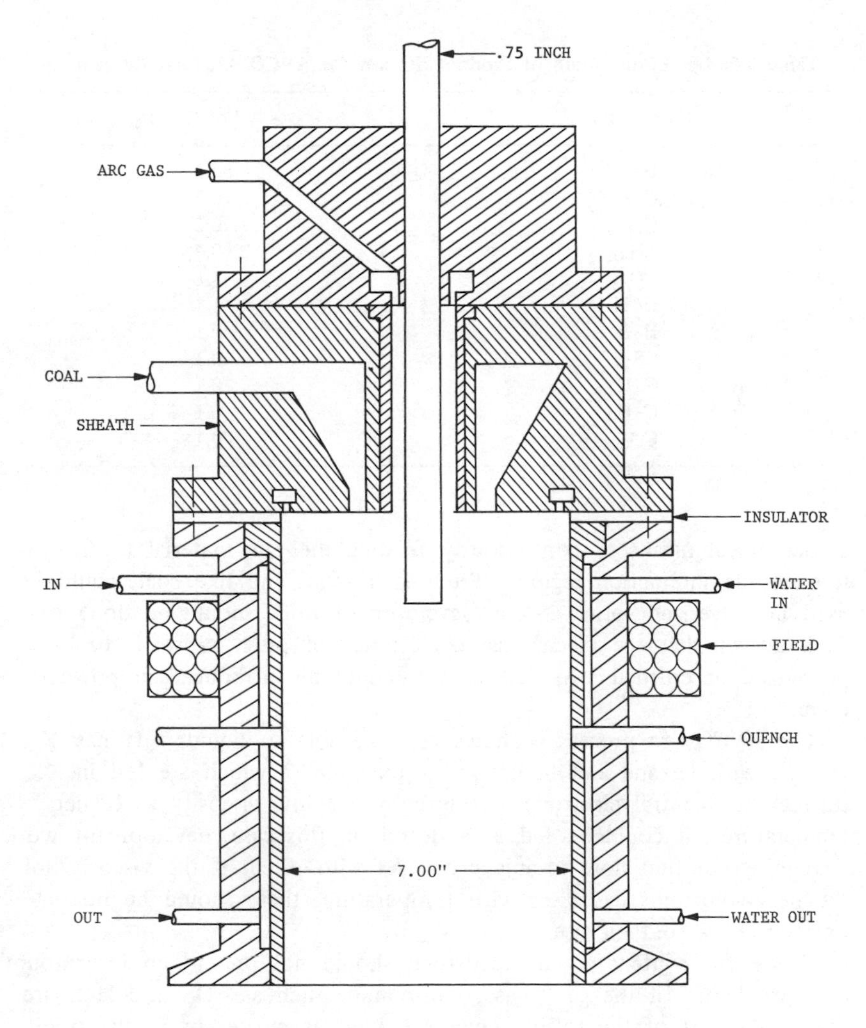

Figure 9-7. One megawatt pilot plant reactor.[11]

Use of Pitch and Heavy Bottoms as Feedstock for the AVCO Process

The AVCO arc process has been developed for use with coal as feedstock. However, pitch and heavy bottoms (solid and liquid residues of oil refineries) are also feedstock candidates. No experiments have been done with these thus far, although it is the opinion of AVCO researchers that technical problems arising from their use are solvable.[16] There is no anticipated problem with the chemistry; the difficulty, however, is finding

Table 9-5. Gas Composition of Product Stream for AVCO Arc-Coal Reactor[11]

Compound	Vol %
H_2	82.0
C_2H_2	10.7
C_2H_4	0.2
CH_4	0.7
CO	4.8
HCN	1.1
H_2S	0.1
CS_2	0.1
C_4H_2	0.1
C_3H_4	0.1
CO_2	0.1

a mechanical means of continuously feeding the raw material to the interelectrode interaction region. Pitch is a solid, and like coal, could be used in pulverized form. Some development work should be done to find out whether the mechanical consistency of pitch allows it to be pulverized or crushed. In that case it could be introduced in powder form.

The AVCO arc process is, however, amenable to liquids. It has worked with hexane and heptane, for example,[16] which are fed in through the central electrode. If pitch has a low viscosity at higher temperatures, it could be fed as a liquid; in this case, development work is required to find how to inject it. As with pitch, if the viscosity of the heavy bottoms decreases with temperature, there should be no serious obstacles to their use.

The sulfur content of the feedstock should not present an insurmountable problem. In the gas phase, compounds such as COS and H_2S are found; the rest of the sulfur remains behind as pyrites or in the original form found in the feedstock. Separation of the sulfur-containing gaseous compounds from the retrieved acetylene is technically feasible. A plant design study of the appropriate equipment is given in a study performed for AVCO by Blaw-Knox Chemical Plants, Inc.[12] The equipment for scrubbing sulfur, however, will add to the capital costs.

Economic Considerations

As indicated in the data in Table 9-6, the manufacture of 1 lb of acetylene by the AVCO arc process requires the following: raw materials–0.1 lb of hydrogen gas and 3.3 lb of coal; electrical energy–4 kW/lb

Table 9-6. Summary of Results Obtained or Expected with Various AVCO Reactors[11]

	Consumable Anode	Arc Heater Hydrogen AHR	Rotating Arc RAR	Combined RAR & NAR	Prototype Reactor R4	1-MW Reactor Design Study	10-MW Plant
Hydrogen Flow Rate (scfm)				24	15-40	500	3,000
Coal Flow Rate (g/min)	200	200	430	650	225-900	6250	75,000
Carbon Yield (%)	5-12	16	18	35	35	35	
C_2H_2 Mol Conc (%)	3-4	8	8	10	15	15	
Energy Consumption (kWh/lb C_2H_2)	30-15	6.8	6.5	4	4	4	4
dc Power (MW)	0.04	0.03	0.08	0.08-0.1	0.05-0.18	1	10
Anode Diameter (in.)				12	2.5	7	
Cathode Diameter (in.)						0.75	
Efficiency (gas enthalpy/ electrical input)		75					
Gas Pressure (mm Hg)		100-150					380
Current (A)		610			560	2000-500	1,250
Voltage (V)		131				(500)2000	8,000
Unit size (lb C_2H_2/yr)							20×10^6

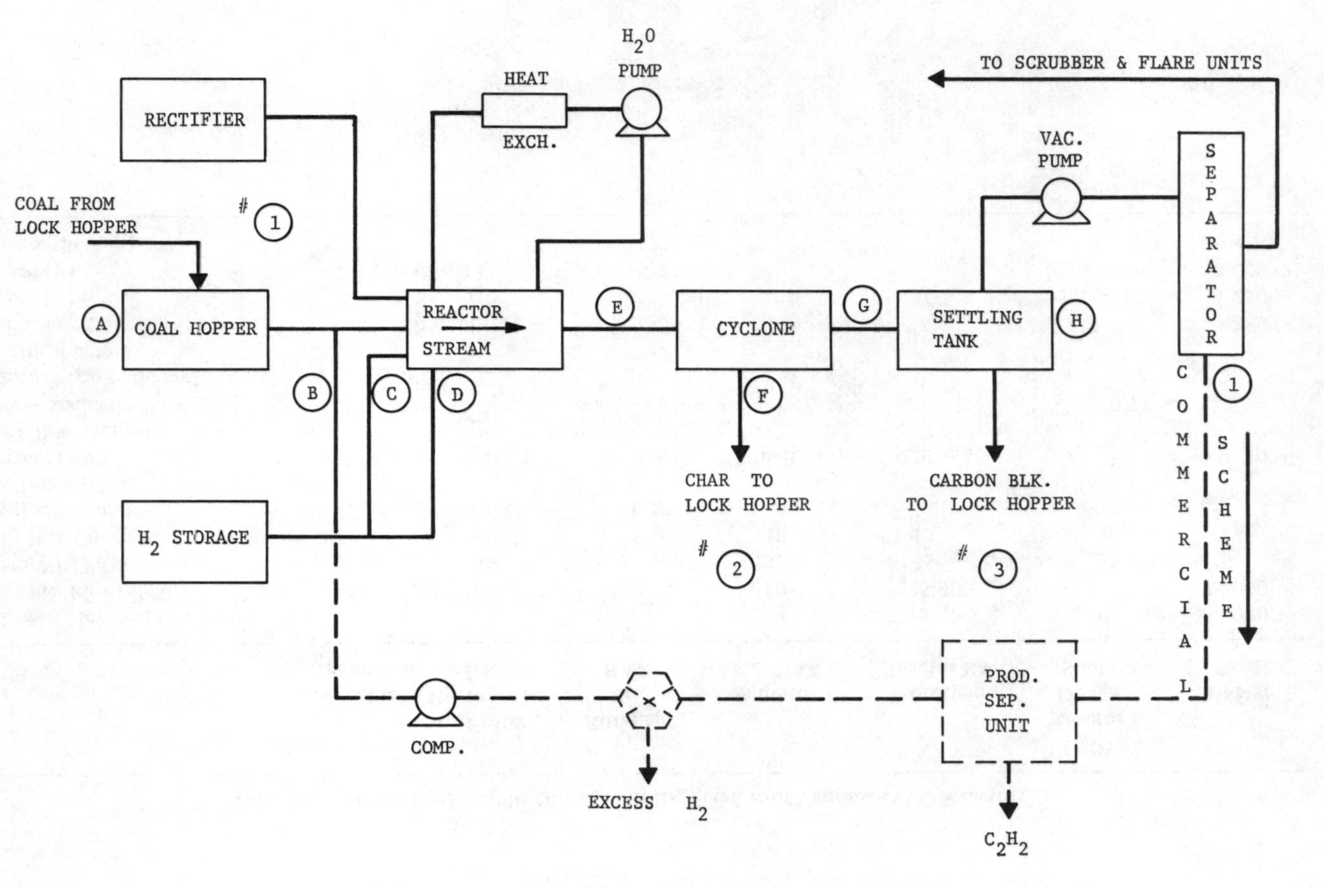

Figure 9-8. AVCO arc-coal process flow diagram.[11]

of C_2H_2. The 0.1 lb of hydrogen gas required in the AVCO process is present in the coal itself. The whole operation produces some excess hydrogen.

Plant costs (capital investments) for various plant types were estimated in the study by Blaw-Knox Chemical Plants, Inc., for AVCO. In the study, a 300-million-lb/hr plant was developed, and product costs projected and compared to costs of other processes. The results of this study are given in Table 9-7. According to the AVCO report, "it should be

Table 9-7. Economic Data on Arc-Coal Process Relative to Other Methods of Producing Acetylene (From Study by Blaw-Knox Chemical Plants, Inc.)[11]

Factored Costs	3.42	2.64	2.31	3.38	3.51
Raw and Process Materials	1.11	9.00	4.42	6.26	2.40
Utilities (net)	4.42	.09	.41	1.18	3.54
G and A, Sales, Research and Interest on Working Capital	1.21	1.20	.81	1.20	1.71
C_2H_2 Recovery Costs	–	–	–	–	2.13
Subtotal	10.16	12.93	7.95	12.02	13.29
By-product Credit	(3.82)	(3.69)	–	(3.70)	(9.00)
Net Production Cost (Excluding Interest on Working Capital)	6.34	9.24	7.95	8.32	4.29
Interest on Capital Investment[a]	1.21	.91	.73	1.09	1.48
Net Production Costs	7.55	10.15	8.68	9.41	5.76

Case Definition:

Base Case	-	The Avco Arc-Coal Process for a 300 million lb/yr acetylene plant using a recycle gas stream quench and taking by-product credit for carbon black, HCN and char.
Case A	-	A partial oxidation plant of the same size using a methane feedstock.
Case B	-	A partial oxidation plant of the same size using a naphtha feed.
Case C	-	A Wulff[b] plant of the same capacity.
Case D	-	The base case except that naphtha is used as the quench and the ethylene produced is recovered and taken as a by-product.

[a]The capital investments for the base case and cases A through D are \$53, 40, 32, 48 and 64 million, respectively.

[b]European-designed plant in which the thermal decomposition of hydrocarbons occurs in a regenerative furnace.

noted that the Blaw-Knox estimate is based on very conservative (7 mils/ kWh) power costs. If improvements on this price can be negotiated for a 100-MW load, the most notable effect would be to additionally reduce the cost of the two Arc-Coa cases."[11]

The Westinghouse Plasma Torch

The development of commercially viable plasma torches for producing acetylene is considered possible;[17] however, the development will offer some extremely interesting technical challenges, as both residual oil and pitch, the two heavy feedstocks under consideration, can have significant amounts of sulfur and mineral matter. The Westinghouse Electric Corporation of the United States has been involved for some time in research and development on plasma heating devices and their industrial applications. According to M.G. Fey of Westinghouse, in a private communication to the MITRE Corporation, a downstream system for the production of acetylene, quite similar to that of the AVCO proposed arc-coal process, is envisioned. After hydrocarbon quenching, the valuable products (acetylene, ethylene and carbon black) would be absorbed from the product stream, the undesirable sulfurous and mineral compounds removed, and the tail gas recirculated back to the arc heaters along with fresh feedstock. Figure 9-9 shows the Westinghouse three-phase Electric Arc Plasma Reactor, a design well suited to the sort of reaction being considered.

Fey cited that Westinghouse had been involved in an extensive test program in cooperation with a large American petrochemical producer to determine the chemical performance of the Westinghouse arc heater for direct high-intensity hydrocarbon pyrolysis. In that test program, three hydrocarbon feedstocks were evaluated: methane, propane, and butane. Complete chemical analyses were made of the arc heater product distribution as functions of the principal variables, *i.e.,* arc power, feedstock flow rate through the arc heater and pressure.

Tests were run on three arc heaters: the MARC 11 (nominal rating–1000 kW); MARC 31 (nominal rating–3000 kW); and an older version of the 300-kW units, the MARC 30. In all cases, the arc heaters were operated on single-phase, 60-Hz ac power, drawn from rotating test generators in the High Power Laboratory located at the Westinghouse-East Pittsburgh Works. From the data gathered, it was possible to determine feedstock conversion levels, selectivity to various products, and specific energy requirements (SERs) for all three feedstocks, and for the several arc heaters tested. The following description of the system was provided by Fey and includes the results obtained in the test program.

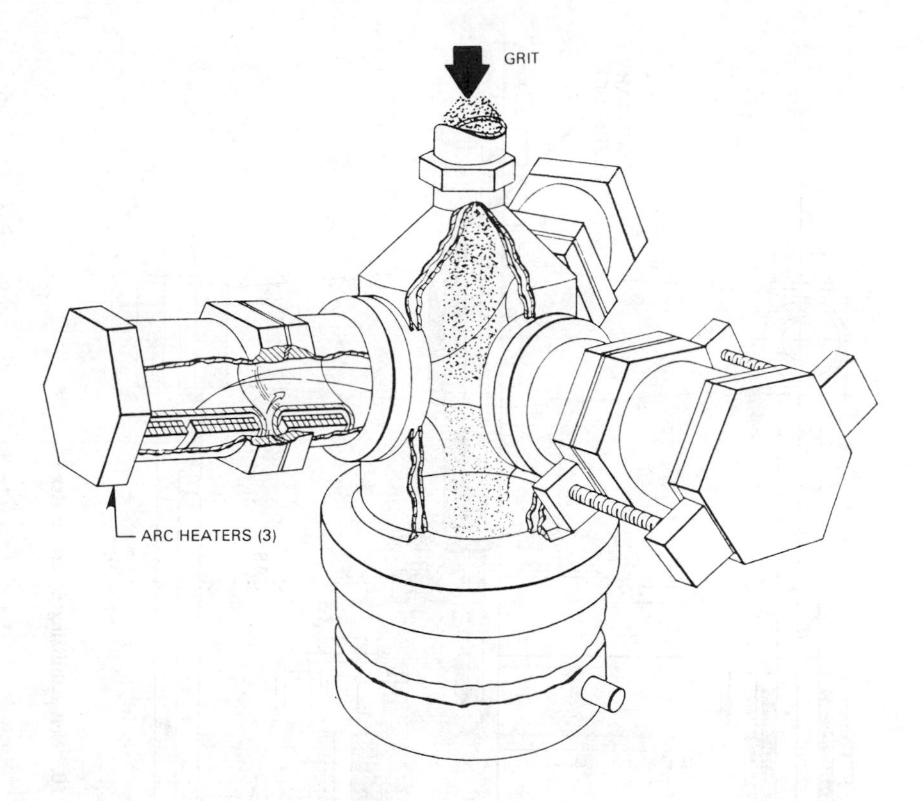

Figure 9-9. Three-phase ac Electric Arc Plasma Reactor.[18]

Description of the Westinghouse System

All three arc heaters tested were single-phase devices of the self-stabilizing configurations shown in Figure 9-10. This patented design employs water-cooled tubular electrodes axially separated by a small (about .040 in.) gap. In the MARC 11 arc heater, the electrodes have a 3.5-in. inside diameter, with an internal volume of about 0.08 cubic foot, while in the MARC 30 (and later version MARC 31) units, the electrodes have a 5.5-in. inside diameter, with an internal volume of about 0.25 cubic foot. A line frequency ac arc is initiated by sparkover in the gap between the electrodes and is blown to the interior of the electrodes by the incoming feedstock gas. Entrance velocities through the electrode gap are on the order of several hundred feet per second. Intimate mixing is achieved between the arc and the feedstock by magnetically induced arc rotation, with speeds on the order of 1000 revolutions per second. The

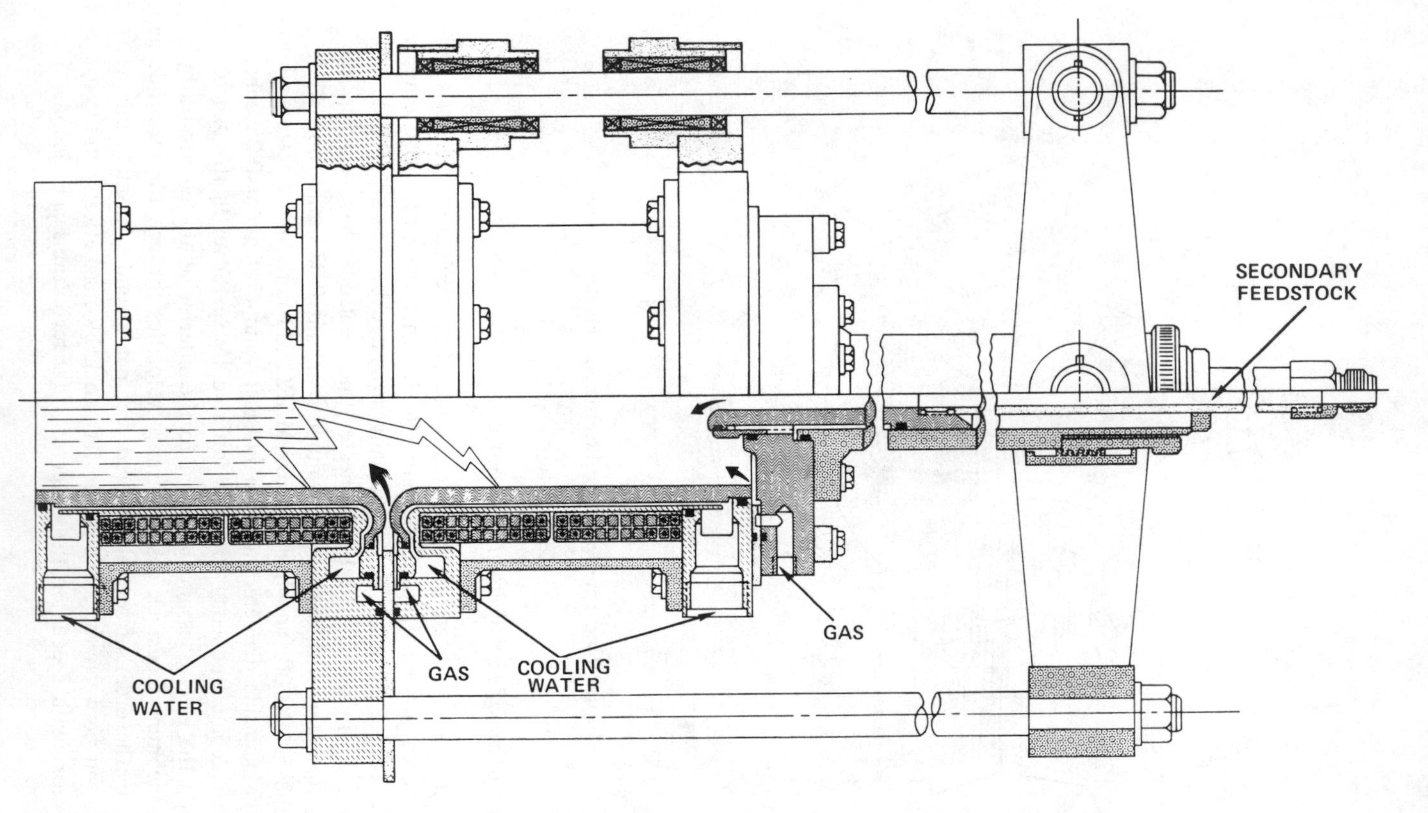

Figure 9-10. Self-stabilizing ac arc heater.[18]

magnetic field arises from opposed solenoid field coils located within each electrode assembly. The field coils are water-cooled, and energized by a low voltage dc power supply (typically a welder).

For an arc heater of this type, several conditions must be satisfied in order to attain commercially satisfactory electrode life at acceptable replacement costs. These conditions include: (1) the arc rotation speed must be high enough to avoid instantaneous destruction of the electrode surface by the arc root; (2) all interior surfaces must be sufficiently cooled for long service operation; (3) the electrode cost must be as low as possible; and (4) the downtime for maintenance must be minimized. In handling hydrocarbons, care must be taken to avoid buildup of solid carbon. The first condition is satisfied by providing a sufficiently strong magnetic field to induce adequate arc rotation and by closely matching its shape to that of the electrode surfaces. A local heat flux in the range of 1 to 5 million Btu/ft^2/hr is experienced by the electrode near the arc roots from the convective and radiative losses of the arc column and heated feedstock. In order to avoid local overheating, cooling water passages must be close to all interior surfaces and coolant velocities must be high for good heat transfer. Electrode costs are minimized by proper material selection and an emphasis on simplicity of electrode design. Copper was chosen for the electrodes because of its high thermal and electrical conductivity and its low cost relative to other candidate materials. Low maintenance cost is also attained by careful design, which yielded simple, inexpensive electrodes that can be replaced in a few minutes. Buildup is avoided by careful design, and by the high pressure gas vibrations resulting from the line frequency power pulsation.

Power was delivered to the arc heater by two 20,000-kva High Power Laboratory test generators driven in tandem by a 3000-hp electric motor. The generators were connected to provide an output voltage which could be varied up to 4000 volts ac, with current limitation provided by air core inductors. In all, seven inductors were available at the facility, three with 60-Hz impedance of 3.3 ohms each, and four of 0.2 ohms each. By connecting the inductors in various series–parallel combinations, arc currents could be easily varied from a few hundred to several thousand amperes.

While the system described above is suitable for factory test purposes, it is, of course, far too complex and expensive for industrial requirements. In the latter case, as with other arc heater systems which have been sold, the arc power supply is a fully engineered package consisting of a transformer, circuit breaker, a tapped current limiting reactor (inductor) and a capacitor bank for power factor correction, as well as a fully interlocked system for monitoring and control.

Methane was supplied to the arc heater from a pressurized tube trailer, as no high pressure natural gas lines were available at the Westinghouse plant with sufficient flow capacity. The tube trailer capacity was about 55,000 standard cubic feet at a pressure of 2300 psig. The gas flow system consisted of a dome-loaded pressure regulating system followed by a sonic flow measuring orifice. Shut-off was provided by pneumatically actuated ball valves. Methane mass flow rate could be varied up to about 1.0 pound per second. The propane and butane were stored in 500-gallon tanks. Liquid was withdrawn from the tanks, pumped through a turbine flow meter at about 250 psig, vaporized in a steam heat exchanger and then piped through a pressure regulator to the arc heater. As with methane, shut-off was provided by pneumatically actuated valves. Propane and butane mass flow rates could also be varied up to about 1.0 pound per second. Cooling water was provided by a 500 psig, 150 gallon per minute, two-stage centrifugal pump. The flow rate to each of the water-cooled components was measured by turbine flow meters, and water inlet and outlet temperatures were monitored by either copper-constant thermocouples or RTD units.

The electrical operating parameters of the arc heater, *i.e.*, arc current and voltage, were monitored using standard current and potential transformers. Arc power was measured using a Lincoln thermal converter, with input signals from the current and potential transformers. The power measurement device was chosen as the most reliable and accurate available after an exhaustive search and evaluation of alternate possibilities. All arc heater operating parameters, *i.e.*, arc current, voltage, power, water flow rates, inlet and outlet temperatures, feedstock flow rates, field coil current, etc., were simultaneously monitored using a 36-channel recording oscillograph.

The cracked gas stream was sampled from several locations along the 4.0-in. inside diameter exit pipe connected to the arc heater exhaust flange. Several water-cooled sample probes were used. A 0.25-in. inside diameter probe was used in tests of the MARC 11, while 0.12-in. inside diameter probes were used in the tests of the MARC 30 and MARC 31. Gas samples were removed from the stream, wall quenched in the probes, from which they flowed through soot removal filters to sample bottles and were exhausted through a vacuum pump. A manifold arrangement with solenoid valves was used to remotely collect multiple samples during individual tests. In this way, several arc heater operating parameters could be evaluated in a single test run. Analysis of the product sample was performed by a mass spectrometer located at the Westinghouse Research Center in nearby Churchill Borough. Analysis was usually performed within one day after tests were run.

After sampling, the arc pyrolized gas stream flowed through 4.0-in. diameter water jacketed steel piping to a combustion stack where it was burned. The steel stack was 36 inches in diameter by approximately 30-feet high, and water-jacketed over the lower 18 feet. Atmospheric air was aspirated into the bottom of the stack and ignition was accomplished by a small oxy-propane vortex igniter.

Results and Economic Considerations

All internal parts in the arc heater and piping system exposed to the product system were water-cooled, and the arc heater interior had essentially no thermal inertia. It was thus possible to start, stop and vary conditions very quickly, allowing only enough time for equilibrium conditions to be established in the sampling system. Prior to starting the arc, the entire gas system was purged with nitrogen. The arc was usually ignited in nitrogen as this procedure minimized any danger of fire. After running on nitrogen for a few seconds, hydrocarbon was admitted into the arc heater through the electrode gap and the nitrogen flow was stopped. Gas flow rate and arc power levels were varied and samples were taken at each condition. In this way the product distribution could be measured at varying arc power levels and gas enthalpies. Tests were terminated when the samples were collected or in the event a sample probe became inoperative due to carbon blockage. Data were reduced from recorded values to determine arc power, arc heater efficiency, feedstock flow rate and enthalpy of the arc-cracked gas stream. The gas analysis data indicated mole fractions of each of the gaseous species in the product stream, and by a hydrogen-carbon mass balance, a complete analysis was determined for each operating condition.

Tests were run using all three arc heaters with arc power varying from 369 kW to 1097 kW for operation of the MARC 11 unit and from 1410 kW to 2833 kW for tests with the MARC 30 and MARC 31 heaters. Table 9-8, Section 1, is a summary listing of typical operating characteristics for the three arc heaters at atmospheric pressure. The mass spectrometric analyses of the product gas streams can be seen in Section 2 for the test conditions shown in Section 1. Section 3 lists the mole fractions of the principal products corrected for carbon (which was produced as soot) as well as the specific energy requirement.

Surprisingly good agreement was achieved for the three heaters over the range of conditions tested. The specific energy requirement is about 4 kWh per pound of acetylene in the enthalpy range of 3000 to 4000 Btu/lb, and the acetylene mole fraction in the product gas stream can be seen to peak at about 13%. The energy requirement is somewhat

Table 9-8. Data on Westinghouse Plasma Torch Tests[17]

Section 1. Arc Heater Operating Characteristics

Test Number	Arc Heater	Sample Time (sec)	Arc Power (kW)	Methane Flow Rate (lb/sec)	Arc Heater Eff. (%)	Arc Heater Enthalpy (Btu/lb)
LH027	M-11	102	1097	0.193	84.8	4560
31Ho25	M-31	67	2675	0.467	73.3	3996
H421	M-30	120	2420	0.453	77.2	3900

Section 2. Spectrometric Analysis of Product Gas Stream from Methane

Test Number	Arc Heater	(sec)	CH_4	H_2	C_2H_2	C_2H_4	C_2H_6	C_3H_4	C_3H_8	C_4H_2	C_6H_6
LH027	M-11	102	27.42	57.17	12.73	0.88	0.72	0.16	0.11	0.51	0.07
31H025	M-31	67	27.60	58.00	13.00	0.50	0.05	0.20	0.01	0.50	0.08
H421	M-30	120	28.9	56.3	13.00	0.70	0.05	0.30	0.02	0.60	0.10

Section 3. Carbon-Corrected Product Analysis and Specific Energy Requirement with Methane

Test Number	Arc Heater	Time (sec)	Mol % in Product				SER (kWh/lb C_2H_2)
			C_2H_2	C	H_2	CH_4	
LH027	M-11	102	11.9	6.5	53.6	25.7	4.5
31HO25	M-31	66	12.0	7.3	53.8	25.6	4.4
H421	M-30	120	12.4	4.4	53.9	27.6	4.1

Section 4. Arc Heater Performance of Propane and Butane Tests

Test Number	Feedstock	kW	Mass Flow Rate (lb/sec)	Heater Eff. (%)	Enthalpy (Btu/lb)
H442-4	C_3H_8	2477	.727	85.3	2758
H445-3	C_4H_{10}	1771	.677	73.9	1834

Section 5. Arc Heater Performance of Methane Plus Butane Tests (Butane Injected into Swirl-Stabilized CH_4 Stream)

				Mass Flow Rate (lb/sec)		Enthalpy (Btu/lb)	
Test Number	Time (sec)	kW	Heater Eff. (%)	CH_4	C_4H_{10}	CH_4	$CH_4 + C_4H_{10}$
H-464	150	1699	71.63	.1093	.3627	10556	2445

Section 6. Product Gas Mass Spectrometric Analyses, in Mol %, of Propane and Butane Tests

Test Number	Feedstock	H_2	CH_4	C_2H_2	C_2H_4	C_2H_6	C_3H_6	C_3H_8	C_4H_2	C_4H_6	C_4H_{10}
H442-4	Propane	46.39	7.06	16.5	5.57	.82	1.37	21.14	.65	.30	.20
H455-3	Butane	55.58	6.77	18.18	3.38	.71	2.52	2.41	.76	.39	9.08

Section 7. Percent Reaction, Concentration in Product and SER from Propane and Butane Tests

Test Number	Feedstock	Reaction (%)	C_2H_2	C	CH_4	SER (kWh/lb C_2H_2)
H442-4	Propane	53.50	14.78	10.42	4.99	4.42
H455-3	Butane	70.39	15.09	17.01	2.81	2.87

Table 9-8. Continued.

Section 8. Product Gas Mass Spectrometric Analyses, in Mol %, of Methane Plus Butane Tests

Number	Time (sec)	H_2	CH_4	C_2H_2	C_2H_4	C_2H_6	C_3H_6	C_3H_8	C_4H_2	C_4H_6	C_4H_{10}
H464	150	57.37	14.44	14.50	2.78	.64	1.39	1.28	.20	.22	7.18

Section 9. Overall Percent Reaction, Concentration in Product and SER from Methane Plus Butane Tests

Test Number	Time (sec)	% Reaction	Mol % of Total Product			kWh/lb C_2H_2
			C_2H_2	C	C_2H_4	
H464	150	59.01	11.23	22.54	2.15	4.57

lower than data reported for other arc equipment (such as that reported by Chemishe Werke Huels–about 5 kWh per pound). The lower energy requirement is attributed to the violent turbulence which exists in the arc heater as a result of the magnetic arc rotation and the line frequency energy pulsations. The units at Huels operate on dc power with gas swirl indicted arc rotation; both these phenomena would tend to reduce turbulence.

The modest amount of soot produced in these tests was examined for surface area and found to be in the range of 100 square meters per gram, suggesting its application as a fairly high quality carbon black.

A test series was undertaken in which we examined the product selectivities at elevated pressure–up to about seven atmospheres. Since the acetylene production was found to decrease at elevated pressures, an argument might be advanced for operating the arc heater at sub-atmospheric pressure, if the increased yield of acetylene could be justified (mindful of the resultant increased compression cost for the acetylene absorption step in the process).

Tests were run in the MARC 30 arc heater at arc power levels ranging from 835 to 2693 kilowatts. The tests were run at atmospheric pressure, and in the first series these heavier feedstocks were injected through the arcing region at flow rates varying from 0.58 to 0.81 pounds per second for propane and from 0.65 to 0.93 pounds per second for butane. In a later series, butane was injected into arc-heated methane streams. In these latter tests, the arc heater power level was varied from 893 to 2894 kilowatts, methane flow ranged from 0.11 to 0.29 pounds per second and the butane quench stream was varied from 0.10 to 0.72 pounds per second. Section 4 of Table 9-8 lists typical arc heater performance characteristics for the propane and butane test series, while Section 5 lists a typical arc heater performance for the methane-butane quench series corresponding to maximum acetylene concentration. It will be noted that the enthalpies are generally much lower than those in the methane tests. This is because of the reduced thermal stability of these heavier molecules, and it is the reason for the low specific energy requirements (3 kWh/lb C_2H_2) reported for the Du Pont process which used these same feedstocks. Section 6 lists the raw analyses of the cracked gas stream leaving the arc heater. The acetylene concentration can be seen to vary up to 16.5 mole percent for the propane pyrolysis and up to 18.2 mole percent for the butane tests. Significant amounts of ethylene (C_2H_4) are also produced. As expected (and can be seen in Section 7) when corrected for carbon, the analyses indicate significant amounts of soot produced in these preliminary tests. It is presently thought that the carbon selectivity can be significantly reduced, and this should be the subject of additional test work at the pilot plant stage. Table 9-8 lists the product gas analyses which were obtained in the tests in which arc heatered methane was quenched with butane. It will be noted that the maximum acetylene concentration attained during this test series was 14.5%. Specific energy requirements are about the same as were achieved for methane pyrolysis.[17]

Figures on the approximate capital costs of the Westinghouse arc heat equipment were also obtained from Fey, and are presented in Table 9-9.

Table 9-9. Approximate Capital Costs of Westinghouse Arc Heaters[17,a]

Westinghouse Arc	Arc Power (MW)	Capital Cost[b]
Single-Phase Unit[c]	1 x 1 = 1	225,000 - 260,000
Three-Phase Array	3 x 2 = 3	350,000
Three-Phase Array	3 x 3 = 9	1,000,000

[a]Includes electrical package, and control and instrumentation.
[b]Approximately 10% more for shipment overseas.
[c]Can be derated down to 0.5 MW.

Other Arc Processes Under Development

The Fluid Convection Cathode Process

This process, presently in the research stage, is being carried out at the Columbia University Chemical Engineering Research Laboratories, under the direction of Professor Samuel Korman. In this system, the gaseous reactants actually form the plasma, but admitting the reactants to the cathode end of the plasma column, using an unusual concentric cone cathode. Much higher temperatures can be achieved, sometimes exceeding 10,000°K. Although the high-intensity arc has been studied for some time, attempts at its use for hydrocarbon synthesis is a recent endeavor. Figure 9-11 shows the concentric cone cathode used for the high-intensity arc in the fluid convection cathode process.[19]

Using the fluid convection cathode, Korman studied the synthesis of hydrocarbons. At temperatures from 10,000-20,000°K, the hydrogen in the feed was found to be monatomic and to impinge on the carbon anode at its sublimation temperature of about 4000°K in the arc crater. Gas

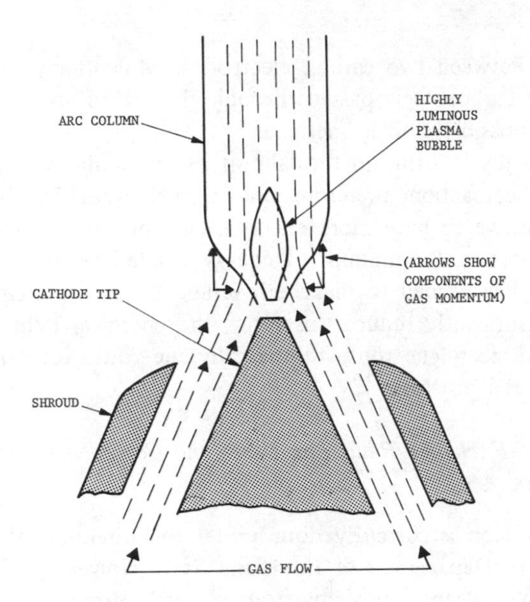

Figure 9-11. Concentric cone cathode for high-intensity arc.[19]

samples drawn from this region during operation were found to contain hydrocarbons, with the composition of the hydrocarbons dependent upon the flow rate of the hydrogen feed to the arc, the temperature and the residence time of the arc effluent in the downstream hot zone. In addition, the composition of the confining wall had an effect. Hydrogen, though always present in stoichiometric excess, causes a transition of the hydrocarbon product ranging from 100% acetylene at low flow rates, to mixtures of acetylene and methane with occasional amounts of propylene. At high flow rates, methane is predominant.

The Arc Heating and Calcium Carbide Process

Research on a new process, an arc heating and calcium carbide system, is currently being done at the Massachusetts Institute of Technology School of Engineering, Cambridge, by R.F. Baddour, J.B. Howard and H.P. Meissner. According to Baddour, the process is expected to produce acetylene from coal by producing calcium carbide with an electric arc; the calcium carbide is then reacted with water when the acetylene is needed.

The feedstock is calcium carbonate $CaCO_3$ and carbon. First plans called for striking an electric arc between a calcium carbonate electrode and a carbon electrode. Now, however, it is envisioned that the arc will be struck between two carbon electrodes, while finely dispersed powder of calcium carbonate is passed through the arc in an atmosphere of hydrogen and possibly a little methane.

The work is just starting in the laboratory; preliminary results on the energetics of the reaction, therefore, are not yet available. It is hoped that this will prove to be a more energy-saving process than other methods. By present processes, the amount of energy needed to make acetylene is roughly twice that needed to make ethylene. If, as is anticipated, this process can significantly reduce this ratio, the chemical industry might again prefer the acetylene route to the ethylene route for a number of chemical synthesis processes.[20]

Investigation of Exchange Reactions Using the Complementary Shock Tube Technique

This investigation is currently done under the direction of R.D. Kern, in the Chemistry Department of Louisiana State University, Systems School of Science, New Orleans. It stems from research supported by a previous National Science Foundation grant, in which the complementary shock tube technique proved very effective in measuring the rates of homogeneous gas-phase reactions by recording the dynamic growth and decay of products, reactants and intermediates. Further applications of this technique are in progress.

The formation of DCl (D is a symbol used for deuterium) from the exchange of CD_4 and HCl is a simple process that takes place amidst the concurrent, but vastly more complex pyrolysis of methane. Studying the exchange will provide insight for an investigation of the more complicated reaction. The exchange of C_2D_2 with HCl is another example of a simple reaction concurrent with acetylene decomposition, and study of this reaction will be useful to the CD_4/HCl exchange, since the major product of methane pyrolysis above 2000°K is acetylene.

For all the reactions mentioned, the measurement and interpretation of the activation energy and time dependence of product formation is related to symmetry concepts and the mechanism by which these processes occur, thereby furnishing information of interest to a wide spectrum of theoretical and experimental chemists.

POTENTIAL APPLICATIONS OF ARC PROCESSES

Arc processes offer a wide range of application in the hydrocarbon field, depending on the feed. For example, in the fluid convection cathode system, when steam is substituted for hydrogen in the arc, the principal hydrocarbon component is methane. As the flow rate increases, acetylene predominates. A solid pulverized petroleum residue fed in entrained argon into the fluid convection cathode arc produces acetylene, predictably, under conditions corresponding to the analogous hydrogen feed.

Dr. Korman believes that the results from his studies thus far indicate that the fluid convection cathode arc may be useful in a number of applications in the gasification of coal and other carbonaceous feeds and in the direct manufacture of hydrocarbons from coal.[4,21]

GENERAL ASSESSMENT OF THE FUTURE

The Acetylene Market

The commercial use of acetylene is predominantly as a chemical feedstock in the plastics and synthetic chemical industry. The production of acetylene in the United States was about 1000 million pounds per year in 1970, dropping sharply since then, as indicated in Table 9-10. The carbide and partial oxidation processes accounting for about 40 and 60%,

Table 9-10. Acetylene and Ethylene Production in the U.S. (x 10^9 lb/yr) (International Trade Commission and U.S. Department of Commerce)

Year	Acetylene			Ethylene
	Total	from Hydrocarbon	from Calcium Carbide	Total
1970	1.07423	0.64454	0.42969	
1971	0.89427			
1972	0.83032			
1973	0.5988			
1974	0.5654	0.28975	0.27565	23.522

respectively. There are no arc processes in operation in the United States at this time, the Montague plant of Du Pont having closed in 1968.

Much of the stagnancy of the acetylene market, despite an ever-growing plastics industry, is due to an increase in the use of ethylene, rather than acetylene, as a feedstock for vinyl chloride. According to the AVCO report, the demand for polyvinyl chloride will increase 9-11%/yr, to 8 billion pounds per year during this current decade. Most of this increase will be used in the construction industry as piping, flooring and siding to relieve the drain on metal and forest resources.[11] The chemical reaction used to fabricate vinyl chloride from acetylene is

$$C_2H_2 + HCl \rightarrow CH_2 = CHCL$$

It involves expensive acetylene, cheap hydrochloric acid and low processing costs.[11]

Ethylene ($CH_2 = CH_2$), instead of acetylene is now used as the raw material for the production of the plastics monomers, such as vinyl chloride and vinyl acetate. The chemical reactions are as follows, and run either by direct chlorination:

$$CH_2{=}CH_2 + Cl_2 \rightarrow CH_2Cl\text{-}CHCl_2 \rightarrow CHCl{=}CH_2 + HCl,$$

or by the oxychlorination process:

$$CH_2{=}CH_2 + HCl + \tfrac{1}{2}O_2 \rightarrow CH_2Cl\text{-}CH_2Cl + H_2O \rightarrow CHCl{=}CH_2 + HCl.$$

The ethylene route involves a cheaper hydrocarbon (ethylene, having a double bond instead of a triple one as acetylene has, is easier to fabricate) but expensive chlorine (or oxidation of HCl) and high processing costs.[11]

Ethylene itself can be manufactured by cracking hydrocarbons in a tubulated heater. The feedstock consists of a mixture containing ethane, naphtha, raffinate, propane, etc. The energy used is given as 6780 Btu (1700 kcal) per pound of ethylene and all by-products, which include fuel gas mixed with hydrogen, some acetylene (2%), butadiene, benzene, toluene, liquid fuel, etc. Of this, ethylene represents 30%, which means that 22,600 Btu (5600 kcal) have to be spent to recover one pound of ethylene. With the present heat rate of fossil fuel power plants of 10,000 Btu/kWh, the equivalent in electrical energy is 2.26 kWh.[22]

Union Carbide, in a review of processes to make acetylene, found that arc processes are not favorable for their company at this time. Some of the reasons were:

1. The cost of electricity is going up;
2. Carbon keeps forming in the process, creating a problem of carbon disposal;
3. There is a need for separating and concentrating the acetylene out of all that is processed. The whole stream produced must be processed, and the equipment is expensive;
4. Union Carbide prefers smaller units of production at more locations. The separation of acetylene becomes an expensive task to perform with small units. Controls, for example, or other costs do not grow when separation equipment size grows;

5. From the standpoint of energy consumption the electric arc process is not very different from the CaC_2 process. It costs about 4000 kWh/ton to make CaC_2. When water is added, CaC_2 produces 32% of its weight in acetylene, and this acetylene is pure. CaC_2 can be shipped easily around the country, transporting 1/3 of its weight in acetylene. By contrast, 200-lb metal cylinders carry about 20 lb of acetylene.[23]

The choice a manufacturer has to make between the acetylene and ethylene processes thus depends on a great number of factors involving the state of the chemical industry at a given place and time. However, the cost of the available feedstocks will be one of the very important factors under consideration. Although the processing of vinyl chloride from ethylene is more expensive than processing from acetylene, the lower original cost of the ethylene more than compensates for the processing costs. A method of producing acetylene more cheaply then has considerable potential for recapturing a large portion of the vinyl market.[11]

Economic Comparison of Different Processes to Produce Acetylene

As stated previously, production of acetylene in the United States is almost exclusively from the calcium carbide and partial oxidation processes. A comparison can be made of these two methods in terms of raw materials used per pound of acetylene produced (Table 9-10).

For calcium carbide-produced acetylene, 0.88 lb of calcium oxide and 0.56 lb of coal are first needed to produce 1 lb of calcium carbide. Energy used in this initial reaction is 3.4×10^3 Btus. The carbon from the calcium carbide (37.5%) goes into the acetylene by hydrolization, virtually without external energy. Requirements for 1 lb of acetylene are then 2.15 lb of calcium oxide and 1.38 lb of coal, and 8.35×10^3 Btus of thermal energy. For the partial oxidation process, 1.85 lb of methane and 8.35×10^3 Btus of energy are needed to produce 1 lb of acetylene.

An examination of the raw material and energy requirements for the processes for producing acetylene by electric means indicates that the AVCO process needs 10% less electrical energy than the calcium carbide route, but requires more than twice as much coal. However, the AVCO process is more convenient than the CaC_2 process, which involved a lot of inert material (calcium), requiring an additional reaction (converting CaC_2 to acetylene), and which presents a disposal problem for the resulting $Ca(OH)_2$. Another advantage of the AVCO process is the hydrogen required in the process if formed in the coal; no external supply of hydrogen is needed, and the process even produces some extra hydrogen, which can be sold. The Du Pont process requires 33% less electrical energy than the carbide route, but uses methane or liquid hydrocarbons. Du Pont and

Table 9-11. Use of Raw Materials and Energy and Capital Investment for Different Processes

	Calcium Carbide	Partial Oxidation	Du Pont Acetylene Arc	AVCO Arc	Westinghouse
Raw Materials (lb C_2H_2)					
CaO (lb)	2.15				
Coal (lb)	1.38			3.3	
CH_4 (lb)		1.85	either 1.5		
C-10 (lb)			or 1.5		
H_2 (lb)				0.1	
Energy (lb C_2H_2)					
Thermal (kcal)		2108			
Electric (kWh)	4.5		3	4	4.1 - 4.5
Plant Cost					
Million $/300 million lb C_2H_2/yr		40-32	48[a]	53-64	20[b]

[a]Wulff Plant
[b]For reactor and electric equipment alone.

AVCO plants were more expensive than partial oxidation plants according to the consistent values and ranges given in the AVCO report;[11] however, the partial oxidation route to acetylene requires natural gas, which is in short supply. Although no more development work has been done on the AVCO process since 1972, this work might be revitalized in the future, and researchers have new ideas on how to improve the process further.[16]

The Westinghouse arc heater shows promise if a feeding mechanism can successfully be developed that allows the use of pitch and heavy bottoms as feedstock. As a matter of fact, the Westinghouse torch has been operated at power levels exceeding 3 MW. When three such torches are operated on three phases simultaneously, the overall power approaches 10 MW. One can thus state that the Westinghouse torch and the Du Pont Reactor are the reactors that have been operated at the highest power level. The range of 5-15 MW is considered to be the suitable power range for units of this sort. It is not felt that units of larger size than indicated by this power range are desired, because of power-switching difficulties that arise above these values.

It is expected that the fluid convection cathode process, which brings the reaction products into a region of the arc which is at a much higher temperature (10,000° K), might give somewhat higher efficiency, both in the use of raw materials and electrical energy. However, it is not yet out of the laboratory stage (with an acetylene output at the grams per hour level) and could require the same number of years of development as the AVCO process required (the AVCO process is now at the 180 lb/hr level of production).

The choice of a process consequently depends on a careful balance of the availability and costs of raw materials, electrical energy, and investment capital with the output and price of the final product (acetylene). This choice will, of course, vary drastically from country to country at different times. Also, the choice is only likely to happen if therc is a sudden or foreseen increase in demand, *i.e.,* the choice is probably only applicable to new capacity or additions to existing capacity.

REFERENCES

1. Miller, S.A., Ed. *Acetylene, Its Properties, Manufacture and Uses,* Vol. 2 (New York: Academic Press, 1966).
2. Hardie, D.W. *Acetylene, Manufacture and Uses* (Fairlawn, New Jersey: Oxford University Press, 1965).
3. Rutledge, T.F. *Acetylene Compounds: Preparation and Substitution Reactions* (Cincinnati, Ohio: Van Nostrand Reinhold, 1968).

4. Baddour, R. and Timing. R.S. *The Applications of Plasmas to Chemical Processing* (Cambridge, Mass: MIT Press, 1967).
5. Gladisch. *Pet. Refiner* (June 1962).
6. Gladisch. *Chem. Eng. Tech.* 41:204 (1969).
7. *Hydrocarbon Proc.* (June 1971).
8. *Chem. Eng. News,* 38 (January 24, 1972).
9. "Arc-Coal Process Development–Final Report, May 1966-April 1972," AVCO Corporation Systems Division, Lowell, Massachusetts.
10. Babcock, J.A. "Design Engineering of an Arc Acetylene Plant," *Chem. Eng. Proc.* 71 (3):90 (March 1975).
11. "Phase I Feasibility Report, AVCO Arc-Coal Process," R&D Report No. 34, AVCO Space Systems Division, Lowell, Massachusetts.
12. "300 mm lb/yr Acetylene Plant–AVCO Arc Coal Process," Blaw-Knox Chemical Plants, Inc. BKC-1661, No. 30 (1971).
13. Schultz, R.A., Du Pont de Nemours. Unpublished Report, February 22, 1968.
14. U.S. Patent 3,320,146, May 16, 1967.
15. Myers, A.F., Du Pont de Nemours & Co. Private communication.
16. Gannon, R.E. AVCO Corp. Private communication.
17. Fey, M.G., Supervising Engineer, Arc Heater Project, Westinghouse Electric Company, East Pittsburgh Divisions. Private communication.
18. Fey, M.G. and F.J. Harvey. "The Role of Plasma Heating Devices in the Electric Energy Economy," Presented to the American Chemical Society, September 9-12, 1974.
19. "High Energy Arcs Used to Make Hydrocarbons," *Chem. Eng. News* 51:36-37 (April 23, 1973).
20. Baddour, R.F. Private communication.
21. Gross, B. *et al. Plasma Technology* (New York: American Elsevier, 1969).
22. Andersen, H., Union Carbide Corp. Private communication.
23. Wiesendanger, Union Carbide Corp. Private communication.

UNCITED REFERENCES

Amman, P.R., R.F. Baddour, M.M. Johnson, T.W. Mix and R.S. Timmins. "Coal Conversion in an Electric Arc to Make Acetylene," *Chem. Eng. Prog.* 52-57 (June 1964).

Brooks, B.W. and R.M. Spearot. *J. Appl. Chem. Biotechnol.* 25:279-288 (April 1975).

Hesp, R.W. and D. Rigby. "Electron Activated Cracking of Gaseous Hydrocarbons, Some Factors Affecting Reactor Design and Rating Up of Operation in The Production of Acetylene from Hydrocarbons," *J. Appl. Chem. Biotechnol.* 24:13-33 (January-February 1974).

Karman, S. *et. al.* "Synthesis of Hydrocarbons," *American Chemical Society, Advances in Chemistry Series, Symposium on Coal Justification,* Dallas, Texas (1973).

Appendix

List of Patents on Electric Means for Producing Acetylene

Process	Patent No.	Year	Title	Author
Du Pont de Nemours Process	3,209,059	1965	Electric Arc Furnace with Outer Cylindrical Electrodes and a Scraping Member with a Reciprocating Motion to Remove Deposits	Fleitz
	3,185,754	1965	Electric Arc Furnace with Central Electrodes and a Cylindrical Electrode and an Arc Rotating Around Central Electrode	Doukas, Reed
	3,179,733	1965	Rotating Electric Furnace in which One Electrode is Wall and a Cooled Scraper Member	Schotte
	3,073,769	1963	Process for Making Acetylene by Pyrolysis of Hydrocarbon in an Electric Arc Furnace	Doukas
	3,375,316	1964	Electric Arc Furnace for the Preparation of Acetylene by Pyrolysis of Hydrocarbons	Harris
	3,347,774	1967	Arc Furnace for Pyrolysis of Hydrocarbon to Acetylene	Myers
	3,320,146	1967	Process of Making Acetylene in an Electric Arc	Neeley
AVCO Process	none			
Other	3,333,927	1967	Electrothermal Process for Producing Acetylene and Carbon Black	Baddour
	3,384,467	1968	Methods and Means for Converting Coal	Amman, Baddour, Mix

10

SECONDARY BATTERIES–RESEARCH AND DEVELOPMENT STATUS*

INTRODUCTION

This chapter emphasizes the advances reported in secondary battery research and development in the United States during 1975 and 1976, and covers the electrochemical couples that have been the subject of the majority of this effort. Included are:

Lead-acid	Pb/PbO_2	Acid electrolyte
Zinc-nickel	Zn/NiOOH	Alkaline electrolyte
Iron-nickel	Fe/NiOOH	Alkaline electrolyte
Iron-air	Fe/O_2 (air)	Alkaline electrolyte
Zinc-chlorine	Zn/Cl_2	Zinc chloride electrolyte
Sodium-sulfur	Na/S	Ceramic electrolyte
Sodium-sulfur	Na/S	Glass electrolyte
Sodium-chloride	$Na/SbCl_3$	Ceramic electrolyte
Lithium-iron sulfide	$Li/FeSor/FeS_2$	Molten salt electrolyte

The two major objectives of the secondary battery programs are (1) load leveling in electrical utilities, and (2) application in the projected increased utilization of battery-powered electric vehicles. Targets in terms of life and cost for the utility batteries are given in Table 10-1. The performance goals given in Table 10-2 for the lithium-aluminum/metal sulfide system can be considered as valid goals for other advanced battery systems.

It is the consensus of battery technologists that for the present, the *Lead-acid* battery is the only one commercially available. Its primary drawbacks are low energy density, relatively poor cycle life on deep cycling, high weight for vehicle applications, and cost per kWh. Research

*By: Ralph Roberts

Table 10-1. Capital Costs for Utility Batteries ($/kWh)[1]

The Target:[a] What Utilities Can Afford	10-yr life	20-yr life
2 hr	25-50 $/kWh	40-80 $/kWh
5 hr	15-35	25-60
10 hr	10-30	20-50
The Expectation: What Developers Can Achieve	10-yr life	10-20 yr life
Lead-Acid	50-80 $/kWh	–
Advanced	–	20-35 $/kWh

[a]The range for each pair of figures reflects wide variation in possible ratios of peak to off-peak energy cost. For peak energy, the crucial variables are generator capital and fuel costs. For off-peak energy, the main variable is fuel cost for the baseload generators used to charge battery storage.

Table 10-2. Performance Goals for Lithium-Aluminum Metal Sulfide Batteries[2]

Battery Goals	Electric Vehicle Propulsion	Off-Peak Energy Storage
Power		
Peak	60 kW[a]	40 MW
Normal	20 kW	10 MW
Voltage, V	140	1000
Specific Energy, Wh/kg	120-160[b]	80-150[b]
Energy Output		
Discharge Time, hr	2	5
Charge Time, hr	5	5-7
Watt-hour Efficiency, %	70	80
Cycle Life	1000	3000
Cost of Capacity, $/kWh	20-30[c]	20-30[c]
Heat Loss through Insulation	150 W	100 kW

[a]Based on the power to accelerate a 1570-kg car from 0-90 kmh in 15 sec.
[b]Includes cell weight only; insulation and supporting weight for battery would add approximately 20% of the weight.
[c]Includes cost of cells, but not battery structure and insulation.

on lead-acid batteries centers on morphological changes in the active material during cycling and the metallurgy and corrosion mechanism of grid materials. The use of grid alloys other than those of lead-antimony is the major change in current battery technology. The use of Pb-Ca and Pb-Ca-Sn for the negative electrode grid to decrease hydrogen evolution during charge is a major trend in battery development. These have been applied to "maintenance-free lead-acid batteries" and are included in the designs for the projected utility lead-acid batteries. A computer study on current distribution in grids has suggested alternate grid structures leading to more uniform current density and possibly improved cycling capabilities of lead-acid batteries.

An interesting approach to lower the specific weight for lead-acid systems is the development of a bipolar cell by Westinghouse. A battery with a specific power of 400 W/kg and specific energy of 27.5 Wh/kg has been constructed. This is an improvement in current capability but below the objective given by Port[3] of 40 Wh/kg.

A comparison of the designs for a 10-hr-discharge 100-MWh battery for utilities is given in Table 10-3. The proposed Gould tank-type battery has the lowest estimated initial and replacement cost but its projected lifetime is less than the other designs proposed.

Table 10-3. Comparison of Results of Ten-Hour Lead-Acid Battery Load Leveling Design Studies[4]

Characteristic	Gould	C & D	ESB
Battery Type	Tank/Paste	Individual/Paste	Submarine/Tubular
Cell Design			
Capacity Ah	64,000	10,000	10,000
Size (1 x w x 1 meters)	2.44 x 0.91 x 1.22		0.55 x 0.55 x 1.43
Weight, kg	5,508	1,201	1,020
Specific Energy Wh/kg	22.7	16.5	19.6
Life			
Cycles	1,750	2,500	2,500
Years	7	10	10
Efficiency, %	80	83	82
Price $/kWh			
Initial	41.6	62.9	65-70
Replacement	21	50	50-55
Price/Cycle			
¢/kWh/cycle	1.6	2.2	2.3
100-MWh Battery			
No. Parallel Strings	2	10	10
No. of Cells	768	5,000	5,250

The effort on lead-acid batteries is continuing with a significant part of this being directly funded by the manufacturers. Continued advances are anticipated in decreased battery weight, increased power and energy density and cycle life.

Among the alkaline batteries, the largest effort is on the *zinc-nickel* battery. Increased cycle life of the zinc electrode and decreased fabrication cost of the nickel electrode are major objectives. In the case of the former, advances have been made by anode shaping[5] and electrolyte stirring.[6] Two approaches to less-expensive cathode construction offer a decrease in nickel electrode cost. One is the pressed electrode[7] and the other a chemically deposited nickel fiber plaque.[8] Cells have been operated with the former. In cells with the shape-modified zinc anode and the pressed nickel cathode, 500 cycles in a 40-Ah cell have been achieved. Cost estimates for such batteries are of the order of $50/kWh. Efforts on their improvement are continuing.[5] It is also planned to demonstrate 300-Ah batteries in electric vehicles.

In the case of the *iron-nickel* battery, few technical details on electrode and cell design of the battery recently reported by Westinghouse have been published.[9] A novel modification in this battery is the use of a circulating electrolyte during charge. The battery configuration now being tested in vehicles has 80 cells, a voltage of 96.5 V and is rated at 16.5 kWh. It has an energy density of 42.7 Wh/kg, including auxiliary hardware and trays.

The Westinghouse *iron-air* cell[10] is less advanced. The air electrode utilizes a tungsten carbide catalyst and is reported to have been operated at current densities as high as 600 mA/cm^2 and for 500 cycles without any deterioration due to atmospheric carbon dioxide. A 1-kWh battery containing 30 cells has been operated. Performance goals for an electric vehicle iron-air battery are energy densities greater than 100 Wh/kg, a power density of greater than 88 W/kg, and a cycle life of more than 1000.

The aqueous *zinc-chlorine* battery[11] is based on a circulating zinc chloride electrolyte which contains the cathode reactant, chlorine. It requires external refrigeration on charge and heating on discharge. This permits the external storage of chlorine as the solid hydrate, $Cl_2.6H_2O$, during charge and its solution during discharge. The cell has low polarization with the associated high electrochemical efficiency during charge and discharge. A 50 kWh battery is reported to weigh 318 kg or 157 Wh/kg. The battery is being evaluated for both utility and vehicular application. During 1977, 100 V/100 kWh modules were to be built, and three of these modules were to be assembled into a 100 V/300 kWh battery.

The principle of the *sodium-sulfur ceramic electrolyte* system is relatively simple. Molten sodium and sulfur at 300-350°C react electrochemically via the transport of sodium ion through a sodium-conducting ceramic electrolyte. The end product is a mixture of sodium-sulfur compounds. The major problems have been in developing reproducible beta or beta″ ceramic electrolyte bodies and low-cost materials compatible with the sulfur cathode. To date, this system has been investigated in single cells. Recently, reported work has shown excellent progress in the preparation and increased lifetime of the ceramic electrolyte. These advances have been made in the programs of the University of Utah working with the Ford Motor Co.[12] and at General Electric.[13] In addition, an alternative ceramic electrolyte, $Na_3Zr_2Si_2PO_{12}$, has been reported by Hong.[14] This has characteristics that promise a better electrolyte than beta alumina. Another important advance has been in the demonstration of cathode structures,[13] which enable high electrochemical efficiency in the reaction of the sulfur. Advances also have been made in seals required to electrically isolate the anode and cathode. The major problem to be overcome is the poor corrosion resistance of the inexpensive metals and alloys to the cathodic reaction products. These are required for containers and current collectors. Alternative approaches being investigated for containers include various electrical conducting ceramic liners and molybdenum and high-temperature resistant polymer coatings. During the period of April 1975 to March 1976, cells operated at General Electric Co. underwent 647 cycles with a total current of 377 Ah/cm^2 of ceramic. The effort at Ford includes the testing of a cell for 6400 cycles and 1060 Ah/cm^2. These results indicate significant advances. The G.E. future effort includes the fabrication of multiple ceramic tube cells. Computer design and manufacturing cost studies also are being conducted with the objective of achieving 100-kWh module batteries for testing under utility conditions in 1981.

Sodium-sulfur glass electrolyte differs from ceramic electrolyte cells in the use of thin hollow sodium ion conducting glass fibers as the electrolyte. Recent advances have been made in glass fiber technology leading to longer living cells.[15] In a 0.5-Ah cell over 1400 deep, discharge-charge cycles have been achieved. Five- and forty-ampere hour cells have been constructed and are being tested. This cell has similar problems with respect to current collectors and cathode containment as the modification with ceramic electrolyte. A cost projection has been made for the base cell which gave a production cost of $23/kWh, assuming the annual manufacture of 10^6 kWh of cells.

Sodium chloride differs from the sodium-sulfur cell in the cathode. The electrochemically active cathodic material is a metal chloride with antimony chloride being preferred. This is suspended in sodium aluminum chloride, which serves along with the beta alumina as the electrolyte. It operates at a lower temperature than the sodium-sulfur cell, 180-250°C. The lower operating temperature has enabled the use of polymeric seals in the experimental cells. However, it has not been established that these materials will be satisfactory for long-term operation. The other major problem is in the current collectors. To date, tungsten and molybdenum have been used; however, these are too expensive for use in commercial batteries. Flat disc and tubular electrolyte cells have been examined. The cycle efficiency is greater than 80%. The disc-type cells have been operated for 640 cycles corresponding to 120 Ah/cm^2 of ceramic electrolyte. An estimate of $20/kWh has been made for the manufacturing costs of a computer-optimized cell design. Development of the cell is continuing and a plan for the construction of a commercial load leveling battery has been formulated.

Lithium iron sulfide is based on the reaction between lithium and iron II or iron IV sulfide using the LiCl-KCl entectic mixture as the electrolyte. The operating temperature is 375-475°C. Lithium alloyed with aluminum or silicon has been used as the electrolyte. Both circular horizontal and prismatic vertical cell structures are being investigated. The major problem areas are materials, especially separators and cathode current collectors and containers. The latter problem is much greater for the FeS_2 cathodes. Among the materials being studied for separators are various ceramics including aluminum and boron nitrides, alumina, beryllia and yttria. Molybdenum has been found to be the best metal for use with FeS_2, whereas this metal and several alloys, especially nickel-chromium and related Inconels, are compatible with FeS. A number of alternative techniques for electrode production are under investigation. Lithium-aluminum anode FeS_2-CoS_2 cathode cells have recently been reported with cycle lives of greater than 360 and close to theoretical ampere hour efficiency. These showed a capacity decline of over 18% during the 360 cycles, and life-testing is continuing. A cost estimate for manufacturing Li-Al/FeS_2 batteries indicates a selling price of $29.16/kWh is achievable in the second or third plant constructed.

Selected characteristics of cells under development are summarized in Table 10-4. All data are not for the same cell or battery, and some of the values are projected objectives that have not been obtained experimentally.

Research and development on *both* the various cell combinations are continuing. Greater emphasis is being placed on cell design and fabrication and manufacturing methods to develop systems that will meet present development performance and cost goals.

BACKGROUND

Secondary battery research and development in the United States is supported by industry, government, and the electric utilities. The governmental agencies include the Energy Research and Development

Table 10-4. Summary of Batteries Reviewed

	Battery Couples					
	Pb-Acid		Zinc-	Iron-	Iron-	Zinc-
	Utility	Auto	Nickel	Nickel	Air	Chlorine
Funding $K 1976						
ERDA	120		134	50	270	162
EPRI	85		--	--	--	143
Organization[a]	Various	ESB	ERC	Westinghouse	Westinghouse	EDA
Cell						
Anode	Pb	Pb	Zn	Fe	Fe	Zn
Electrolyte	H_2SO_4	H_2SO_4	KOH	KOH	KOH	$ZnCl_2$
Cathode	PbO_2	PbO_2	NiOOH	NiOOH	O_2 (Air)	Cl_2
Energy Density, Wh/kg	21	24 (1-hr rate)	33	44-55 (2-hr rate)	>100	88[b]
Power Density	--	--	--	--	>88[c]	70[c]
Capacity, Ah	10,000 (10-hr rate)	--	300	180	--	--
Cycles	2500[b]	400	500	1500	500	>500
Cost, Estimated $/kWh	$37	$40	$50	--	--	$25

Table 10-4 (continued)

	Battery Couples						
	Sodium-Sulfur			Sodium-Chloride	Lithium-Sulfur		
Funding \$K 1976							
ERDA	3350		536	--		3505	
EPRI	1875		--	305		250	
Organization[a]	Ford	General Electric	Dow	ESB	Argonne	General Motors	Atomics Int'l
Cell							
Anode	Na	Na	Na	Na	LiAl	LiAl	LiSi
Electrolyte	β Al_2O_3	β Al_2O_3	Glass	β Al_2O_3	KCl, LiCl	KCl, LiCl	KCl, LiCl
Cathode	S	S	S	$SbCl_3$,	FeS, FeS_2	FeS_2	FeS
Energy Density, Wh/kg	0.2 Wh/cm^2[c]	--	--	--	140	115	85
Power Density	0.4 W/cm^2[c]	--	--	.04 W/cm^2[c]	80	7.6	--
Capacity, Ah	--	--	0.5	0.8	--	--	--
Cycles	9000	647	$>$1400	$>$640	$>$360	445	240
Cost, Estimated \$/kWh	--	--	\$23	\$20	\$29	--	--

[a]Source of data.
[b]Program goal.
[c]Area of β alumina.

Administration (ERDA), the Department of Defense (DOD) and the National Aeronautics and Space Administration (NASA). The funding by the electric utilities is primarily through the Electric Power Research Institute (EPRI). The government and EPRI are the major sources of support for the more advanced, not yet commercial systems. Work on the lead-acid, nickel-iron, nickel-cadmium, and zinc-silver batteries is supported by both groups. Improvement of the lead-acid and nickel-cadmium batteries is receiving proportionately more direct industrial support than the other two combinations. Very little work on the iron-nickel systems is being done by the organizations contacted. Zinc-silver secondary battery research and development is largely under the auspices of DOD and NASA. This is due primarily to the high cost of this battery and the limited cycle life of those developed to date. Because of these considerations, zinc-silver batteries will not be discussed.

The information included is based on publications during the past two years and papers presented at various meetings, especially the Inter-Society Energy Conversion Engineering Conference of 1975 and 1976 and the Electrochemical Society meetings of 1976. Contact was also made with the major research and development centers, both government and industry, to make certain the latest available information was included. The breadth of the current secondary battery activities has made it impractical within the limitations of this chapter to discuss these programs with the responsible scientists and engineers. It has also been impractical to undertake a full literature review, especially of the patents issued. For the most part, only literature since January 1975 will be cited. Earlier information is included where required for understanding. Original references to the earlier information have not been cited but are included under Uncited References.

The current increase in activity on secondary batteries has been due to the fuel shortage. Various studies indicate that secondary batteries offer the potential of energy, and especially petroleum-derived fuel savings, in the evolution of the future national energy technology. Two applications have received most emphasis—load leveling in electric utilities and the propulsion of electric vehicles. Prevalence in the United States of a large number of two-car families, approximately one-third of those having automobiles, is considered to offer a market for electrically powered vehicles. There have been several studies of the required battery characteristics for these applications.

Secondary Batteries for Utility Systems

Secondary batteries have been given much consideration for application to load leveling for electric utilities. Some of the advantages are:[1,16]

- mitigation of oil shortage and import problems;
- uniquely versatile and rapid in their response to electric system needs;
- valuable for system regulation;
- battery output can exactly match incremental changes in load without efficiency penalty;
- modular construction for easy installation and lower capital cost;
- shorter lead time for construction;
- few, if any, site restrictions;
- potential savings in transmission costs;
- increased reliability resulting from modularity; and
- virtually free of environmental problems such as pollutants, heat dissipation and noise.

The major drawback to current batteries is that they do not meet required performance, life and cost standards.

Figure 10-1 shows an estimate by Birk[1] of the relative cost of various storage and base-load systems for power plants. It is to be noted that advanced batteries are estimated to have a cost advantage for an annual operation of approximately 1000 hr, *i.e.*, for peaking. To have potential for greater annual use it will be necessary to reduce the energy cost.

Table 10-5[1] outlines various candidate systems being considered by EPRI and the current targets for their development.

Secondary Batteries for Electric Vehicles

The other major application for storage batteries motivating the recent increased research and development activities on batteries is for use in electric vehicles. Estimates such as the one shown in Figure 10-2[16] indicate that starting with coal as the energy source, the electric vehicle would have an overall energy efficiency greater than the internal combustion engine. Some of the potential advantages for electric vehicles, as given by Yao and Birk[16] are:

- mitigation of oil shortage and import problems;
- conservation and efficient use of available energy resources;
- air pollution abatement; and
- lower operating and maintenance costs.

Among the disadvantages of the electric vehicle are:

- low vehicle range with available and most projected advanced secondary batteries;
- low cycle life of most secondary batteries; and
- time required for recharging.

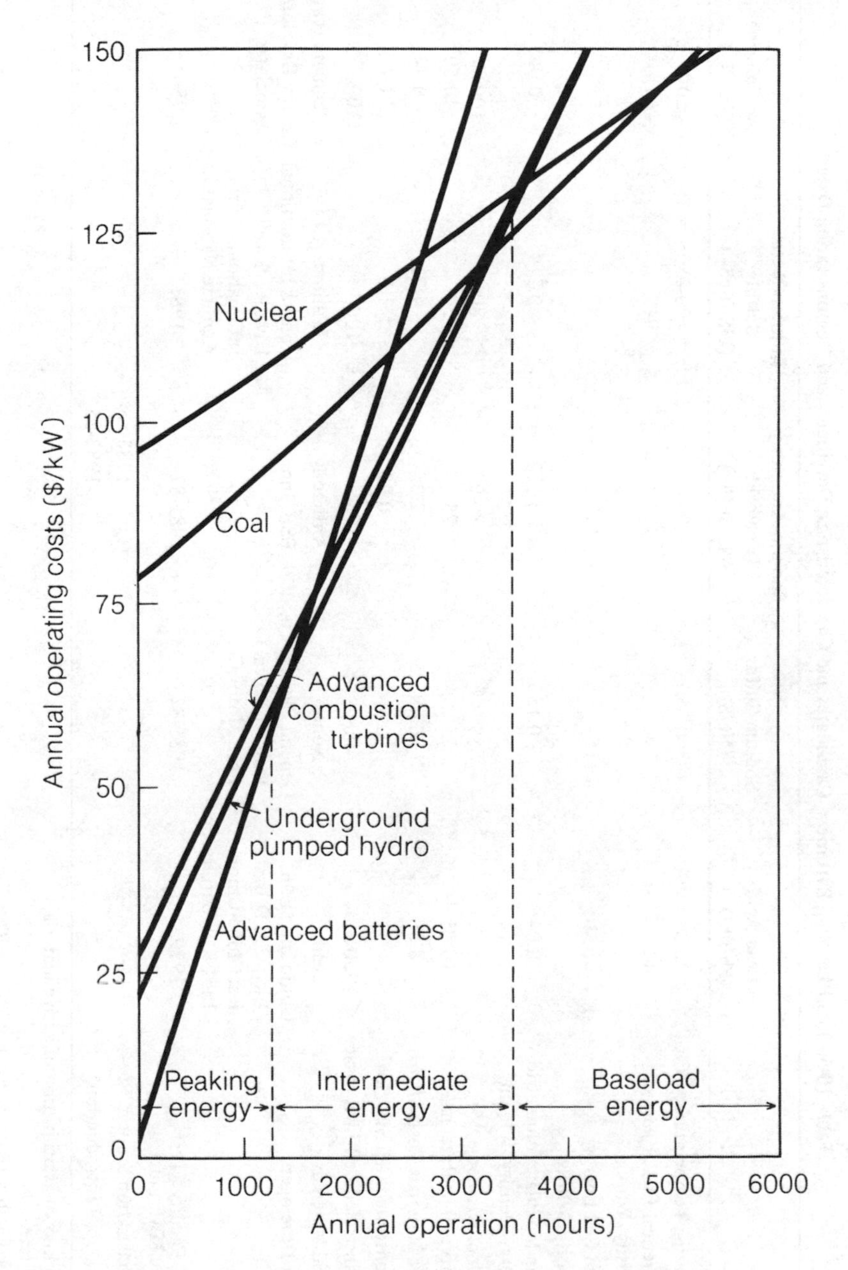

Figure 10-1. Relative cost of different modes of power generation.[1]

Table 10-5. Load Leveling Batteries: Candidates and Characteristics; Developers and Demonstration Dates[1]

	Lead-Acid (Pb/PbO_2)	Sodium-Sulfur (Na/S)	Sodium-Antimony Trichloride ($Na/SbCl_3$)	Lithium-Metal Sulfide ($LiSi/FeS_2$)	Zinc-Chlorine (Zn/Cl_2)
Operating Temperature, °C	20-30	300-350	200	400-450	50
Theoretical Cell Energy Density, Wh/kg	242	794	772	949	464
Design Cell Energy Density, Wh/kg	20	154	110	187	55
Design Modular Volumetric Energy Density, Wh/cm^3	0.046	0.153	0.122	0.214	0.043
Depth of Discharge,[a] %	25	85	80-90	80	100
Density, 10-hr Rate, mA/cm^2	10-15	75	25	30	40-50[b]
Active Materials Cost, $/kWh	8.50	0.49	2.35	4.27	0.74
Demonstrated Cell Size, kWh	<20	0.5	0.02	1.0	1.7
Demonstrated Cell Life, cycles	<2000	400	175	1000	100
Critical Materials	Lead	None	Antimony	Lithium	Ruthenium (catalyst)
Major Developers	Gould Inc.; ESB, Inc.; C & D Batteries; Globe-Union, Inc., K-W Battery Co.	General Electric Co.; Dow Chemical Co.; Ford Motor Co.	ESB, Inc.	Atomics International Div., Rockwell International Corp.; Argonne National Lab.	Energy Development Associates
BEST Facility[c] Test (5-10 MWh unit)	1979	1981-82	1981-82	1981	1980
Demonstration Station			1983-85		
Commercial Introduction			1983-85		

[a]Also known as utilization of active materials.
[b]5-hr rate.
[c]BEST Facility (Battery Energy Storage Test Facility).

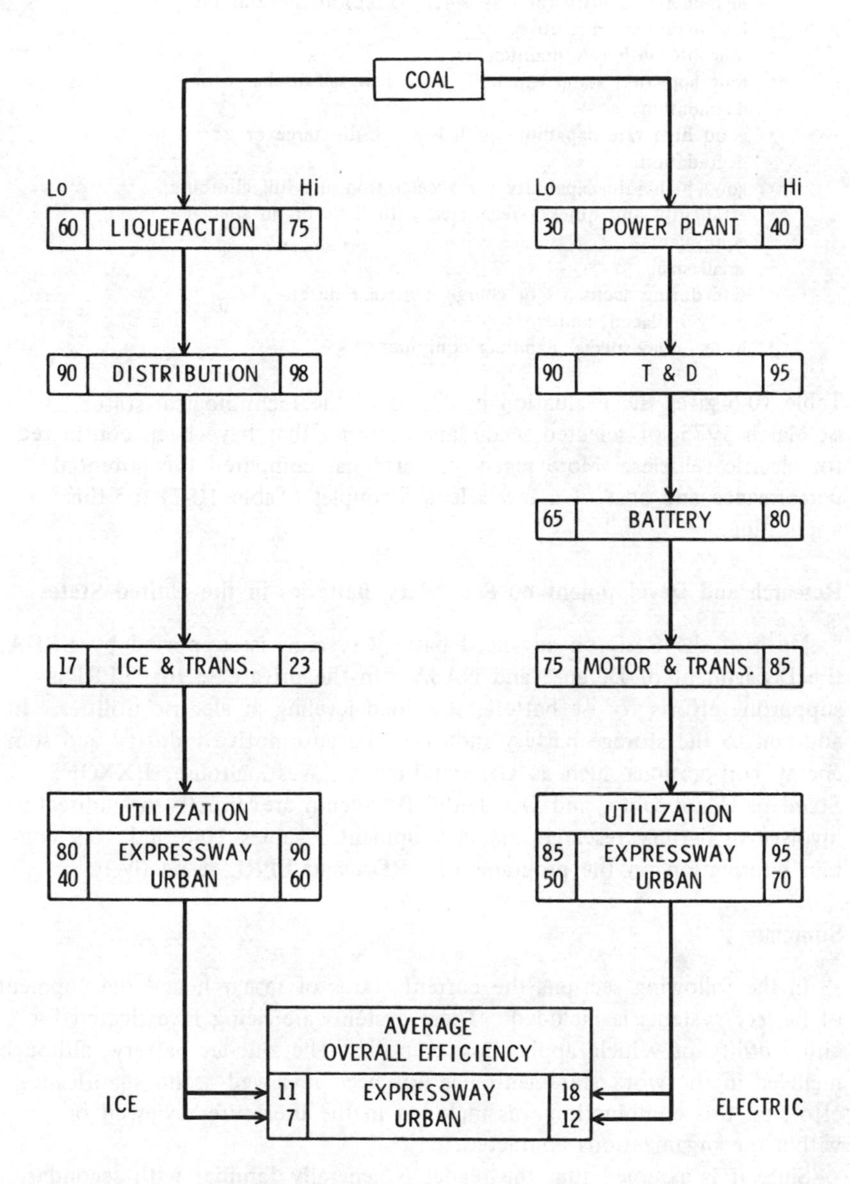

Figure 10-2. Comparison of energy conversion efficiencies: electric car vs synthetic gasoline-powered car.[16]

Gross[17] has described some of the technological requirements for electric vehicle batteries:

- high-energy density (at least 44.1 Wh/kg) at the 2-hr rate;
- low manufacturing cost;
- long life with low maintenance;
- long activated stand capability with low self-discharge or degradation;
- good high rate capability with low self-discharge or degradation;
- good high rate capability for acceleration and hill climbing;
- efficiently and quickly recharged with little or no special equipment;
- small size;
- safe during accidents or charge control failure;
- easily replaced; and
- little or no special handling equipment.

Table 10-6 gives the evaluation by Gross of the technological status, as of March 1975, of selected secondary batteries that have been considered for electric vehicles. More recently, Port[3] has compared the potential performance and cost of a few selected couples (Table 10-7) for this application.

Research and Development on Secondary Batteries in the United States

Much of the R&D on advanced battery systems is supported by ERDA, the Department of Defense and NASA. In the private sector, EPRI is supporting efforts to use batteries for load leveling in electric utilities. In addition to the storage battery industry, the automotive industry and some energy corporations such as General Electric, Westinghouse, EXXON, Standard Oil of Ohio, and Occidental Petroleum are directly or indirectly involved in battery research and development. Tables 10-8 and 10-9 contain information on the programs of ERDA and EPRI, respectively.

Summary

In the following sections the current status of research and development of battery systems is included. These systems are being investigated for either utility or vehicle application or both. The zinc-air battery, although included in the work statement, has not been reviewed as no significant effort on this combination was included in the literature reviewed or within the organizations contacted.

Since it is assumed that the reader is generally familiar with secondary battery systems, no discussion of general considerations or historical

Table 10-6. Evaluation of Selected Candidate Secondary Batteries[17]

Battery	Major Factors Affecting Possible Use	Short-Term Prospects (0-5 yr)	Long-Term Prospects (5-15 yr)
Lead-Acid	Low cost; better energy density and life needed	Excellent	Good
Nickel-Zinc	Need improved cycle life, better oxygen transport	Fair	Good
Zinc-Bromine	Need separator for long-activated stand	Poor	Fair
Nickel-Iron	Uncertain cost	Good	Good
Nickel-Cadmium	High cost, limited cadmium supply	Poor	Very poor
Nickel-Hydrogen	Cost and hydrogen tank weight	Good	Good
Zinc-Chlorine Hydrate	Weight, cost and reliability	Poor	Fair
Zinc-Air	Weight, cost and life	Poor	Poor
Aluminum-Air	Rechargeable aluminum electrode	Poor	Fair
Iron-Air	System complexity, cost, life	Fair	Good
Lead-Air	Weight and life	Very poor	Very poor
Sodium-Sulfur	Solid separator, life, cost	Fair	Good
Sodium-Phosphorus/Sulfur	Solid separator, life, cost	Poor	Fair
Sodium-Selenium	Limited selenium supply, cost, improved energy density	Very poor	Poor
Lithium-Sulfur	Improved sulfur electrode, life	Poor	Good
Lithium-Chlorine	Improved energy density	Poor	Good
Lithium-Tellurium Tetrachloride	Improved energy density	Poor	Good
Lithium-Selenium	Limited selenium supply, cost, improved energy density	Poor	Fair
Aluminum-Chlorine	Improved aluminum electrode	Poor	Good
Lithium-Phosphorus/Sulfur	Improved sulfur electrode, life	Poor	Good
Magnesium-Chlorine	Requires more basic research	Very poor	Poor
Calcium/Barium-Chlorine	Requires more basic research	Very poor	Poor
Calcium-Nickel Fluoride	Improved solid electrolyte doping, better electrode geometry	Very poor	Fair
Lithium-Sulfur Dioxide	Cost, high rate capability	Poor	Poor
Lithium-Lamellar Structure	Cost, high rate capability	Poor	Poor
Antimony Redox	Requires more research and development	Poor	Fair
Potassium-Sulfur	Solid separator, life, cost	Fair	Good
Lithium-Bromine	Requires more research and development	Poor	Good

Table 10-7. Comparison of Energy Sources for Electric Vehicles[3]

	Lead-Acid EV-106[a]	Lead-Acid Improved	Nickel-Zinc	Zinc-Chlorine	Sodium-Sulfur
Specific Energy, Wh/k at 1-hr Rate	24	40	60	110	125
Cycle Life	400	800	1000	350	500
Initial Cost/kWh	\$40	\$45	\$100	\$60	\$50
Lifetime Cost/kWh	10¢	5.6¢	10¢	17¢	10¢
Probability	100%	85%	50%	25%	10%

[a]ESB golf cart battery.

background is included. For overall guidance, Figure 10-3, which gives the relationships between the theoretical gram equivalent weight of the electrochemical couple (weight per Faraday) and the theoretical specific energy, is included.[18] The underlined couples are included in this review.

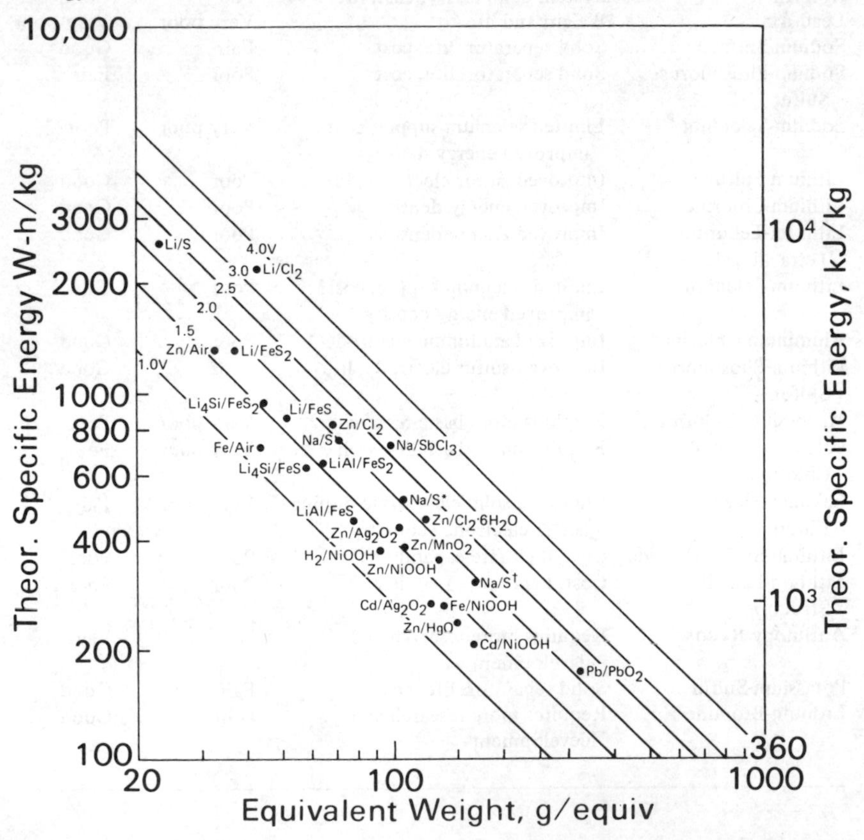

Figure 10-3. Theoretical specific energy of a number of electrochemical couples.[18]

Table 10-8. ERDA Secondary Battery Program

Organization	Investigator	Subject	1976 Contractual Amount ($)
		Lead-Acid Batteries	
Argonne National Laboratory	N. P. Yao	Program Management	
Subcontractors			
Gould Inc.	D. Douglas	Load Leveling Battery Design	20,000
ESB	T. Ferrell	Near Term Electric Vehicles	
C&D		Near Term Electric Vehicles	100,000
Gould		Near Term Electric Vehicles	
Globe Union		Near Term Electric Vehicles	
		Nickel-Zinc Batteries	
Argonne National Laboratory	N. P. Yao	Program Management	
Subcontractors			
Gould Inc.		Near Term Electric Vehicle Batteries	
Eagle Picher		Near Term Electric Vehicle Batteries	
Energy Research		Near Term Electric Vehicle Batteries	100,000
Yardney		Near Term Electric Vehicle Batteries	
		Nickel-Iron Batteries	
Argonne National Laboratory	N. P. Yao	Program Management	
Subcontractors			
Westinghouse	J. Brown	Near Term Electric Vehicles	
Eagle Picher		Near Term Electric Vehicles	50,000
		REDOX Systems	
Lewis Research Center, NASA	L. Thaller	REDOX Battery Systems for Load Leveling	380,000 (Includes subcontracts)
Subcontractors			
Giner, Inc.	J. Giner	Characterization of REDOX Couples	60,000
EXXON	G. Ciprios	Assessment of REDOX Systems for Load Leveling Applications	120,000
Ionics	R. Lauver	REDOX Membrane Development	90,000
		Other Aqueous Batteries and Supporting Studies	
General Electric	F. Will	Zinc-Bromine Battery Research and Development	375,000 (20 months)
EDA	P. Symons	Zinc-Chlorine Battery Development	620,000
Westinghouse	E. Buzzelli	Development of the Iron-Air Battery	270,000
Lockheed	H. Halberstadt	Lithium/Water/Air System Studies	200,000
Linfield College	D. Hamby	Zinc-Electrode Investigations	34,000
Lawrence Livermore Laboratory	J. Cooper	Lithium/Water/Air, Analog Studies, Recycle Studies	160,000
Lawrence Berkeley Laboratory	C. Tobias J. Newman	Fundamental Electrochemical Investigations	200,000

Table 10-8. ERDA Secondary Battery Program (continued)

Organization	Investigator	Subject	1976 Contractual Amount ($)
		Sodium-Sulfur Batteries	
Dow	C. Levine	Hollow-Fiber Sodium-Sulfur Battery Development	536,000
MIT (Lincoln Lab)	J. Kafalas	Characterization and Testing of New Fast Sodium Ion Transport Electrolytes	100,000
Franklin Institute	P. Morgon	Beta-Alumina Stability	50,000
Ford	S. Weiner	Sodium-Sulfur Battery Development	3,200,000 (Includes subcontracts)
Subcontractors			
Univ. of Utah	R. Gordon	Beta-Alumina Separator Pilot Plant	1,000,000 (approx.)
Aeronutronics		Ce.. Fabrication and Testing	1,000,000 (approx.)
		Lithium Sulfur Batteries	
Argonne National Laboratory	P. Nelson	Li/S Battery Development	3,460,000 (Includes subcontracts)
Subcontractors			
Atomics International	C. Sudar	Li/S Battery Research and Development	365,000
Gould	R. Rubisko	Li/S Cell Fabrication and Development	286,000
Eagle Picher	R. Hudson	Li/S Cell Fabrication and Development	326,000
Catalyst Research	D. Stakem	Li/S Cell Fabrication and Development	99,000
Carborundum	R. Hamilton	BN Separator Development	140,000
Carborundum	R. Hamilton	BN Separator Supply	120,000
Fiber Materials	J. Cook	BN Separator Development	6,000
Zircar, Inc.	B. Hamling	Y_2O_3 Separator Development	5,000
Argonne National Laboratory	P. Nelson	Li/S Auto Battery Mobile Test	825,000
Subcontractors			
Goulton, Ind.	J. Cox	Battery Charging/Voltage Equalization Circuitry	18,000
		Other Battery Systems and Supporting Research and Development	
EIC	B. Brummer	Nonaqueous Li/S and Na/S Battery Systems	150,000
Electrochimica	M. Eisenberg	Li/S Spinel Cathodes	65,000 (9 months)
Univ. of Tennessee	G. Manatov	Investigation of $Al/AlCl_3$ Cathode Battery Systems	60,000
Univ. of Florida	H. Laitinen	Characterization of Molten Salt Electrolytes	38,000 (18 months)
Univ. of Florida	R. Waler	Development of BN Paper Separators	45,000
Argonne National Laboratory	P. Nelson	New High-Energy Battery System Studies	200,000
Argonne National Laboratory	N. P. Yao	National Automobile Battery Test Facility	200,000

Table 10-9. EPRI Battery Program

Organization	Subject	1976 Contractual Amount ($)
Bechtel Corp.	Engineering Design and Cost Estimates of Alternative Storage Concepts for the Zinc/Chlorine Load Leveling Battery	120,000 (9 months)
Arthur D. Little, Inc.	Capital Cost of Advanced Batteries for Utility Energy Storage—Assessment and Methodology Study	55,000 (12 months)
ESB, Inc.	Development of Moderate-Temperature Molten Salt Load Leveling Battery	305,000 (12 months)
Atomics International	Lithium-Sulfur Battery	250,000 (12 months)
General Electric	Sodium-Sulfur Battery	2,500,000 (32 months)
Energy Development Associates	Zinc-Chlorine Battery	523,000 (12 months)
Cornell University	Improved Beta-Alumina Electrolytes for Advanced Batteries	120,000 (24 months)
Public Service Electric and Gas	Battery Energy Storage Test (BEST) Facility	3,400,000 (36 months)
ESB, Inc.	Application of Lead-Acid Batteries for Energy Storage on Electric Utilities: Design Study and Cost Projection	85,000 (24 months)
Gould, Inc.	Assessment of the Zinc-Bromine Battery for Utility Loading	275,000 (18 months)
CGE (France)	Improvements of Electrolyte and Seal Technology for Sodium-Sulfur and Sodium-Antimony Trichloride Load Leveling Battery	90,000 (12 months)
Battelle Research Center (Geneva)	Exploration of REDOX Batteries for Utility Energy Storage	35,000 (12 months)

THE LEAD-ACID BATTERY

The recent popularization of the "sealed" lead-acid batteries is another factor that has led to increased research and development on this battery system. Since much of this work is being conducted by the battery industry, many of the details on the scope and results of these efforts are proprietary.

Problem Areas

The technological shortcomings of current lead-acid batteries have recently been discussed by Simon and Caulder,[19] Cook[20] and Cairns.[18] These problems include:

- alternate battery systems that promise improved energy densities;
- low weight fraction of reactive materials in the total battery structure;
- decreased grid corrosion to help improve cycle life;
- improved utilization of active materials, especially at high rates of drain;
- decreased gassing of lead-acid batteries to enable truly sealed batteries;
- cohesion and adhesion of the lead dioxide;
- further decrease in internal resistance through improved separators;
- effect of depth of discharge and recharging rate on battery lifetime.

Research and Development

The increased emphasis on lead-acid batteries as the short-term electrochemical system for utility load leveling and electric vehicles has resulted in increased activity in research and development. Research is focusing on studies on morphology of active materials and alternate lead alloys for grids. Development areas include alternate cell configurations, adaptation of new grid materials, improved separators, methods for control of gas evolution, lower weight containers, and improved battery hardware, such as intercell connectors, seals and leads.

Results of Recent Research Activities

Active Materials. Simon and Caulder have reviewed some of the more recent studies pertaining to understanding the morphological changes in the active material and the relationship of these changes to material availability and cycle life.[19] One of the causes of the loss of capacity with cycle life can be traced, in some cases, to purely physical changes. It has been observed that growth in crystal size leads to material unavailability. The larger crystals, due to passivation or more complex factors, such as change in lattice spacing due to absorbed impurities that may deactivate the crystal, decrease the area of active material.

Two reported properties[19] of the positive active PbO_2 appear to be of significance as guides for better material utilization and prolonged capacity retention. The first of these was the discovery that part of the PbO_2 formed initially was electrochemically inactive in subsequent charge-discharge cycles. It also was observed that the portion of inactive material increased with increasing cycle life.

Differences in the thermal decomposition of electrochemically and chemically prepared PbO_2 have been observed. However, the active material in the positive plates after long cycling gave decomposition curves similar to those of the less-active chemically prepared material. The electrochemically prepared material also showed an exothermic peak at 180°C, which was not shown by the chemically prepared materials. Mass spectrometry showed that water was evolved. Nuclear magnetic resonance studies showed that the electrochemically prepared PbO_2 contained more water than did the chemically prepared material and that obtained from a battery after a large number of cycles. The electrochemical PbO_2 also showed a difference in the structure of the nuclear magnetic resonance hydrogen spectrum. This material contained some hydrogen with a 200 nm separation as well as with the 160 nm separation in the chemical and cycled PbO_2.

It also has been observed that on cycling, the lead dioxide, which was initially compact and dense, became porous and open in structure.[21] This change took place gradually, and eventually this structure disappeared. It was replaced by a very loosely connected, nondescript structure that resembled neither the original nor porous material. This helps explain the loss of active material on cycling. This work has only been carried out with lead-antimony grids and is being continued with other grid compositions.

Grid Materials. The increasing interest in maintenance-free "sealed" lead-acid batteries for automobiles has led to the greater use of lead-calcium grids. According to Simon and Caulder, these grids do not stand up well to deep discharge-charge cycling.[19] However, the automobile battery is presently operated at almost constant potential, making the use of the lead-calcium grid feasible for this application. It should be noted that these alloys are also being considered for other lead-acid battery applications, especially for the negative electrode grid.

According to Rao and Mao, the grids for the maintenance-free battery should resist corrosion during overcharge and also must be free of impurities leading to low-hydrogen overvoltage to minimize gas evolution.[22] This has led to the adoption of Pb-Ca-Sn alloys for this application. This has also led to the trend of decreasing the antimony content in Pb-Sb alloys to around 3% Sb with the addition of other alloying elements.

Rao and Mao report that in the Pb-Ca-Sn system, composition control of calcium and tin during grid casting is essential to obtain corrosion resistance. For alloys with a calcium content in the range of 0.07-0.09%, limiting the tin content has been found essential to obtain the characteristics needed for processing. In comparative tests of batteries with the

Pb-Ca-Sn grid (composition not specified), water was lost at a lower rate than with the low Sb-Pb grids. In addition, this lower rate persists throughout the cycle life, whereas in the case of Sb-Pb it increases by almost a factor of three during cycle life.

Studies by Devitt and Myers[23] on the inclusion of 7-420 ppm of bismuth in a Pb-Ca-Sn grid containing 0.07% Ca and 0.7% Sn showed that the presence of bismuth did not affect battery performance. The studies with bismuth were carried out, as this is an impurity occurring in primary lead.

Metallurgical studies on various aged lead alloys containing calcium and tin have been reported by Prengaman.[24] His results showed that for lead-calcium alloys, the rate of aging and ultimate strength level are determined by the calcium content and rate of cooling. He also noted that the ternary alloys, Pb-Ca-Sn, aged more slowly than the binary, Pb-Ca. Based on these studies, he concluded that where ability to handle is of prime importance, the composition chosen should be such that Pb_3Ca or mixed Pb_3Ca-Sn_3Ca precipitates are present. The former provides initial strengthening for handling. The Sn_3Ca produces better mechanical properties and improved creep resistance.

Prengaman also noted that the grain structures of the Pb-Ca-Sn alloys contribute to the corrosion behavior. Recrystallized structures, with or without tin, corrode at a much faster rate than the nonrecrystallized structures. This is attributed to the finer grain structure of the recrystallized alloys as well as to the condition of the grain boundary.

Further studies on the corrosion of lead-5% antimony grids with emphasis on the effect of heat treatment on anodic corrosion have also been reported recently by Marshall and Tiedemann.[25] The attack on the as-cast material was found to change from interdendritic to a more uniform mode, as the overpotential was increased. In the heat-treated samples the interdendritic attack was decreased, especially at the higher overpotentials, as was the corrosion rate. However, at high potentials the heat-treated alloys corroded intergranularly. This mode of corrosion was not observed in the as-cast materials, thus heat treatment should improve the lifetime of the grid provided that high overvoltages are avoided during the charge cycle.

The trend toward the use of alloys other than that of lead-antimony for grid materials has already been noted. The use of these grids is being planned for batteries other than the current automotive low-maintenance ones. In recent design studies for utility-type lead-acid batteries, these alloys have been incorporated as grid materials. The ESB[4] study includes the lead-calcium alloy for the negative grid and a lead-arsenic-antimony one for the positive grid. C&D Batteries[26] propose the same negative grid with a lead-antimony positive. Most of the proposed designs for the utility batteries are conventional, using flat plate electrodes. The only exception is the ESB design which utilizes tubular-positive plates. These are similar to the positive plates developed for submarine batteries. Although in the past this type of construction increased battery costs, it is reported that improvements in processing and equipment have increased productivity significantly. A comparison of various proposed designs for electric utility batteries for 10 hr of operation is given in Table 10-3.[27]

Towle and Kaulkis cite some results of studies by Tiedemann and coworkers of Globe Union on a computerized model of the current distribution on a lead-acid grid.[28] The computations have shown good agreement with experimental measurements. They show that grid resistance can be affected by modifications in thickness, spacing or angle of grid wires, location of the current collection tab and overall height-to-width ratio. An automotive battery grid design resulting from this computation is shown in Figure 10-4. The five benefits expected from the application of this technique to a utility cell are:

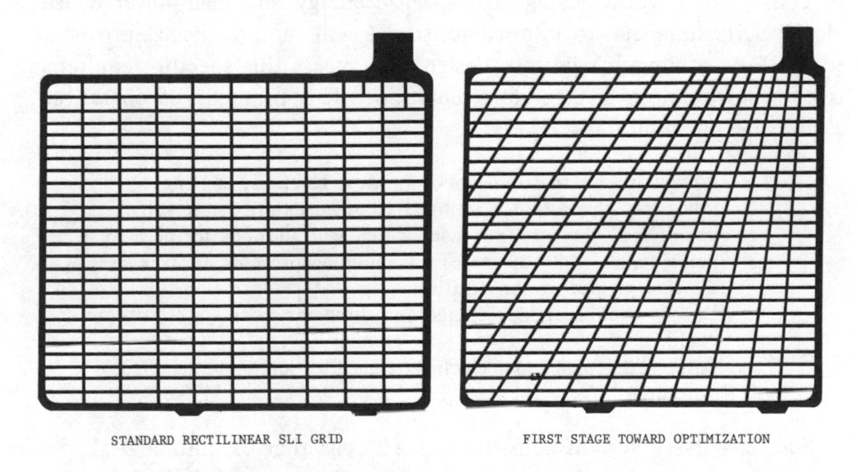

Figure 10-4. Computer grid design for lead-acid batteries.[28]

1. significant improvement in high-rate discharge performance;
2. minimum cost for grids;
3. improved utilization of active material leading to lower cost;
4. improved cycle life through greater uniformity in "working" of the active material; and
5. reduced tendency to concentrate grid corrosion near the tab, thus improving cycle life.

A bipolar lead-acid battery has been proposed by Kassekert, Isenberg and Brown[29] for use as an electric vehicle battery. The bipolar batteries employ flat sheet electrodes that are pasted on one side with positive active material and on the other side with negative active material. The flat sheet, called the current collector, is an electronic conductor that is dense to prevent electrolyte passage from one side to the other. Electrolyte leaks would cause self-discharge. Many bipolar electrodes, stacked together with separators and sealed around the periphery, form a battery pile. In this construction principle, current flow is normal to the electrode orientation which results in reduced electric resistance of the current collector. In traditional batteries, much resistance is introduced by current flow along grid-type current collectors, which cause cell heating, especially at high-power surges. The bipolar construction also allows for uniform active material utilization. Major problems in the engineering of bipolar lead-acid batteries are effecting a controlled electrode separation, sealing of the periphery and reliable operation of many cells in the series connection.

The bipolar battery concept is often referred to as the ideal principle of construction for achieving high specific energy and high-power density. However, heating due to internal resistance will require special provision for battery cooling. A battery design that meets this specific requirement is one that employs electrolyte circulation. This principle of operation allows the possibility for:

- minimizing internal resistance, by close electrode spacing (necessary electrolyte is pumped from an external reservoir);
- use of low concentration acid, which will allow better negative performance while the necessary pumping prevents an excessive dilution of acid at the positive;
- battery cooling at high charge and discharge rates via electrolyte cooling;
- water maintenance and electrolyte purification to be handled centrally.

Such a battery system consisting of 12 cells that exhibits a peak power capability of over 440 W/kg and a specific energy of 27.5 Wh/kg at the two-hour rate, has been constructed. A simulated comparison of this battery with the EV 106 golf cart battery (Table 10-7) for use in a commuter vehicle is shown in Table 10-10.

One of the major weight savings in the lead-acid battery system has been through reduced case weight. Substitution of plastics for the hard-rubber casing is the reason. Work is also in progress on improved separators to decrease internal resistance, weight and cost.

Table 10-10. Comparison of Bipolar and Conventional Lead-Acid Batteries[29]

	EV-106	Bipolar (Simulated)
Top Speed	60 mph (96.54 kmh)	67 mph (107.80 kmh)
0-40 mph (0-64.36 kmh)	14 sec	11 sec
0-50 mph (0-80.45 kmh)	25 sec	19 sec
Range-Commuting Cycle	76 mi (122.28 km)	105 mi (168.95 km)
Range-Constant 30 mph (48 kW/h)	29 mi (46.66 km)	41 mi (65.97 km)

Planned Activities

Tables 10-8 and 10-9 include the current activities on improved lead-acid batteries in the ERDA and EPRI programs. In view of the current emphasis on improved energy efficiency in utilities and the passage of the Electric Vehicle Research and Demonstration Act by the U.S. Congress, an increased level of effort on lead-acid battery development can be predicted with a high degree of certainty. The directions to be taken are for increased lifetime, especially for batteries with deep cycling, decreased cost and maintenance. To achieve this, continued effort to understand in great detail the morphology of active materials and how such problems of crystal growth, active material loss and grid corrosion can be overcome, is needed. In addition, development work along the lines discussed is expected to continue. Proof of projected capabilities of one or more of the other systems to be discussed in the following sections of this report will change the emphasis on lead-acid batteries. Unless there is an unanticipated rapid advance in other battery systems, this increased effort on lead-acid batteries can be expected to last until the early to mid-1980s.

ALKALINE ELECTROLYTE BATTERIES

Several alkaline electrolyte batteries have been the subject of recent investigations. These include combinations of iron or zinc anodes with nickel oxide or air cathodes. Of these, the major effort in the United States is on the zinc-nickel couple, with lesser efforts on iron-nickel and iron-air. Little recent work has been reported with respect to studies in the United States on the zinc-air system, and this couple will not be reviewed. All of the cells using these couples have a potassium hydroxide solution as the electrolyte. In some instances, other hydroxides such as that of lithium are added. These batteries are primarily being considered for electric vehicle applications.

Problem Areas

It is best to discuss problem areas in terms of the particular electrode. The zinc electrode has received most attention, and Gross[17] has summarized some of the problem areas. In the case of the zinc-nickel system, cycle life is limited by the zinc electrode. Due to the solubility of the zinc oxide, zinc dendrites develop during charge, causing shorts. This solubility also causes redistribution of active material leading to a shape change with a resulting capacity loss. The charging of zinc in zinc-air batteries, *in situ,* has not been successful. In addition to the problems cited above for the zinc electrode, the air electrode becomes passivated under short-term peak power loads.

The iron electrode, due to its insolubility, does not show the recycling problems characteristic of zinc. The iron-nickel (Edison) batteries have demonstrated very high cycle life and are very rugged. The use factor of the iron electrode has been low. In addition, because of its low hydrogen overvoltage it has a relatively low charging efficiency with considerable hydrogen evolution during charge. The classic iron-nickel battery also has poor high discharge rate characteristics.

On the cathodic side, the nickel-oxide electrodes are very satisfactory for long-life. Higher coefficients of use of active material are desired, however, and methods for improving performance and decreasing the cost of manufacture of the nickel oxide cathodes are being investigated.

Improved air cathodes are an important problem in the case of metal-air batteries just as in the case of fuel cell development. High-rate air electrodes that do not require noble metal catalysts are a research and development objective. No catalyst has been discovered that operates close to the full theoretical potential of the oxygen electrode. This lower than reversible operating potential also leads to a larger potential difference between the charge and discharge potential. This lowers the energy efficiency of the charge-discharge cycle. Carbon dioxide in the atmosphere can also lead to carbonation of the alkaline electrolyte with attendant performance decrease. During *in-situ* charging of air batteries, carbon- or silver-bearing electrodes generally have not demonstrated long life.[17]

Recent Research

Most advances have been primarily developmental. With the exception of the oxygen electrode, the other three electrodes show good electrochemical reversibility.

Bennion and his collaborators have been investigating the modeling of the porous zinc electrode. In one of the latest reports[30] on this effort, the model and its results are summarized. The reaction mechanism used is a dissolution-precipitation one:

$$Zn + 4\ OH^- \rightarrow Zn(OH^-)_4^{=} \qquad (a)$$

$$Zn(OH^-)_4^{=} \rightleftarrows ZnO + H_2O + 2\ (OH^-) \qquad (b)$$

Based on the mathematical treatment of this mechanism, together with the required electrochemical and transport equations, the following description of the discharge and charge processes was postulated:

"At the beginning of discharge in a fresh cell, the zinc dissolution reaction takes place near the front of the electrode decreasing the conductivity of the matrix, and the solution becomes supersaturated with zincate ions. The electrolyte concentration, KOH, falls in the reaction region, decreasing both solution electrical conductivity and zinc oxide solubility. The reaction shifts toward the front face due to the decrease in solution conductivity and toward the backing plate due to the decrease in matrix conductivity. When the matrix conductivity is much greater than the solution conductivity, the latter effects are negligible. On further discharging, as the reaction product, ZnO, precipitates on the solid surface from the supersaturated solution, it begins to cover up the remaining zinc surface, decreasing active zinc surface. This small active zinc surface and the resulting high resistance of the charge transfer reaction shift the reaction toward the depth of the electrode with increased potential differences.

"During the charging half cycle, the electrolyte concentration, KOH, increases and the solution is undersaturated with zincate. The reaction, depending on the previous discharged state and the previous active zinc surface, takes place inside the electrode where the sum of the resistances of solution, matrix, and charge transfer reaction are minimized. Active zinc surface and matrix conductivity increase as charge proceeds. However, decrease in active surface for reaction B and depleted zincate concentration in the solution, eventually limits the charge transfer reaction, causing the cell potential to rise."[30]

A somewhat different and more empirical approach to the problem of reversibility and shape change of the zinc electrode has been used by

Charkey.[5] This is based on a nucleation site concept. Due to the solubility of the zinc hydroxide or oxide formed during the discharge, a portion of the electrode is lost. Among the factors that influence the dissolution rate are the electrolyte concentration at the edges of the electrode pack, cell geometry and discharge rate. A greater dissolution rate is expected at the edges where the excess electrolyte volume exists. On the subsequent charge, the zinc in solution tends to diffuse and plate back on the high-area portions of the electrode, *i.e.*, the electrode center. Thus, the movement of the zinc will be away from the edges and toward the center of the electrode.

Recent studies of Pickett and Maloy[31] on microelectrodes containing both nickel and cobalt oxides, prepared by electrooxidation of the coprecipitated hydroxides, have shown that the presence of cobalt gives improved electrochemical characteristics when compared with a pure nickel oxide electrode. The difference between charge and discharge potential was lowered from 150 to 75 mV with 10% cobalt hydroxide in the nickel oxide.

Recent Developments

Although development leading to improved electrode performance may be transferable from one battery system to the other, it is best to summarize recent developments in terms of specific couples.

Zinc-Nickel Battery Developments

Greatest activity has been on the improvement of the zinc electrode. McCoy[6] has reported on a zinc-nickel oxide cell using a revolving thin-apertured shutter-separator. This leads to a zinc electrode that does not suffer from shape change, retaining its performance on repeated deep discharges. This is attributed to shutter rotation, which agitates the electrolyte and prevents dendrite formation and unequal zinc distribution. The cell design is detailed in the reference. The negative electrode is a thin copper sheet coated with zinc. The cell is placed into operation in the discharged state. The potassium hydroxide electrolyte, containing sufficient zinc oxide to furnish the desired capacity (1.5 g ZnO/Ah), is added to the cell containing the electrodes and separator. The shutter is operated for a brief period to disperse the zinc oxide, and the cell is then charged. These batteries were tested at the 5-Ah rate with a depth of discharge of 65% of the nickel oxide capacity. The initial test, 291 cycles, was terminated due to the growth of zinc deposits. By redesigning the shutter, improved performance was obtained, and no edge growth was observed after 50 cycles. Further work is in progress to improve cycle life and energy density with the objective of attaining 77 Wh/kg.

In addition to the development of a zinc electrode based on the considerations described, Charkey[7] has investigated more conventional zinc-nickel cell structures than that of McCoy. Various methods for varying the zinc electrode have been studied which have led to: (1) improving the current distribution in the negative by using solid foil collectors; (2) minimizing electrolyte (zinc) distribution through the use of potassium titanate additives and separators; and (3) wrapping the negative electrode in the separator to minimize shape change. Details of zinc electrode shape based on the model described[5] have not been published.

Variations in the structure of the nickel-oxide electrode also have been investigated by Charkey.[7] Three types of structures were included: (1) sintered nickel plaques alternately impregnated with nickel nitrate followed by cathodization in potassium hydroxide; (2) sintered nickel plaques in which nickel hydroxide is electrochemically deposited; and (3) nonsintered nickel electrodes made by blending $Ni(OH)_2$, graphite and a binder, and rolling or pressing the mixture on a nickel screen.

The nonsintered nickel electrode showed lower capacity per unit area than the others, but higher capacity per gram except at the high rates of discharge—one half hour and faster.

Charkey also has used inorganic and combined inorganic-organic separators. The composition of the inorganic separators has not been published; however, they show comparable conducting properties with the organic ones and are more resistant to attack by the electrolyte, especially at higher temperatures of operation.

In zinc-silver cells using the shape-modified zinc electrode,[5] more than 500 deep discharge cycles at the 5-hr rate in 40-Ah cells have been achieved. The cause of failure was shorting because of separator degradation. These improved electrodes are being tested in 300-Ah batteries with nonsintered nickel electrodes. Estimates for batteries based on the modified electrodes indicate that a $50/kWh battery is a distinct possibility.[5] This includes present material and labor costs.

A different approach in attempting to obtain improved performance of the zinc-nickel system was reported by Donnel.[32] Both the nickel and zinc electrodes are enveloped by separators. For the nickel electrode, a nonwoven nylon was used and for the zinc, an inorganic material (composition not specified). The interelectrode separator was made from a nonwoven polypropylene. The cells constructed were rated at a 100-Ah capacity. After a limited number of cycles, the cells were taken apart to observe changes. Zinc sludge was observed at the bottom of the electrode envelope and distortion of the envelopes was noted in some cells. Other cells had a zinc sludge at the bottom of the cell stack. The erosion of the negative material was primarily across the top and along the edges of the electrode. This study also included the construction of 300-Ah batteries, which are to be tested at the Lewis Research Center of NASA.

The discharge characteristics of a 20-Ah zinc-nickel battery being developed for aircraft[9] are shown in Figure 10-5. It shows good performance down to -10°C at drain rates of 200 A. The specific design of the battery was not detailed.

Advanced nickel-oxide technology for improved nickel-oxide cathodes has been reported by Miller and Brown.[8] The nickel electrode matrix

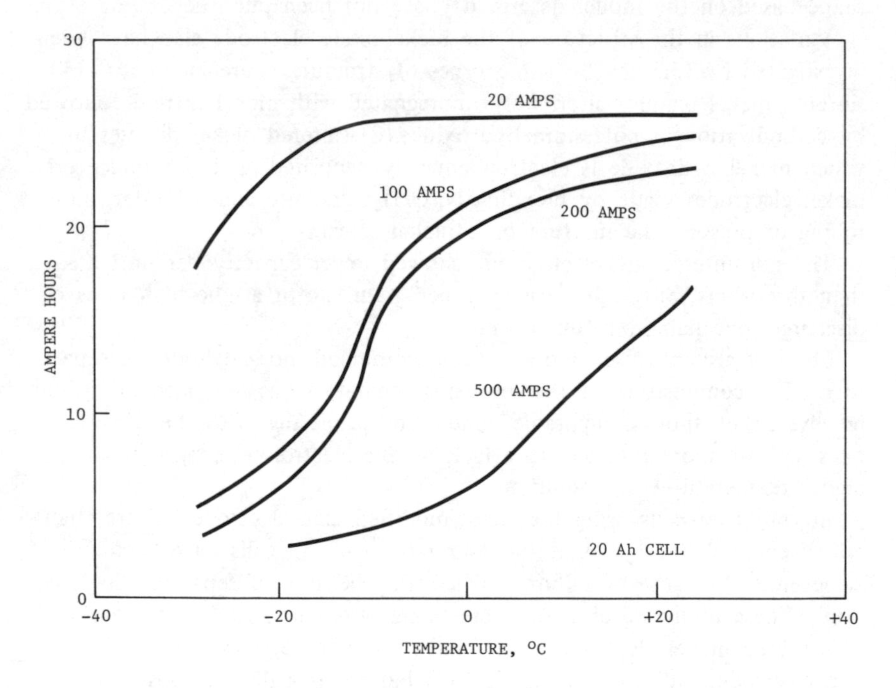

Figure 10-5. Zinc-nickel battery capacity vs temperature.[9]

is prepared by uniformly chemically depositing nickel from nickel carbonyl on an organic felt. The organic material is removed, leaving a pure nickel matrix, which requires approximately 50% less nickel and has a surface area at least an order of magnitude greater than sintered nickel plaques. Nickel electrode matrices up to 0.65-cm thick have been made. These matrices are well suited for loading with active material. They can be impregnated by either chemical or electrochemical deposition of nickel hydroxide or oxyhydroxide. This type of matrix electrode with electrochemically deposited active material showed much greater resistance to distortion during cycling than a conventional chemically impregnated sintered plaque electrode. The test conditions were deep discharge followed by high overcharge.

An automated pilot plant production facility for the electrochemically impregnated, nickel vapor deposited matrix electrode has been operated. Based on this work, it has been estimated that when manufactured on a large scale, this type of electrode will significantly reduce the cost of the nickel-oxide electrode relative to the cost of sintered plaque electrodes.

The Lewis Research Center of NASA recently reported on the performance of a nickel-zinc battery in an electric vehicle.[33] It was reported to provide almost twice the range, 95.6 km compared with 47.3 km of a lead-acid battery occupying the same space in the vehicle. It also delivered almost twice as many start and stops (190 vs 99).

Planned Activities in Nickel-Zinc Batteries

The nickel-zinc battery is being given serious consideration as the follow-on power source to lead-acid batteries for electric vehicles. Work will be continuing on improved cycle life, especially for the zinc electrode. This will not only involve alternate geometries but also improved separator materials. To what extent dynamic anodes or moving electrolytes will be incorporated to improve battery life is highly dependent on the weight and cost of the auxiliary equipment required for dynamic systems and their energy consumption. The development of less-expensive, higher performance nickel electrodes, such as the nonsintered types,[7,8] should help decrease the cost of the system. The rate of testing large batteries for electric vehicle applications both in the laboratory and in vehicles is expected to increase.

The Nickel-Iron Battery

Effort on this battery, some aspects of which have been discussed here, is primarily centered at the Westinghouse Research Laboratory.[34] There is also a related effort in progress at Eagle Picher, but this has not been described in detail in the available literature.

Recent Research and Development

Research and development on the nickel electrode reported earlier also apply to the nickel-iron battery. The referenced report gives no details on the method of fabrication of the iron electrode used at Westinghouse.[34] Its charge-discharge characteristics are shown in Figure 10-6. The discharge curve indicates that this proceeds in two stages, the first

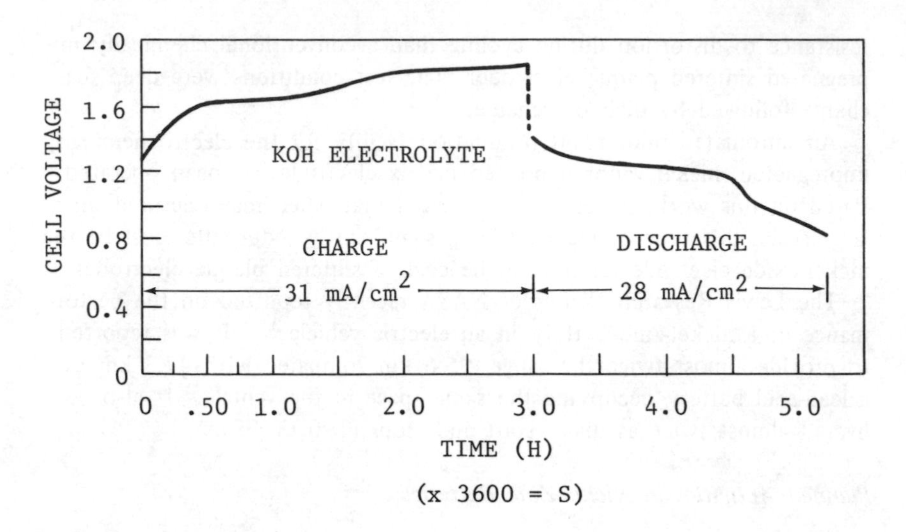

Figure 10-6. Charge-discharge characteristics of the new iron-nickel cell.[34]

to form $Fe(OH)_2$ and the second to form $Fe(OH)_3$. The performance of an 80-cell, 96.5-V, 16.5-kWh battery is shown in Figure 10-7. The circulating electrolyte battery was developed for an electric vehicle test battery. The advantages claimed for circulating the electrolyte during charge are:

- improved energy per unit volume;
- equal electrolyte concentration in each cell;
- proper liquid level in each cell;
- low maintenance time;
- faster recharge capability with thermal control; and
- safe handling of gases on charge.

The electrolyte circulation is cut off during discharge. The 16.5-kWh battery, with auxiliary hardware and tray, weighed 386 kg. Batteries based on this configuration are being tested in various electric vehicles.

Iron-Air Batteries

Buzzeli recently reported on the Westinghouse iron-air battery,[10] primarily considered for the electric vehicle application. The unit cell of the battery consists of two bifunctional, rechargeable air electrodes and one central iron electrode. The air electrode is made from carbon with teflon as the hydrophobic agent. The catalyst, 2 mg/cm^2, is tungsten carbide. No details of the method for iron electrode construction are included in the available literature. The iron electrode is designed to give a 4-hr

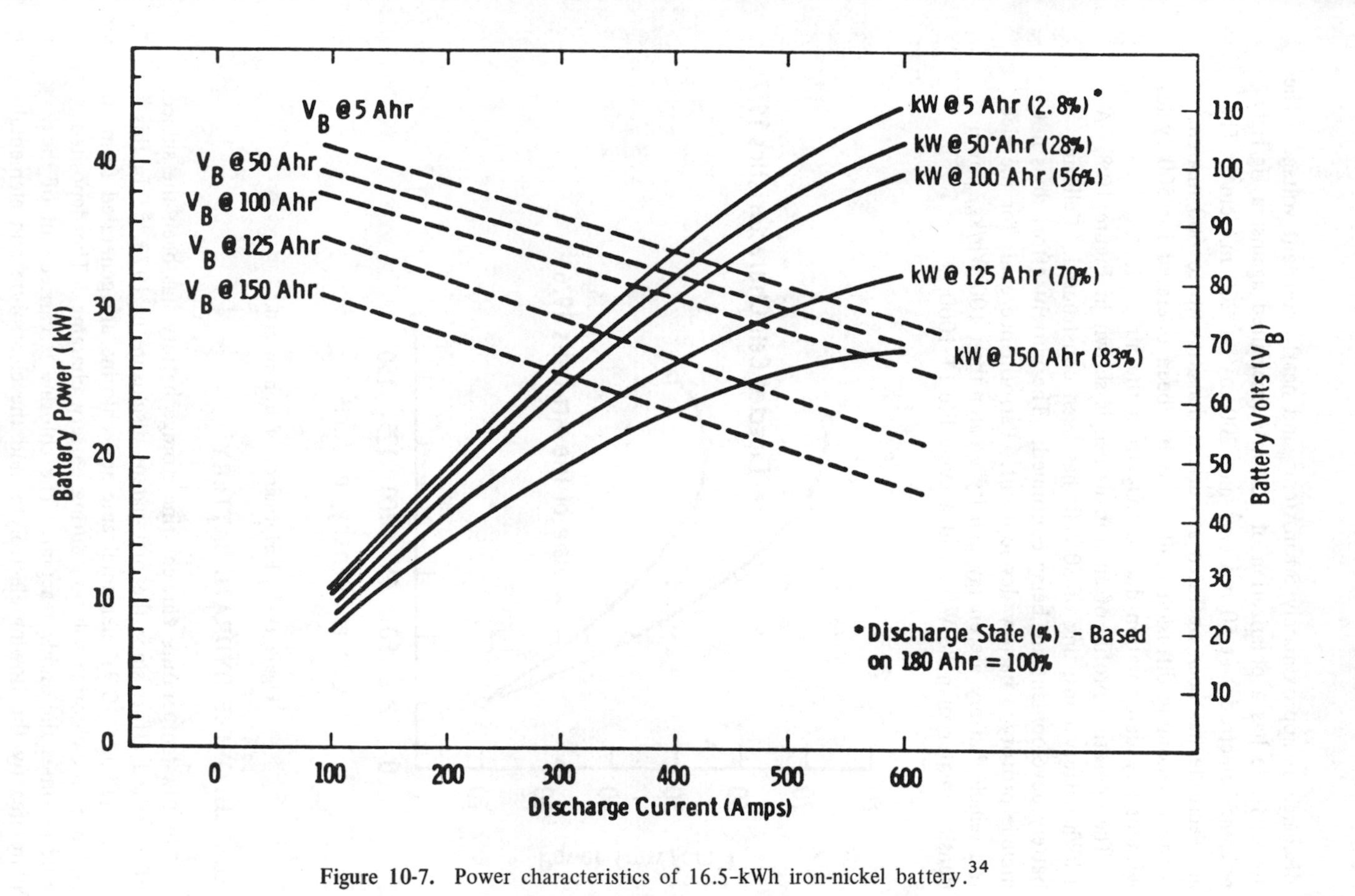

Figure 10-7. Power characteristics of 16.5-kWh iron-nickel battery.[34]

discharge at approximately 50 mA/cm^2 and nearly constant voltage. The air electrode has a polarization of 100 mV (measured against a Hg/HgO reference electrode) at 50 mA/cm^2 and 200 mV at 200 mA/cm^2. The air electrode has been operated at current densities up to 600 mA/cm^2 without becoming diffusion limited. It has been operated for 500 cycles without any deterioration due to CO_2 from the air.

The operating profile of an iron-air cell is shown in Figure 10-8. A 1-kWh battery consisting of 30 cells has been constructed. Cell and battery development are being continued. The current effort does not include prototype battery development. Performance goals for the electric vehicle battery are energy density greater than 100 Wh/kg, power density greater than 88 W/kg and a cycle life of 1000.

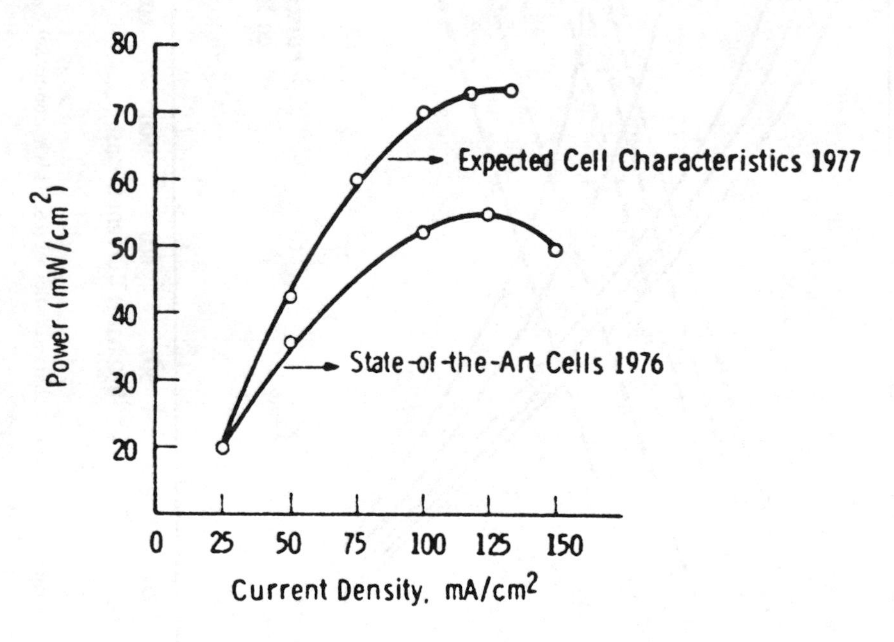

Figure 10-8. Performance of iron-air cell.[10]

ZINC-CHLORINE HYDRATE BATTERY

The low-temperature zinc-chlorine storage battery has been the subject of a limited effort since the first patent disclosure in 1973.[35] In this battery (Figure 10-9), elemental zinc and chlorine are generated from an aqueous zinc-chloride solution during battery charging. The deposited zinc becomes the anode reactant. The chlorine is carried out of the plate area by the flowing electrolyte, and the electrolyte and elemental

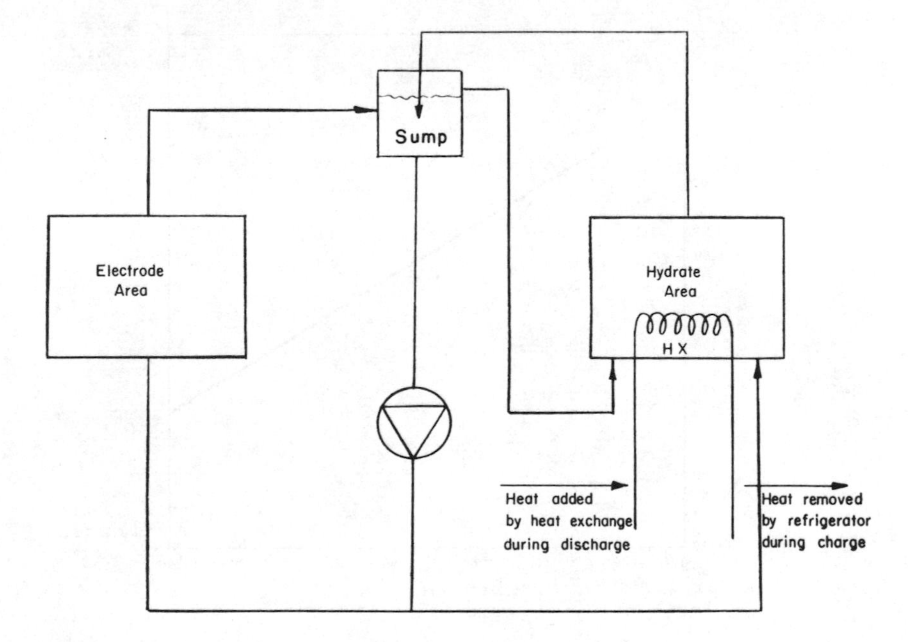

Figure 10-9. Schematic of a zinc-chlorine hydrate battery.[11]

chlorine are cooled outside of the battery with the formation of chlorine hydrate ($Cl_2.6H_2O$), a yellow solid. The chlorine hydrate is stored externally to the cell and the electrolyte recirculated. During discharge of the battery, chlorine is carried by the flowing electrolyte stream to the cathode. The end product of the cell reaction is zinc chloride. The voltage current curve for a single cell is shown in Figure 10-10.[11]

Both the zinc and chlorine electrodes operate with little polarization, as shown by the linear current voltage curve. The circulation of the electrolyte during charge should increase the cycle life of the zinc electrode as this inhibits dendrite formation during zinc deposition. A concern for this system is possible leakage of the chlorine-containing electrolyte in the case of faulty seals, pipe corrosion or accidental rupture.

As shown in Figure 10-9, the total battery system includes the cell, an electrolyte circulating system, a refrigeration system for separating out the chlorine hexahydrate during charge, a sump for storing the solid and a heat source for redissolving the chlorine during the charge cycle. The weight details for a 50 kWh battery are given in Table 10-11; this is 318 kg for the total system or 157 Wh/kg.

Work is continuing on this system to develop batteries applicable to utility load leveling and vehicle propulsion. Studies also are underway to evaluate alternative methods of handling the chlorine during charge and

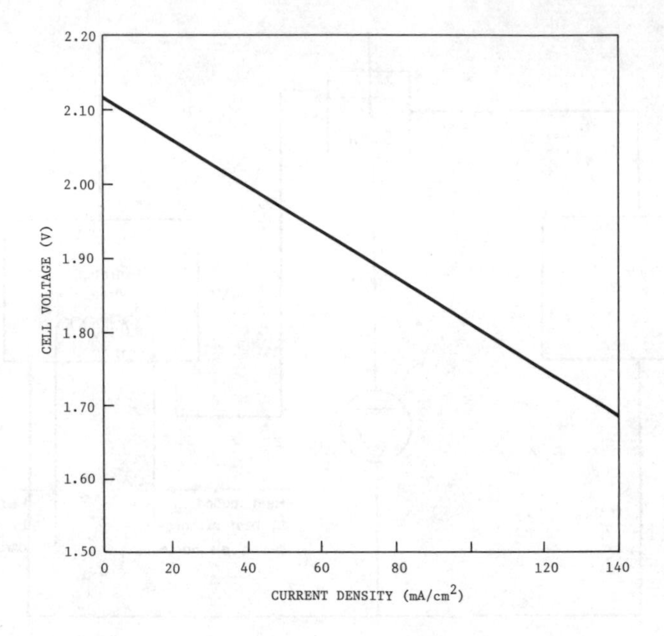

Figure 10-10. Discharge curve for a zinc-chlorine hydrate cell.[11]

discharge. During 1977, 100-V/100-kWh modules are to be built, and three of these modules assembled into a 100-V/300-kWh battery.

Table 10-11. Weight of Zinc-Chlorine Hydrate 50 kWh Battery[11]

	kg
50-kWh Battery Electrolyte, Electrodes, Frames and Hydrate Hx[a]	237
0.86 kWh for Pump	4
Pump and Motor	4
Sump and Main Plumbing	4
Refrigerator Compressor	14
Condenser and Fan	9
Insulation	2
Case	38
Brackets	4
Total	316

[a]Hx (heat exchanger).

SOLID ELECTROLYTE SECONDARY BATTERIES

Two different electrochemical systems that use solid sodium ion-conducting solid electrolytes are currently under investigation. The first is the sodium-sulfur couple with beta alumina[36] or glass[37] electrolyte. The other is the sodium-metal chloride[38] couple with beta alumina and sodium chloroaluminate ($NaAlCl_4$) electrolyte.

The schematic of a sodium-sulfur cell with ceramic electrolyte is given in Figure 10-11.[39] To maintain the sodium and sulfur, together with reaction products, in the liquid phase, the cell must be operated at above 270°C; the usual range of operation is 300-375°C. During discharge, sodium ion is transported through the sodium ion conducting ceramic. There it reacts with the sulfur electrode. The cathode process is quite complex–sodium polysulfides forming that are not soluble in the sodium leading to a two-liquid system in the cathode. A single liquid phase of sodium polysulfides forms when the discharge is approximately 60% complete. The ceramic electrolyte is a thin-walled cylindrical tube or a thin flat plate of beta or beta alumina. The first has a mole ratio of Al_2O_3: Na_2O of between 9 and 11; the second has a ratio of approximately 5.

The chemistry of the glass fiber electrolyte cell is the same as that of the beta alumina cells. The major difference is the use of very thin-wall sodium-conducting hollow glass fibers for the electrolyte. The configuration of this cell is shown in Figure 10-12.[15]

The sodium metal halide cell is shown schematically in Figure 10-13.[40] This cell operates at a lower temperature, 180-250°C, than the sodium-sulfur cell. For the system illustrated, the products of the discharge reaction are sodium chloride and antimony. On charge, these revert to sodium and antimony trichloride.

Problem Areas

For all these systems, the electrolyte is the most critical component. In the case of the beta alumina electrolytes, the production of a reproducible long-lived ceramic body has been found to be sensitive to a number of variables:[16]

- composition;
- surface area;
- powder preparation (synthetic powders);
- additives (impurities);
- fluxes and binders;
- powder fabrication (pressing, extrusion, electrophoretic deposition);
- conditions of powder fabrication; and
- sintering conditions.

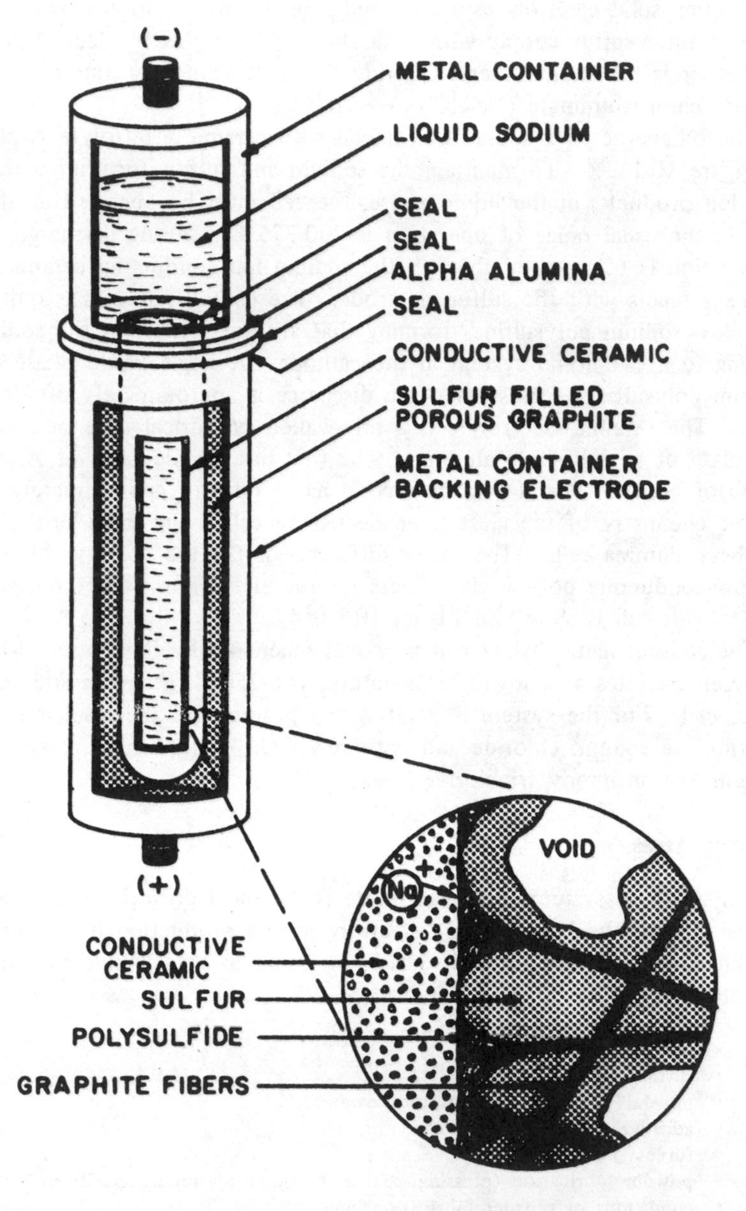

Figure 10-11. Schematic of the Ford sodium-sulfur cell.[39]

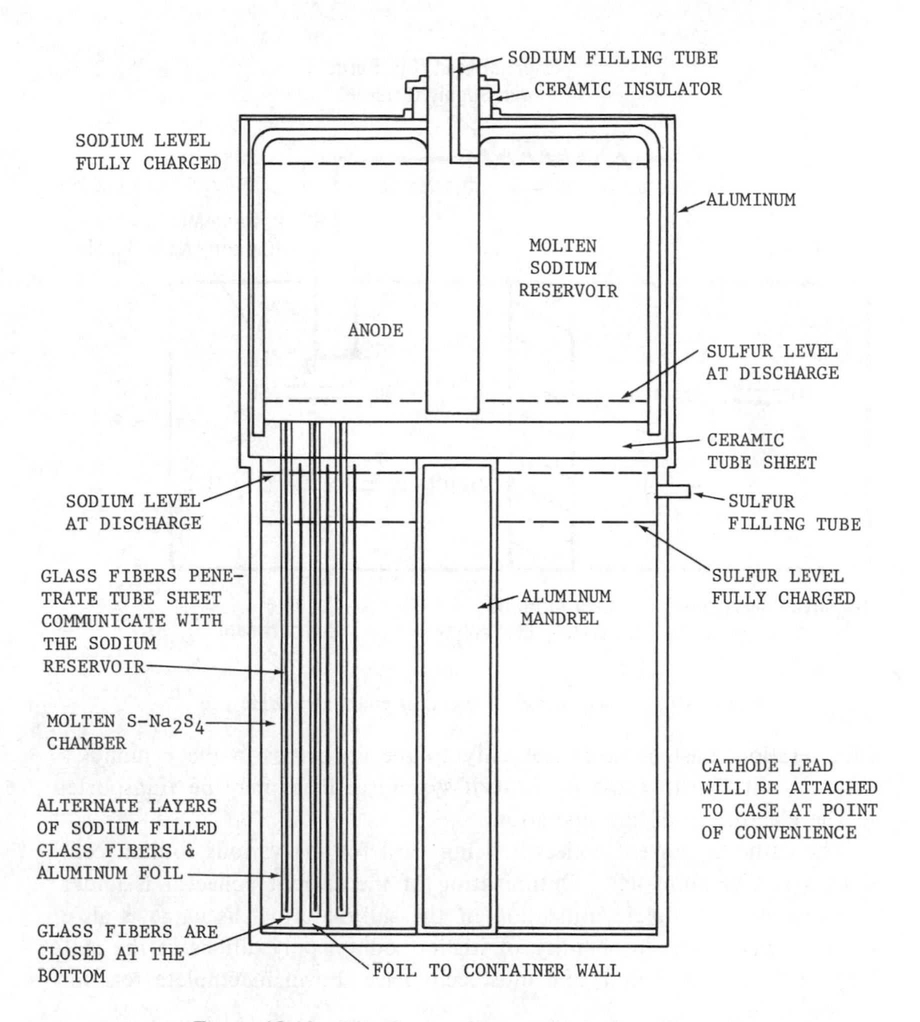

Figure 10-12. The Dow sodium-sulfur hollow-fiber cell.

There is a tendency for the sodium to accumulate in grain boundaries of the ceramic electrolyte, causing it to weaken. There also is a tendency for the electrolyte to lose sodium as a result of prolonged heating—this decreases its conductivity.

Doped as well as pure ceramic electrolyte bodies are being investigated. Most of this effort has centered on MgO as the dopant. There are conflicting claims as to the relative merits of high-purity ceramic electrolyte bodies vis-a-vis doped systems. The beta aluminas are generally moisture sensitive, so that it is necessary to process and handle them in an atmosphere with very low humidity or in a dry box. The lifetime of the beta aluminas has been found to be sensitive to various impurities; thus,

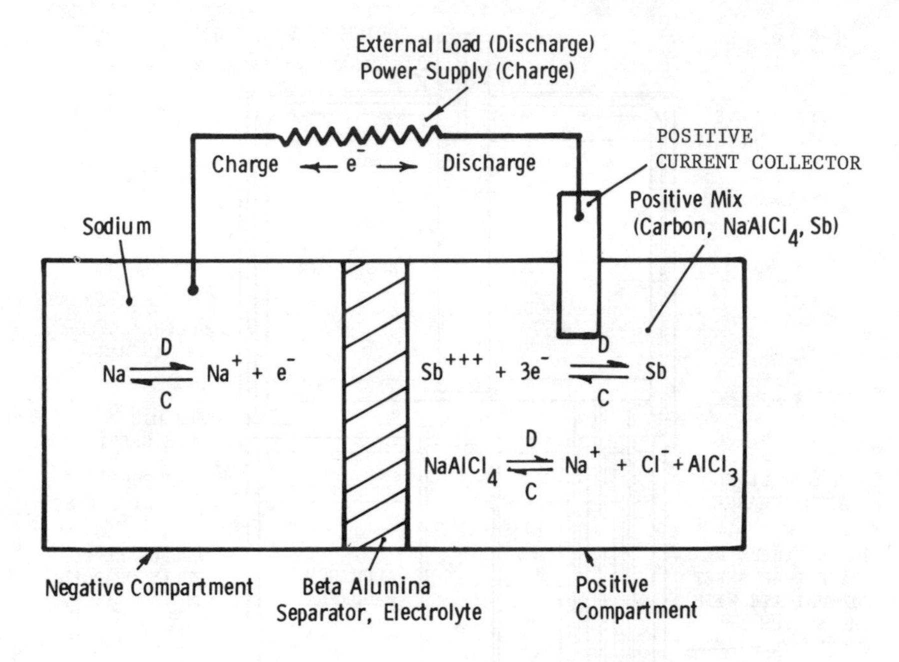

Figure 10-13. Schematic of the ESB sodium-chloride cell.[40]

consideration must be given not only to the impurities in the ceramic material but also to those in the cell system as these may be transported to the electrolyte during operation.

The cathodic current collector being used for the various sodium cell designs is a graphite felt. Optimization of the current collector is under investigation. Complete utilization of the sulfur during discharge is also being studied. The insolubility of higher sodium polysulfides in the sulfur leads to phase separation, and most cells have shown incomplete reaction of the sulfur.

There also is the question of the compatibility of the sulfur with the container. According to Yao and Birk,[16] no organization in the United States had demonstrated the suitability of an economic material as a cathode current collector and container. Work has been conducted on containers coated with carbon, metal, metal sulfide, or oxide and semiconductor materials. Their acceptability for long-lived batteries is yet to be demonstrated.

Another problem is the seals between the ceramic electrolyte and the container and between the electrolyte and cathodic current collector. These both must be resistant to attack by the sulfur and sodium and be able to withstand temperature cycling. In the case of the ESB cell, sealants are less of a problem because of their operability at lower temperatures—as low at 180°C.[38]

In the case of the Dow cell, long-lived glass electrolytes are required. Although the glasses used to date have good cycling life at low depth of discharge (10-25% of capacity), their cycling life at deep depths of discharge (80-100% of capacity) is much shorter. Failure of the glass fibers also occurs at the seals with the tube sheet (Figure 10-12).

A major problem area for the ESB chloride system also is the b″ alumina electrolyte and its reproducibility. The seal problems also have not been completely solved. At the high range of the operating temperature, 250°C, the vapor pressure of the cathode mix is approximately 15 atm. This requires a heavy container if this operating temperature is used. Seal problems also become more difficult at these temperatures.

Recent Research

Electrolyte Research

Hong reported recently on studies of ionic conduction in solids.[14] A general structural relationship, which explains some of the geometric and transport criteria for high ionic conductivity in solids, has been developed. Based on these concepts, various structures have been prepared and studied.[41] Compounds of the type $Na_{1+X}Zr_2Si_XP_{3-X}O_{12}$, where $0<X<3$, were included. All these compounds can be reversibly ion exchanged with Li^+, Ag^+ and K^+ in molten salts. At 300°C, the compound with $x = 2$ has a resistivity of 3 ohm-cm, approximately the same as the best of the beta″ aluminas. Some other characteristics of $Na_3Zr_2Si_2PO_{12}$(I) related to its potential as an alternate solid electrolyte are:

- I does not become hydrated so that it should not have the moisture sensitivity of beta alumina;
- In beta alumina, the Na^+ ions are constrained to two-dimensional motion: in I they move in three dimensions;
- The thermal expansion of I is isotropic, whereas for beta aluminas it is anisotropic; and
- Both compositions are stable in sodium.

I can be processed into a ceramic body at approximately 1200°C, compared with 1500°C or higher required for beta aluminas. To date the use of these compositions in electrochemical cells has not been reported.

The mechanism of sodium ion transport and some of the factors that limit conductivity and stability in beta aluminas have been the subject of a cooperative effort between the General Electric Co.,[13] the State University of New York in Albany and the Brookhaven National

Laboratory. Based on nuclear magnetic resonance studies, it has been suggested that the mechanism for sodium transport is thermally activated, hopping between the three different sites for sodium in beta alumina. Above 120°C, all sodium ions participate in the motion.

Neutron diffraction studies have shown large concentrations of Frenkel defects in the spinel block structure of beta alumina and a formula for the defect derived. This shows the presence of interstitial oxygen ions in the conduction plane, occupying positions that can block sodium ion transport in some of the channels. These studies also have shown that when magnesium ions are added to stabilize beta″ alumina, it substitutes in only one of the four types of aluminum sites. This ion substitution appears to reduce local strain, leading to stabilization.

As a guide to improved ceramic electrolytes, the General Electric Co. (GE) developed a model of the electric properties of the ceramic electrolyte.[42] Based on this model, it was concluded that the electric fields at the grain boundaries must be large, about 10^5 V/cm. This leads to impurity concentration at the grain boundaries, resulting in the elimination at GE of additives that had been used to improve the densification of the ceramic bodies.[43]

The Sulfur Electrode

The Ford effort on the sodium-sulfur battery has included an extension of earlier studies on the sulfur electrode. It had been shown by Cleaver *et al.*[44] that in a potassium-thiocyanate melt, at low concentrations the sulfur/polysulfide electrode undergoes a one electron change. The more recent Ford studies[45] in the same melt show that the reaction becomes much more complex at high concentrations of the reactants. Unlike the earlier observations, these showed that the sulfur deposition and stripping reactions are dependent on potential sweep rate. Based on the shape and position of the reaction peaks on anodic and cathodic potential sweeps, it has been concluded that the high concentration reaction is a two-electron process. In addition, evidence for the participation of chemical rather than purely electrochemical reactions between the sulfur and the polysulfide ion has been obtained. The formation of sulfur on the electrode surface during charge, which reacts with the polysulfide ion, has been suggested as an important step in controlling the rate of charge of the battery.

Work at Ford also has shown that in nonmetallic systems, the addition of 0.1% FeS to the sulfur electrode improves its performance;[45] an increase in capacity was noted in cells to which the addition was made. The evidence indicated that this improvement was due to an increase in electrode wetting, due to the presence of the metal sulfide.

Recent Developments

Ceramic Electrolyte Fabrication

Research on new electrolytes and the relationship between composition and electrolyte preparation were discussed in the section on electrolyte research. Both General Electric and the Utah-Ford efforts have been concerned with ceramic fabrication. The three main features of the General Electric process are:[13]

- use of a commercial alumina, rather than a high-purity alumina, as the powder source;
- forming by electrophoretic deposition; and
- sintering by the stoker or pass-through method.

The beta alumina used is purchased from Alcoa (the Aluminum Company of America). In the electrophoretic deposition process the powder particles are placed in suspension in n-amyl alcohol.[46] The particles are given a negative charge concurrent with milling of the powder to reduce the particle size to about one micron. The material is deposited on a mandrel, using fields of 700 V/cm. The soda (Na_2O) concentration in the final ceramic is 8.4-9.6%.[13]

After forming, the green deposits are removed from the mandrels and are sintered in an oxidizing atmosphere near 1700°C. Using the stoker method, the sintering furnace is preheated to the sintering temperature. A train of sintering boats containing the green ware is passed through the furnace, enabling continuous production of the electrolyte tubes. The furnace atmosphere is dry oxygen. The resulting tubes have not shown degradation after hundreds of deep cycles. Tubes of 1- and 2-cm in diameter and 15-cm long have been fabricated.

The process developed by Utah[12] and Ford[45] involves pressing the powder mixture isostatically to form the green body. Since it was observed that low-temperature sintering processes did not form the fully converted beta″ alumina, a seeding technique (addition of beta″ alumina crystals to the powder) was developed. By this means, a 2-5 minute sintering time at 1600°C has been shown to produce a high-density beta″ alumina. Sodium evaporation during sintering was eliminated by encapsulation of the green body in beta or beta″ aluminas. The encapsulant can be reused. These ceramic electrolytes contained 8.7-8.8% Na_2O and 0.7% Li_2O, the remainder being Al_2O_3.

Sodium-Sulfur Cell Development

Given the ceramic body, the major concerns in cell fabrication are cathodes that operate into the two-phase region, permitting high utilization of the cathode materials and seals. The groups at both Ford[45] and General Electric[13] have studied the effect of voids in the graphite felt cathode current collector on cell performance. The results at Ford indicated that with complete contact between the graphite felt and the ceramic body, the two-phase reaction did not extend to the active material contained in the voids. However, when this contact was discontinuous, the two-phase reaction extended into the voids. This increase in reactivity was attributed to the maintenance of a parallel ionic path. It also was noted that the addition of FeS to the sulfur was more effective in increasing the cyclic capacity in cells with cathodes having uncovered ceramic electrolyte areas than in those having complete contact.

The group at General Electric has also studied various cathode fabrication techniques and designs.[13] The preferred fabrication method is the production of a continuous fiber-bonded electrode. From four to six holes are drilled into the cylindrical carbon-sulfur cathode body. This has led to cells that utilize up to 90% of the cell capacity. This improvement is attributed to a gradient of the electronic resistance inside the sulfur electrode. The resistance increases toward the beta alumina surface and prevents the preferential formation of sulfur layers close to the beta alumina during charging.

The development of improved low-cost cathode containment materials is also being investigated under both these programs. Molybdenum has been found to be satisfactory but is too expensive for use in cells for mass production. Table 10-12 gives the results of corrosion tests on various materials conducted at Ford.[45] The corrosion of low-cost metals has led to attempts to use coatings to permit their use. At GE phenolic resin coatings proved effective on aluminum, but their conductivity must be improved before they can be considered as a possible solution.[39] Molybdenum coatings to protect less-expensive metals were unsatisfactory.

Studies at the University of Utah have centered on electronic conducting liners for the metal container.[12] These materials are fabricated into metal-clad cylinders. Several electronic conducting ceramics have shown excellent resistance to molten sodium polysulfide. The most promising of those examined is polycrystalline rutile doped with 1% tantalum cation. This material, after sintering at 1400°C for three hours, has an electrical conductivity at 350°C of 5.5 $(\text{ohm-cm})^{-1}$. It also possesses excellent thermal shock resistance. It is planned to examine this material in containers having an external wall of nickel or aluminum. Both these metals

Table 10-12. Materials Screened in Sodium Tetrasulfide at 400°C[45]

Group I	**Extensively Corroded Materials** TaB_2, ZrC, VN, VC, NbB_2, $ZrSi_2$, $TiSi_2$, CrB_2, ZrN, CrC, $CaTiO_3$ (+ 0.3% Fe_2O_3), anodized Ti, Ti
Group II	**Corrosion Protection from Sulfide or Oxide Film Formation** Zr, $CrSi_2$, TiC, Zr B_2, TiN, Si, TaC, TaN, AISI 446 stainless steel, Inconel 600 and 601, Mo
Group III	**Intrinsically Corrosion-Resistant Materials** Cr_2O_3, MoS_2, ZrO_2, $La_{0.84}Sr_{0.16}CrO_3$, TiO_2, single crystal and polycrystalline doped with Ta); $SrTiO_3$, $CaTiO_3$ (+ 3.0% Fe_2O_3) Polyphenylene (Pph), Polyphenylene-graphite composites, oxidized stainless steels, NiO

have been shown not to degrade mechanically or electrically in air temperatures of 500°C.

A successful laboratory cell with a mechanical seal has been developed by GE.[43] The use of the mechanical seal permits flexibility in the use of the materials for the sulfur containment compartment and has been used to investigate their behavior. The use of glass ceramic seals for the beta″ electrolyte containing Li_2O has been investigated at Ford, involving both sealing techniques and sealing materials.[45] The latter has included the development of new glasses. Satisfactory seals of alpha to beta″ alumina have been achieved with a sodium-calcium aluminoborosilicate glass.

Table 10-13 gives a summary of the advances made in cycle life of cells up to mid-March 1976. The marked improvement of cycle life together with increased utilization of active material shows the considerable progress that has been made in single-cell performance. Ford[45] has reported that its oldest laboratory cell had, as of the end of June 1976, passed over 1600 Ah/cm^2 in discharge and completed 9000 cycles. However, data on cells utilizing improved ceramics and cathode construction comparable to that in Table 10-13 was not included in the referenced report.

The improvements in ceramic electrolyte reproducibility and cell performance has led GE to undertake the design and development of cells containing several ceramic electrolyte tubes.[13] As many as 45 tubular electrolyte bodies have been built into a single alpha alumina header. Computer programs for calculating the performance of alternate cell and battery designs have been developed. In addition, failure analysis and the effect of various types of failures on battery performance and safety are under investigation.

Table 10-13. Comparison of Past and Present Performance of GE Sodium-Sulfur Cells[13]

	No. of Cells	Current Density in mA/cm^2	No. of Cells with more than 50 Ah/cm^2	No. of Cells with more than 100 Ah/cm^2	No. of Cells with more than 200 Ah/cm^2	Best Life	
						Ah/cm^2	No. of Cycles
August 9, 1973 to March 15, 1975	104	50	21	3	0	120	100
		50	21	4	2 2 still cycling	May 12, 1976 288 still cycling	260
April 29, 1975 to March 17, 1976	112						
		100	16	10	5 4 still cycling	May 12, 1975 377 still cycling	647

It should be noted that the attempt by TRW to develop flat plate cells was not successful.[47] Much of the difficulty was in obtaining satisfactory seals, and this effort has been discontinued.

The efforts at GE have recently been expanded and tests of multiple electrolyte tube cells are to be undertaken. The work will continue to emphasize increase in battery life through improvement of components, particularly seals. The longer range objective is to utilize the cells in 100-kWh module batteries. By 1981, it is hoped to test 50 such modules in a 5-MWh system at the Battery Energy Storage Test (BEST) Facility.

The studies at Ford are to continue evaluating various components and cell fabrication technology. Current results on a selected cell and program goals are shown in Table 10-14. The work at Utah on ceramic body fabrication is to be scaled-up to prepilot plant and pilot plant production of various-sized ceramic electrolyte tubes.

Table 10-14. Test Cell Results and Goal of the Ford Program[45]

Variable	Goal	Results (Cell E17)
Power Density, W/cm^2	0.7	0.4
Energy Density, Wh/cm^2	0.2	0.2
Utilization of Reagents, %	25	30
Electrical Efficiency, %	70	62
Capacity, Ah/cm^2	0.1	0.1
Discharge Time, hr	0.3	0.6
Durability, Ah/cm^2	250	1060
Cycle Life	1000	6400

Glass Electrolyte Cell Status

The interaction between the electrolyte fibers and the "tube sheet" that holds them in place is critical to this cell.[37] This tube sheet is made by applying a paste of powdered low-melting glass. The liquid of the paste is evaporated, and the powdered glass is fused to a solid disc. The low-melting glass is from the $B_2O_3:Na_2O$ system. The glasses used contain from 4-8% Na_2O.

The fibers are made from a borate glass which contains small amounts of additives. Excess sodium ion is added in the form of NaCl to give improved conductivity. For improved lifetime, the trend is toward walls thicker than 10 μ. The increase to 15 μ extended the lifetime by a factor of four. Although excellent life-times were achieved in cells with low depth of discharge, deep discharge markedly shortened the cycle life. These failures occurred immediately below the tube sheet, at scratches in the fiber, or at the sealed ends of the fiber. Other studies have shown that the presence of oxygen or calcium in the sodium also markedly decreased fiber life. These observations have led to improved fiber production, avoiding scratches and other defects. Calcium in the sodium and magnesium in the aluminum foil cathode current collectors caused an increase in cell resistance with cycling. These effects can be avoided by the use of calcium-free sodium and magnesium-free aluminum.

Aluminum has been used for the cathode container. However, because of the possibility of high cell temperatures (600-700°C) that can lead to aluminum foil failure, stainless steel has been examined and 316 stainless steel did not adversely affect cell performance. Figure 10-14 shows the charge-discharge performance of a 3000-fiber 5-Ah cell. A 40-Ah cell with molybdenum-coated aluminum foil in the cathode has also been operated.

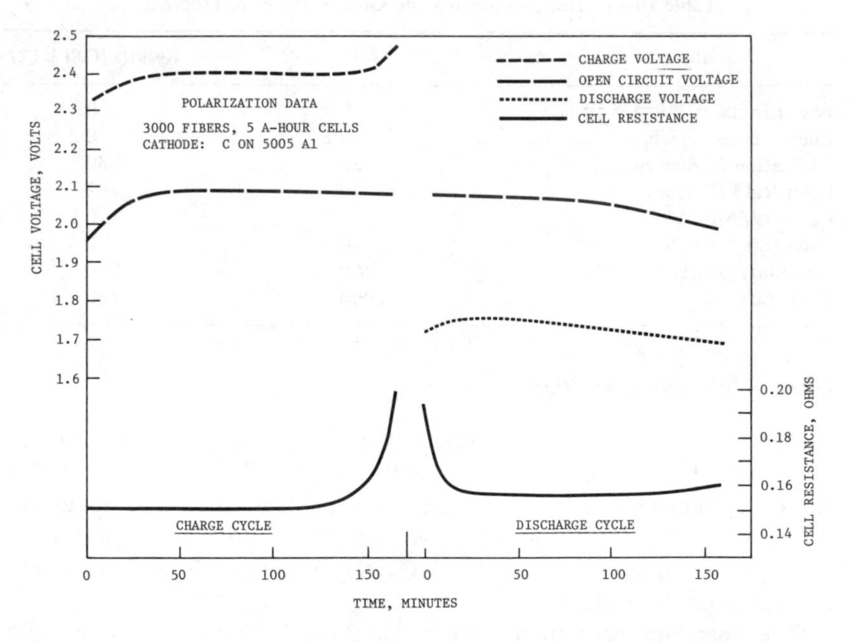

Figure 10-14. Performance of a glass fiber cell.[15]

Over 1400 deep charge-discharge cycles have been achieved on the 0.5-Ah cell. The larger 5-Ah cell has been cycled 190 times with no change in characteristics. At lesser depths of discharge, longer lifetimes are achieved. A 0.5-Ah cell operated at 25% depth of discharge showed a cycle life of over 1800 in 82 days of operation.

The estimated cost for the production of the Dow glass fiber cell is approximately $23/kWh for bare cell production. This is based on an annual production of 10^6 kWh/yr. Several areas for possible cost reduction, especially in labor costs, have been indicated.

Future directions under this program have not been defined. It is anticipated that efforts will continue on fiber technology with emphasis on avoiding flaws in the fibers. Studies on the effects of impurities on performance, cell materials and cell fabrication will probably continue. As these technologies are improved, more extensive cell testing and cell design studies will be incorporated into the program.

The Sodium-Chloride Battery

In addition to the use of antimony III chloride as the cathodic material, copper II, iron III and nickel II chlorides have been examined.[38] However, antimony chloride is preferred. Sulfur, to be discussed below, also has been used. Since no solid-state reactions are involved, high cathodic efficiency is not a problem.

Two types of cells have been operated—one with a flat disc electrolyte and the other with cylindrical electrolyte tubes. In the case of the latter, the $SbCl_3$ and $NaAlCl_4$ are inside the ceramic body, and the sodium is external to it. A mild steel container suffices for retaining the sodium and as the anode current collector. To obtain conductivity at the cathode, graphite powder is added. Tungsten and molybdenum have been used as positive current collectors. The end of the current collector has a brush-like structure to maximize contact with the cathode mixture. The cell has an open circuit voltage of slightly above 3.0 V.

To date, the problem of seals, a major one in the case of the sodium-sulfur battery, has been met with polymeric materials. A properly cured silicone rubber makes an excellent liquid-tight seal for sodium at 200°C. Prototype seals have been tested in sodium for more than 13,500 hr without leaks or failures. Teflon has been used for the cathode seal. It is satisfactory when kept out of contact with the cathode mix. As alternatives to Teflon, glass to ceramic seals are being investigated.

The electrolyte used for this work is a beta alumina with the composition 90% Al_2O_3, 80% Na_2O and 2.0% MgO. It is isostatically pressed and then sintered. The conductivity of this material is less than that of the lithiated ceramic being investigated under the Ford-Utah program. Thus, the substitution of the latter may lead to improved cell performance.

Both disc and tubular electrolyte cells have been tested. Figure 10-15 shows the current-voltage relationship for this cell. Its discharge to charge efficiency is greater than 80%. Cells have undergone 640 cycles

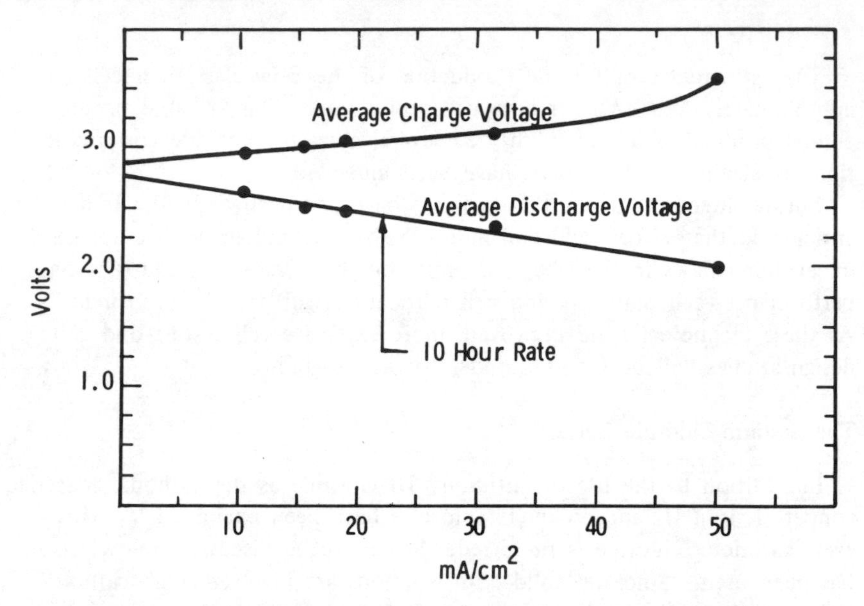

Figure 10-15. Performance of a disc electrolyte $Na/SbCl_3$ cell.[40]

at 100% depth of discharge. This corresponds to 120 Ah/cm^2 of beta alumina in one direction. The testing of tubular cells has been less extensive. These have had a design capacity of 5.3 Ah. The cells had an average discharge voltage of 2.5 and a discharge charge efficiency of over 80%. Up to 160 cycles and 15 Ah/cm^2 of beta alumina were achieved in the longest-lived cell, reported in December 1975. Some cells failed due to cracking of the beta alumina ceramic. Loss of capacity was observed in other cells and was attributed to: (1) dewetting of sodium at the beta alumina surface; (2) positive mix separation; and (3) poor current collector geometry. These difficulties are reported to have been overcome in later cell designs.

Sulfur has also been examined as an alternative to antimony chloride for the cathode reactant. With sulfur the postulated overall reaction is

$$6Na + 2AlCl_3 + 3S \rightarrow 6NaCl + Al_2S_3$$

This has an open circuit voltage of 2.67 compared with 2.08 for the Ford-type cell. These show similar charge-discharge efficiencies to the antimony chloride cells; however, they showed greater degradation in performance during cycling.

A cost optimization for various cell designs based on the use of $SbCl_3$ has been conducted. The following configuration gave the lowest design cost:

Tube height:	35 cm
Tube OD:	4.65 cm
Tube ID:	4.25 cm
Mix:	0.5 cm
Current collector:	Molybdenum
kWh/tube:	0.230
Discharge-charge efficiency:	85% (at 10-hr rate)

This configuration resulted in an estimated manufacturing cost of $20/kWh (1975 dollars) and is optimum for a charging cost of 15 mils/kWh.

Research and development on this cell is continuing under ESB, EPRI and utility funding. A plan aimed at the construction of a commercial load-leveling battery has been formulated.

LITHIUM-SULFUR SYSTEMS–GENERAL CHARACTERISTICS

A program to develop lithium-sulfur cells was initiated in 1968 at the Argonne National Laboratory (ANL). Other groups followed, and Table 10-15 shows the preferred approaches to this system.[16,48] These cells

Table 10-15. Options for the Lithium-Sulfide Systems

Component	AI[16]	ANL[16]	GM[48]
Positive Electrode	FeS_2	FeS_2, FeS[a]	FeS_2[b]
Separator	BeO and Al_2O_3 (rigid)	BN (flexible)	BN cloth
Negative Electrode	Li-Si	Li-Al	Li-Al, Li-Si
Electrolyte	LiCl-KCl	LiCl-KCl	LiCl-KCl
Positive Current Collector	Carbon felt	Molybdenum	Graphite and molybdenum
Positive Lead	Molybdenum	Molybdenum	Molybdenum
Design Electrode	Prismatic	Prismatic	Horizontal
Orientation (present laboratory cell size)	(100 Wh)	(200 Wh)	
Feedthrough	Compression	Compression	Compression
Case	SS (floating potential)	Iron (negative electrode potential	Stainless steel

[a]Additives of Cu_2S, CoS_2, and Co_2S_3 are also utilized.
[b]CoS_2 additive in some cells.
AI = Atomics International Division, Rockwell International.
ANL = Argonne National Laboratory.
GM = General Motors.

operate with an electrolyte of a LiCl-KCl eutectic at a temperature of 375-475°C. Originally, sulfur was used as the positive reactant; however, its high vapor pressure and the solubility of reaction intermediates (lithium polysulfides and hypersulfide) eliminated this approach. As alternative positive active materials, various metal sulfides have been evaluated and currently primary consideration is being given to FeS and FeS_2. Cells with the latter have a 20-25% greater specific energy than those with FeS-positive electrodes. To increase the cycle life of the anode, aluminum-lithium[2] and silicon-lithium anodes are under investigation in place of metallic lithium. These changes decrease the theoretical voltage and increase the weight of material per unit of energy output.

The schematic of an Al-Li/FeS_2 cell is shown in Figure 10-16.[48] The discharge curve for a cell with an FeS_2 cathode is shown in Figure 10-17. It is to be noted that the charge and discharge curves show three plateaus, and the discharge takes place over a wide voltage range; for the discharge

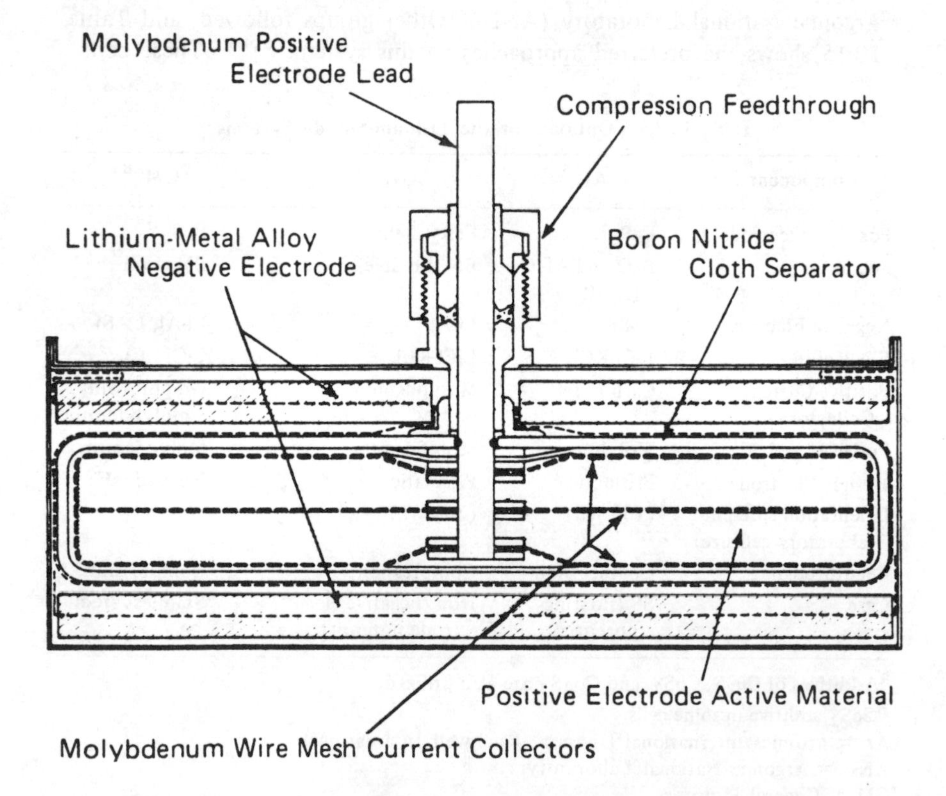

Figure 10-16. Schematic of a lithium aluminum/iron sulfide cell.[20]

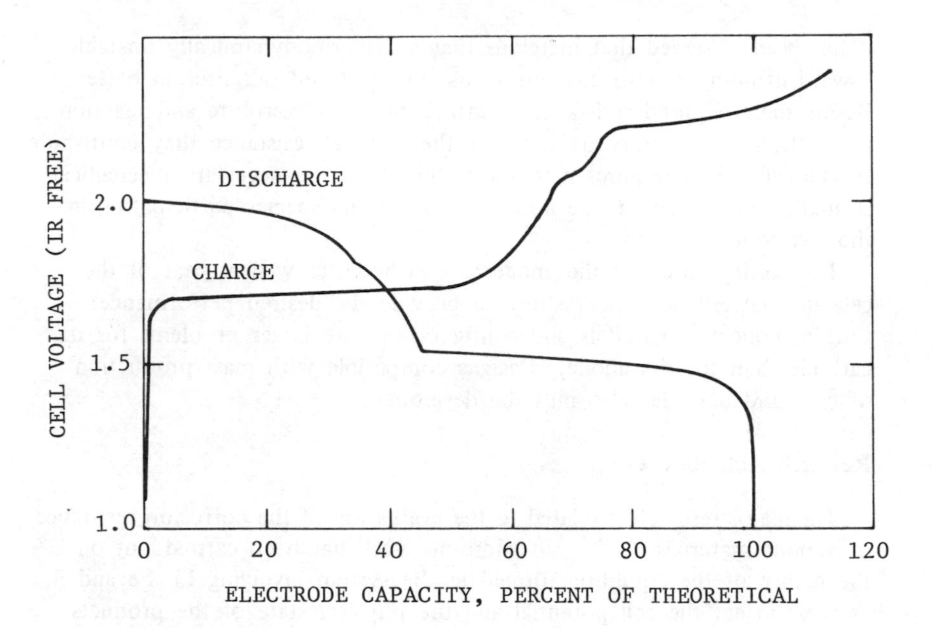

Figure 10-17. Voltage-capacity curve for a Li/FeS_2 cell.[48]

this is from 2.0 to 1.5 V. Good cycle life has been demonstrated for the metal-sulfide electrode. By 1975, ANL had achieved greater than 150 Wh/kg at the 10-hr rate in 200 Wh compact-sealed cells.[16] Performance goals for the lithium alloy-metal sulfide systems are given in Table 10-2.[2]

Problem Areas for the Li/S Cell

As in the other high-temperature cells, materials are the major problem for these systems. As in the case of the Na/S system, these are more severe at thc cathode than at the anode. Although satisfactory metal current collectors have been found for FeS cathodes, this electrode undergoes swelling during recycling. The cause of this swelling and methods for its mitigation are under study. By contrast, the greater reactivity of FeS_2 has made the solution of the current collector problem more difficult for this cathode. Although molybdenum has been found to be satisfactory, less expensive current collector materials or designs that make the cost of the molybdenum used compatible with program cost objectives must be developed.

The separator represents another important materials problem. Several ceramic materials such as aluminum nitride, boron nitride, alumina, beryllia and yttria are being examined as possible separator materials. In general, it has been observed that materials that are thermodynamically unstable toward lithium are also unstable in use in the fused salt lithium batteries. Beside their required resistance to attack by the electrolyte and reaction products, it is necessary to minimize the internal resistance they contribute to the cell. This requires membranes that have the necessary mechanical strength and porosity to permit transport of the species participating in the electrode reactions.

The configuration of the anode and cathode, as well as that of the cell and battery, are under study to provide the desired performance. Current collector materials and configurations are larger problems for the cathode than for the anode. Designs compatible with mass production of cells and batteries also must be developed.

Research Activities

The major research is related to the evaluation of the corrosion resistance of various materials.[49-51] In addition, work has been carried out on the nature of the equilibria formed in the system involving Li, Fe and S, as these affect the cell potential and the physical state of the products of the reaction. Sharma has recently reported on the phase diagram of the lithium sulfide-ferrous sulfide system (Figure 10-18). Two of its main features are: (1) a compound, $Li_2S.FeS$ melting over the range $885 \pm 5°C$ to Li_2S and a 50.5 mol % FeS melt was found; and (2) a eutectic between $Li_2S.FeS$ and FeS at 63 mol % FeS, and melting at $858 \pm 3°C$ was observed. X-ray studies on products of the interactions of FeS and FeS_2 with Li_2S have shown the existence of compounds such as $Li_2S.FeS$, $2\ Li_2S.Fe_3S_4$ and $Li_2S.Fe_2S_3$. Their relationship to the discharge process has not been completely defined.

An improved, but still tentative, phase diagram for the lithium-aluminum system has been determined at ANL as published diagrams were considered to be in error. This revised diagram is shown in Figure 10-19. A major difference is in the region of the 335°C isotherm. This corresponds to the formation of Li_9Al_4.

The swelling of the FeS electrode during cell operation has been attributed to the formation of $K_2Fe_7S_8$.[51] In related experiments using $FeS_{1.5}$ rather than FeS as the cathodic reactant, Sudar and co-workers examined cell performance using an electrolyte with only lithium cations.[49] The electrode showed a higher specific capacity in the mixture of lithium salts than in the KCl-LiCl eutectic. Based on these studies, it was concluded that potassium ions have an adverse effect on reactant utilization.

Sudar and co-workers[49] also have determined the IR-free overpotential of the FeS electrode as a function of the state of charge. These

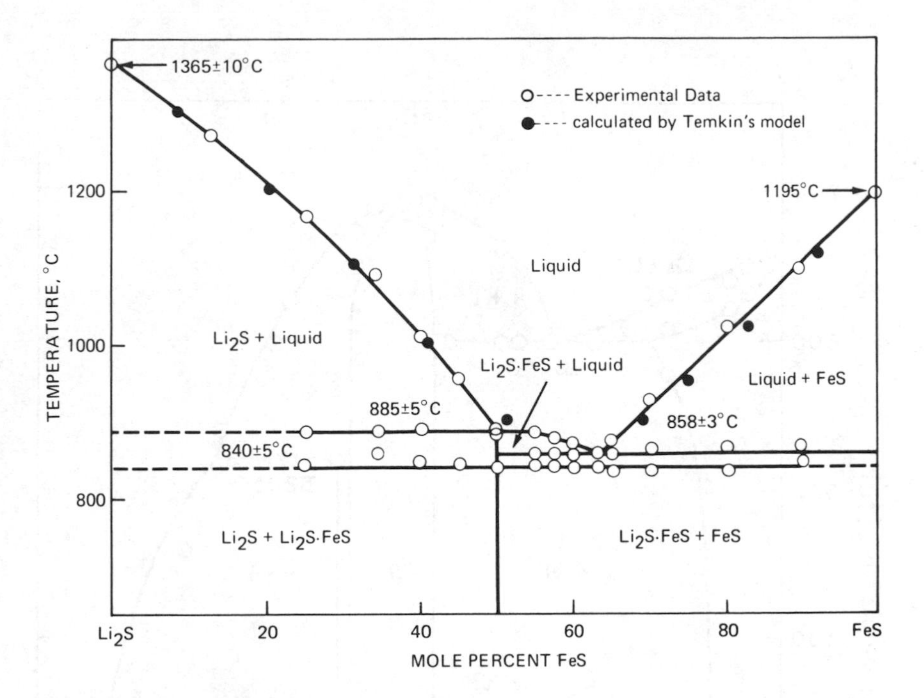

Figure 10-18. Li_2S-FeS phase diagram.[50]

observations, together with voltage relaxation studies, suggest that the electrode mechanism varies to some extent with current density. Based on the observed relatively long voltage relaxation times, diffusion appears to play an important role in the electrode mechanism.

Cairns[18] has summarized the results of various studies on ceramic and metallic materials for Li/FeS or FeS_2 cells. Table 10-16 shows the best ceramics for use as feedthroughs for the cathode and as separators. In related studies at ANL, it was reported that Y_2O_3 felt, prepared by impregnating a rayon felt with YCl_3 and converting this to fibers of Y_2O_3 by controlled pyrolysis, had the greatest resistance to attack of the separator materials studied.[2] The investigation also included Y_2O_3 with 10% asbestos and BN felts. The resistance of Y_2O_3 to attack by lithium in the presence of KCl-LiCl eutectic also has been confirmed by the work at Rockwell International.[49]

The summary of the studies on compatibility of materials for the positive electrode (Table 10-17) shows the difficulty in obtaining low-cost materials for cathode current collectors and containers. This problem is similar to that of current collectors for the sodium-sulfur cell. In related investigations at ANL, several materials exhibited suitable corrosion

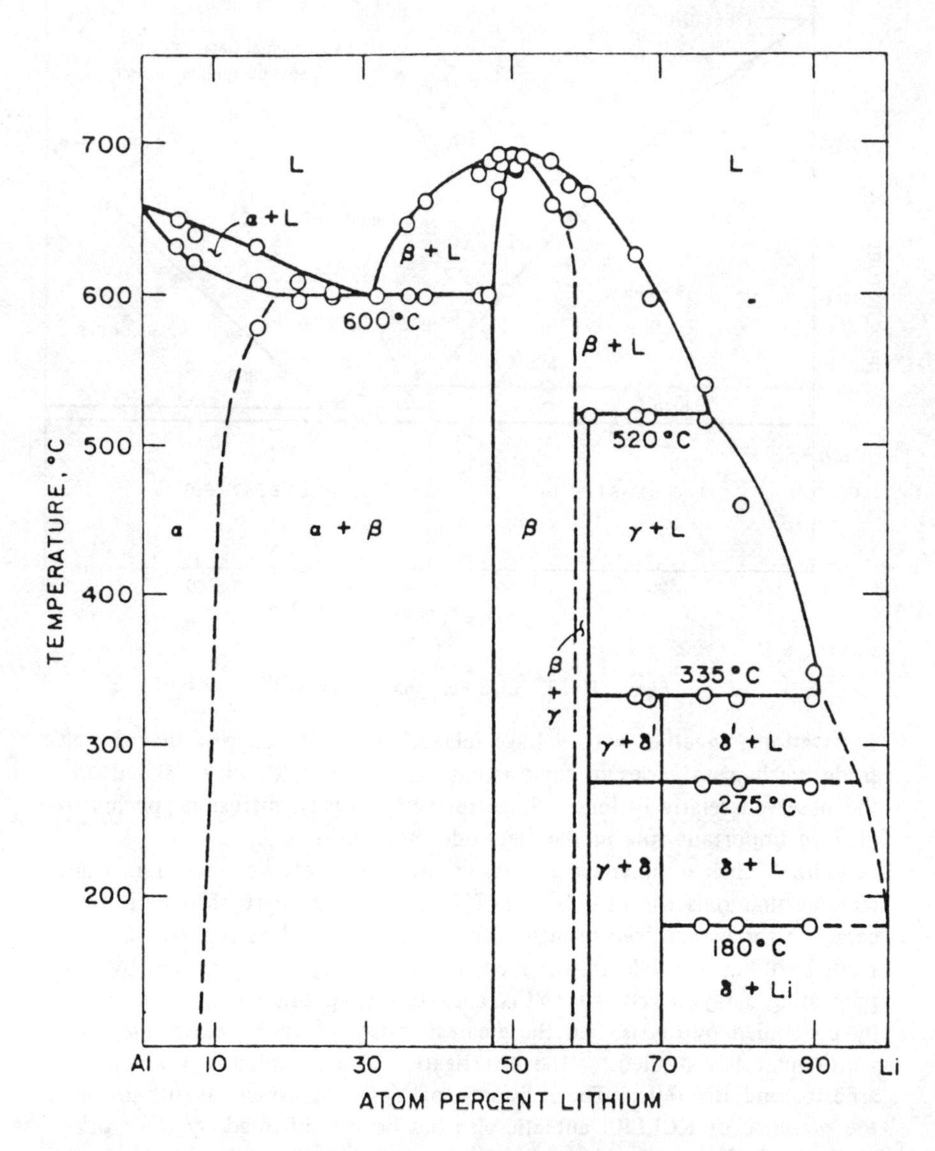

Figure 10-19. Atom percent lithium.

resistance to FeS.[2] These results are shown in Table 10-18 and again emphasize the difficulty of finding a metallic material compatible with FeS_2.

Table 10-16. Best Ceramic Materials for Li/FeS_2 Cells[18]

Feedthroughs		Separators	
AlN	- Molten Li, 400°C, 10,000 hr	BN	- Cloth, paper, purified. Molten Li, 400°C, 2500 hr
BN	- Purified–Molten Li, 400°C, 1300 hr	ZrO_2	- Cloth. In contact with positive only
$CaZrO_3$	- Stoichiometric, High-Purity, Promising.	Y_2O_3	- Cloth. Not very strong
BeO	- High-Purity, Hot-Pressed, Molten Li, 400°C, 1000 hr		

Lithium-Sulfur Cell Developments

Cell developments have followed two alternatives based on the use of either Li-Si or Li-Al as the electrode material. The first of these is being developed by Rockwell International,[49] and the latter by ANL[2] and General Motors.[48] Emphasis is being placed on the FeS_2 cathode because of the materials problems this presents. FeS is also being considered in spite of its poorer electrochemical characteristics.

The compact cell developed by General Motors[48] and shown in Figure 10-17 has been designed in two configurations. These are summarized in Table 10-19. The cell with Si-Li had a higher specific energy (130 Wh/kg compared with 115.5 Wh/kg) due to the lower weight of anode active

Table 10-17. Compatibility of Prospective Positive Electrode Materials in Lithium/Iron Sulfide Cells[18]

Material	Positive Electrode	Results of Corrosion Tests or Cell Tests
Mo	FeS_2, FeS	Little or no attack on properly prepared material
W	FeS_2	
C, Graphite	FeS_2, FeS	
TiN (Coating on Fe)	FeS_2	
FeB (Coating on Fe)		
Fe	FeS	Moderate attack or dissolution
Ni	FeS	
Fe	FeS_2	Severe attack or dissolution
Ni	FeS_2	
Cu	FeS	
Nichrome	FeS_2	
Nb	FeS_2	

Table 10-18. Results of Corrosion Tests in FeS and FeS_2 Environments at 450°C[2]

	Corrosion Rate[a] (mμ/yr)			
	FeS + LiCl-KCl		FeS_2 + LiCl-KCl	
Material	500 hr	1000 hr	500 hr	1000 hr
Molybdenum	2.0	4.1	+1.8	+0.20
Niobium	+30	+22	2900	–
Nickel	+17	3.0	> 6600	–
Armco Iron	470	450	–	–
Inconel 617 (Ni-22Cr-12.5Co-9Mo-1Al)	1.8	2.7	890	1280
Inconel 625 (Ni-22Cr-9Mo-5Fe-3.6Nb)	1.7	3.6	890	930
Inconel 706 (Ni-36Fe-16Cr-3Nb-1.8Ti)	5.3	3.9	4300	2300
Inconel 718 (Ni-19Cr-18Fe-5Nb-3Mo-0.8Ti-0.5Al)	6.6	3.4	2400	–
Incoloy 825 (Ni-28Fe-22Cr-3Mo-2.2Cu-0.8Ti)	2.2	1.9	3140	–
Hastelloy B (Ni-28Mo-5Fe-1Cr)	1.6	3.4	220	680
Hastelloy C (Ni-17Mo-16Cr-5Fe-4W)	1.5	0.33	370	970
304 SS (Fe-19Cr-10Ni)	4.7	2.2	> 5000	–

[a]Numerical values preceded by + represent formation rates of reaction layers. Values preceded by > represent corrosion rates based on the initial weight of the sample. A value of ⩾80 mμ/yr is considered acceptable.

material and better packaging. The presence of CoS_2 improved cell performance in the low voltage part of the discharge. Both cells showed good current efficiencies, greater than 90%, and in the early cycles, cycle efficiencies of more than 70%. Cell 1 operated for 445 cycles with marked degradation in performance. Cell 2 was only operated for 77 cycles with little degradation, but testing was discontinued due to an equipment failure, which caused overcharging.

Argonne National Laboratory and associated contractors (Table 10-8) are investigating various electrode and cell configurations. Both FeS and FeS_2 are included in the cathodic reactants and Li-Al as the anode. Several promising types of positive electrodes have been designed for prismatic cells, including:[52]

- electrodes having porous current collector structures into which the active material is loaded by vibrating;
- electrodes in which a mixture of the active material and electrolyte is hot-pressed onto a metal current collector; and

Table 10-19. Design Summary of General Motors Lithium/Iron Sulfide Cells[48]

	Cell 1	Cell 2
Diameter, cm	11.43	11.43
Volume, cm^3	390.9	320.1
Weight, kg	1.010	0.823
Positive Electrode		
Area, cm^2	163	163
Thickness, cm	1.55	1.65
Theoretical Capacity, Ah	115	126.5
Active Materials	FeS_2	$FeS_2 + CoS_2$
Current Collectors	Graphite cloth & Mo expanded metal	
Negative Electrodes		
Total Area, cm^2	197	197
Thickness, cm	0.60	0.50
Theoretical Capacity, Ah	115	115
Active Materials	55 a/o Li-Al	80 a/o Li-Si
Current Collectors	Stainless steel screen	Ni metal foam and expanded metal

- electrodes formed from a mixture of active material and carbon cement to give a carbon-bonded, porous structure that is easily penetrated by electrolyte.

It has been found that the addition of CoS_2 to the FeS_2 provides additional electrical conductivity and decreases the amount of molybdenum required for current collection.

Similarly, various techniques have been or are being studied for the preparation of the anodes:

- hot-pressed lithium-aluminum alloy with and without aluminum wires;
- powdered Li-Al loaded into a porous metal structure by vibration;
- electrochemical formation using a plaque of compressed aluminum wires; and
- cast lithium-aluminum electrodes.

Various separator materials are being used. In some cell designs, zirconium oxide fabric is used as a retainer for the active materials of the anode and cathode and boron nitride cloth as the separator between anode and cathode. Cells with Y_2O_3 separators were not reported in the referenced reports.

Cells with electrochemically formed Li-Al anodes and CoS_2 containing FeS_2 positives were continuing under test after more than 360 cycles and between 3000 and 4800 hours of operation. They show close to theoretical ampere hour efficiency with capacity declines of 8% or more over the cycles tested.

The development status of the Rockwell International (Atomics International Division) development has been described by McCoy and Heredy.[53] The anode is based on the use of lithium-silicon and cathodes of either FeS or FeS_2. The use of $FeS_{1.5}$, which had been studied earlier, has been dropped as it gave no advantages in performance or materials compatibility when compared with FeS_2. The electrode potentials of the various lithium silicon phases involved in the anode discharge are shown in Figure 10-20. The starting alloy for cells being tested in Li_5Si.

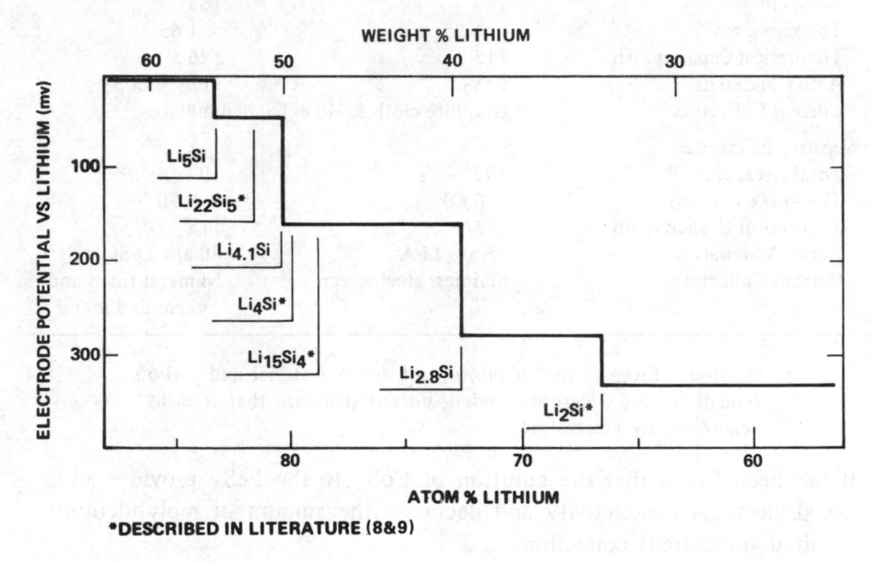

Figure 10-20. Electrode potentials for lithium-silicon alloy compositions.[53]

The negative electrode is based on a metallic honeycomb support. The powdered alloy is added to the support and then covered with 100-mesh stainless steel screens. Electrodes with areas up to 130 cm² have been built. Cathodes of FeS have been primarily used to simplify material selection. With this material, a honeycomb metallic cathode structure filled with an FeS-carbon mixture and covered by a stainless steel wire cloth to retain the active material has been developed. In the compact cells, BN cloth or porous Si_3N_4 separators were used initially and later cells used a composite oxide and nitride separator. The cells consist of two cathodes and one dual-faced anode with electrode areas of up to 130 cm². Cells with the composite separator were reported to have been cycled for months with good overall performance. It has recently been announced that a program to build a 1-kWh cell has been initiated.[54] No details of the cell design were included. However, the demonstration battery had three pairs of electrodes, operated at 400°C, and had an energy density of approximately 85 Wh/kg.

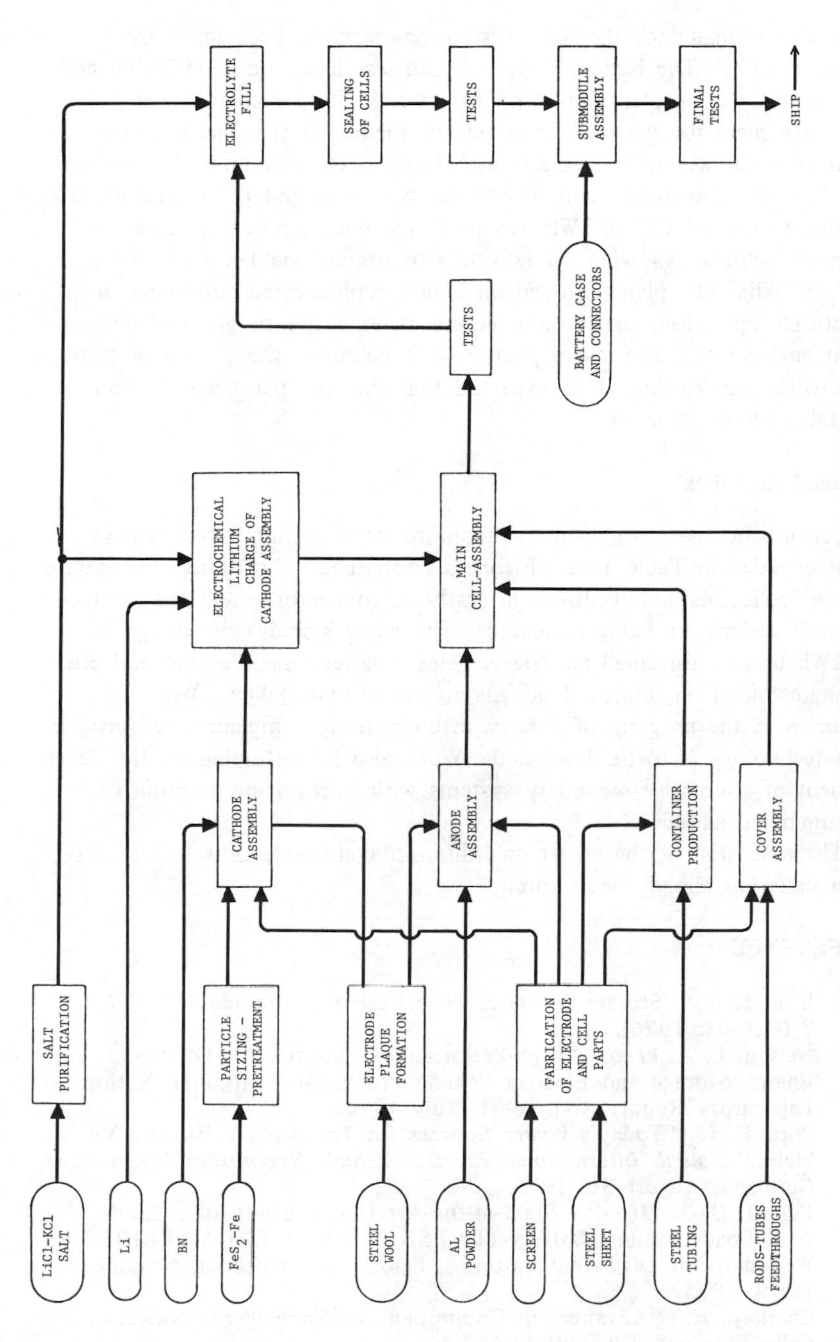

Figure 10-21. Flow sheet for a Li-Al/FeS_2 battery manufacturing plant.[55]

A cost estimate for the ANL system has recently been made by Towle *et al.*[55] The battery design chosen was based on a 0.92-kWh cell. The manufacturing plant was sized to produce 5000 such cells per day. The flow sheet for the plant is shown in Figure 10-21. The cells are assumed to be assembled for sale in battery cases or submodules containing 24 cells. The total plant investment was estimated to be $12,500,000. A selling price of $29.16/kWh was projected with a return of 25% on invested capital. Allowing for recycling of lithium yielded a net price of $27.33/kWh. The plant is based on highly sophisticated automated manufacturing operations that would require an extensive program of development engineering. The above plant cost is based on the second or third one to be constructed. It is expected that the first plant will be considerably more expensive.

Planned Activities

Performance goals for lithium-aluminum metal sulfide batteries have been included in Table 10-2. Effort is continuing on materials for cathode current collectors and feedthrough (cathode connector). Various electrode and cell designs are being evaluated. The latter includes the design of a 30-kWh battery for small electric vehicles. Battery manufacture and cost estimates based on improved designs are to be undertaken. With the inclusion in the program of battery manufacturing companies, cell production technology is to be developed. Work also is continuing on the development of alternative secondary systems with calcium, magnesium, or sodium-based anodes.

The extension of the effort on lithium-silicon anode cells to larger 1-kWh units has already been noted.

REFERENCES

1. Birk, J. R. "Storage Batteries: The Case and Candidates," *EPRI J.* 7 (October 1976).
2. Nelson, P. A. *et al.* "High-Performance Batteries for Off-Peak Energy Storage and Electric Vehicle Propulsion," Argonne National Laboratory Report ANL-76-81 (July 1976).
3. Port, F. J. "Today's Power Sources for Tomorrow's Electric Vehicles," *Fourth International Electric Vehicle Symposium,* Dusseldorf, Germany, August 31, 1976.
4. Ferrell, D. T., Jr. "A State-of-the-Art Design Study of a Lead-Acid Load-Levelling Battery-The ESB Concept," ERDA/EPRI/ILZRO Workshop on Lead-Acid Batteries, Palo Alto, California, November 18-19, 1975.
5. Charkey, A. "Advances in Component Technology for Nickel-Zinc Cells," *loc. cit.* 17(769076):452.

6. McCoy, L. R. "Zinc-Nickel Oxide Secondary Battery," *loc. cit.* 2 (759169):1131.
7. Charkey, A. "Evolutionary Developments in Nickel-Zinc Cell Technology," *loc. cit.* 2(759168):1126.
8. Miller, L. E. and R. A. Brown. "Nickel Battery Systems for Electric Vehicles," Eagle Picher Industries, Inc., Joplin, Missouri (April 1976).
9. Bishop, W. S., D. L. Call, J. K. Wilson and R. A. Brown. "Development of Nickel-Zinc Batteries for Aircraft," report from U.S. Air Force Aero-Propulsion Laboratory and Eagle Picher Industries, Joplin, Missouri.
10. Buzzeli, E. S. *loc. cit.* 9:147.
11. Symons, P. C. "Performance of Zinc-Chloride Batteries," *Electric Vehicle Symposium,* Washington, D.C., February 19-20, 1974.
12. Gordon, R. S. "Ceramics in High-Performance Batteries," *Symposium on Ceramics in the Service of Man,* Carnegie Institution, Washington, D.C., June 7-9, 1976.
13. Chatterji, D. "Development of Sodium-Sulfur Batteries for Utility Application," Annual Report EPRI EM-266 (December 1976).
14. Hong, H. Y. P. "New Solid Electrolytes," *American Chemical Society Meeting,* New York, April, 1976.
15. Levine, C. A. "Sodium-Sulfur Battery System," ERDA Report No. Ey-66-C-02-2565, 00-4 (November 1, 1976).
16. Yao, N. P. and J. R. Birk, "Battery Energy Storage for Utility Load Levelling and Electric Vehicles: A Review of Advanced Secondary Batteries," *Tenth Intersociety Energy Conversion Engineering Conference,* Paper no. 759166 (1975), p. 1107.
17. Gross, S. *Energy Conv.* 15:95 (1976).
18. Cairns, E. J. "Materials Problems in Rechargeable Batteries," General Motors Research Laboratories Research Publication GMR-2170 (June 8, 1976).
19. Simon, A. C. and S. M. Caulder. "The Lead-Acid Battery," in *Proceedings of the Symposium and Workshop on Advanced Battery Research and Design,* Argonne National Laboratory (March 1976), p. A-34.
20. Cook, A. R. "A Review of Battery R&D," *American Power Conference,* Chicago, Illinois, May 1, 1974.
21. Simon, A. C., S. M. Caulder and J. T. Stemmle. *J. Electrochem. Soc.* 122:461 (1975).
22. Rao, P. and G. W. Mao. "Extended Abstracts," *150th Meeting, Electrochemical Society* (1976), p. 14.
23. Devitt, J. L. and M. Myers. *J. Electrochem. Soc.* 123:1769 (1976).
24. Prengaman, R. D. *loc. cit.* 9:31.
25. Marshall, D. and W. Tiedemann. *loc. cit.* 10:1849.
26. Gallwitz, W. *et al.* "Lead-Acid Battery in Peak Power Demands," ILZRO Project LE-239, Final report for period May 12, 1975 to November 12, 1975.
27. Yao, N. P. Second Workshop on Lead-Acid Batteries, Washington, D.C., December 9-10, 1976.
28. Towle, W. L. and U. A. Kaulkis. "High-Performance Lead-Acid Utility Battery," Annual Meeting American Institute Chemical Engineers, 1976.

29. Kassekert, D. W., A. O. Isenberg and J. T. Brown. "High-Power Density Bipolar Lead-Acid Battery for Electric Vehicle Propulsion," *11th IECEC,* paper no. 769070, September 12-17, 1976.
30. Sunu, W. G. and D. L. Bennion. "Theoretical Analysis of Zinc Porous Electrodes," *loc. cit.*9:61.
31. Pickett, D. F., Jr. and J. T. Maloy. "Effect of Electrochemically Coprecipitated Cobalt Hydroxide in Nickel Hydroxide Electrodes," *loc. cit.* 9:136.
32. Donnel, C. P., III. "Development and Fabrication of Large Vented Nickel-Zinc Cells," Final Report NASA CR-134967 (December 1975).
33. News Release, NASA, Lewis Research Center, September 10, 1976.
34. Birge, J. *et al.* "Performance Characteristics of a New Iron-Nickel Cell and Battery for Electric Vehicles," *Power Sources Conference,* Brighton, England (September 1976).
35. Symons, P. C. "Process for Electrical Energy Using Solid Halogen Hydrates," U.S. Patent 3, 713, 888, January 20, 1973.
36. Kummer, J. T. and N. Weber. *Soc. Automotive Eng. Trans.* 76: 1003 (1968).
37. Levine, C. A. *loc. cit.* 2(759095):621.
38. Werth, J. T. U.S. Patent 3, 877, 984, April 15, 1975.
39. Weiner, S. A. "The Sodium/Sulfur Battery: A Progress Report," *loc. cit.* 6:B-219.
40. Werth, J. "Sodium Chloride Battery Development Program for Load Leveling," Electric Power Research Institute, Project 109, Report EPRI EM-230 (December 1975).
41. Hong, H. Y. P. *Mat. Res. Bull.* 11:173-182 (1976).
42. Powers, R. W. and S. P. Mitoff. *J. Electrochem. Soc.* 112:226 (1975).
43. Bush, J. B. "Sodium-Sulfur Battery Development for Bulk Power Storage," EPRI 128-2 (September 1975).
44. Cleaver, B., A. J. Davies and D. J. Scheffrin. *Electrochem. Acta* 18:747 (1973).
45. Weiner, S. A. and R. P. Fisher. "Research on Electrodes and Electrolyte for the Ford Sodium-Sulfur Battery," Annual Report for period June 30, 1975 to June 29, 1976 under contract NSF-C805 (AER-73-07199) (July 1976).
46. Powers, R. W. *J. Electrochem. Soc.* 122:490 (1975).
47. Silverman, H. P., L. S. Marcoux and E. T. Seo. "Development Program for Solid Electrolyte Batteries," Final Report EPRI EM-226 (September 1976).
48. Dunning, J. S., T. G. Bradley and E. J. Zeitner. "Development of Compact Lithium/Iron Disulfide Electrochemical Cells," General Motors Research Laboratory, GMR-2168 (June 10, 1976).
49. Sudar, S. *et al.* "Development of Lithium-Metal Sulfide Batteries for Load Levelling," EPRI EM-166 (May 1976).
50. Sharma, R. A. *J. Electrochem. Soc.* 123:448 (1976).
51. Cairns, E. J. and J. S. Dunning. "High-Temperature Batteries," General Motors Corp. GMR-2065, (January 23, 1976).
52. Nelson, P. A. *et al.* "Development of Lithium/Metal Sulfide Batteries at Argonne National Laboratory," Summary Report for 1975, ANL 76-45 (May 1976).

53. McCoy, L. R. and L. A. Heredy. "Development Status of Lithium-Silicon-Iron Sulfide Load-Levelling Batteries," *11th IECEC*, paper no. 769082 (1976), p. 485.
54. *Chem Eng News,* 7 (November 15, 1976).
55. Towle, W. L., J. E. A. Graae, A. A. Chilenskas and R. O. Ivins. "Cost Estimate for the Commercial Manufacture of Lithium/Iron Sulfide Cells for Load Leveling," ANL 76-12 (March 1976).

11

COMPRESSED AIR IN U.S. INDUSTRY*

INTRODUCTION

The use of compressed air in the U.S. has been established for at least a century as a stable ingredient of its industrial development. There is a great variety of specific applications spread rather evenly throughout all the major industrial sectors of the national economy. The equipment is generally installed as a local in-plant utility to provide a common source of stored energy used to operate tools, material transport systems and other machinery. Compressed air is also used directly as a reactant or drying agent in chemical and various other manufacturing processes.

The reasons for using compressed air, rather than alternative means, are generally associated with convenience and safety and unique physical dynamics rather than the efficient use of energy. In fact, a substantial thermodynamic penalty is paid for using it. The advantages are so strong, however, that even the recent rise in energy costs has so far not discouraged its use. At best, in some industrial plants, older steam-driven compressed air generators have been returned to continuous operation to save on higher electricity charges.

Of all the newest applications discussed in the literature or by individuals associated with compressed air equipment and its use, only the following stand out as novel and promising enough to greatly expand the importance of compressed air in U.S. industry:

1. pneumatic logic and fluidic control devices as major elements in automated assembly and other manufacturing processes;
2. large volumes of compressed air (generally placed underground) as large-scale energy storage systems for off-peak energy use. Special measures must be taken to reduce energy loss over the entire thermodynamic cycle; and

---

*By: Norman Lord

3. large volumes of compressed air used for low-level combustion in a method for *in situ* underground coal gasification.

The first two developments are in their infancy even though their commercial utility has been demonstrated. The last is still an experimental procedure that shows great commercial promise. None of these developments has therefore been established well enough to indicate its alternate economic impact on the market for compressed air equipment. We have in compressed air equipment a well-established, mature industry. The companies engaged in producing compressed air equipment are generally old, financially sound and growing in pace with the U.S. economy. The likelihood is small that any major changes will occur in the near future.

COMPRESSED AIR SYSTEMS

As a local plant utility for distributing ready access to a compact store of energy, compressed air systems offer the following advantages over direct use of electricity, steam, and other energy forms:

- safety;
- convenience;
- compact, low weight equipment;
- low capital, operational and maintenance costs of equipment;
- ease of control; and
- simplicity of use.

The drawbacks of compressed air are primarily the energy loss in its thermodynamic cycle, leakage of air and some difficulty in controlling the operating temperature of machinery driven by compressed air. Any compressed air system includes the compressor itself, unless, as a rare instance, compressed air is supplied by an outside independent utility, and a variety of separate systems that are associated with conditioning the air or are being operated by the controlled release of the air pressure. The major categories of their equipment are shown in the two-part Table 11-1, presenting separate breakdowns of compressor types and associated compressed air equipment.

Air dryers, coolers and purifiers are used with all types of air compressors, being scaled appropriately to the air flow capacity and maximum pressure of the system. The positive displacement compressors are used for the low range of capacities up to 20,000 ft^3 of air per minute, while the dynamic compressors handle flow rates up to 1,000,000 ft^3 of air per minute. The highest maximum pressures of 60,000 $lb/in.^2$ are available with reciprocating, double-acting compressors. Pneumatic tools,

Table 11-1. Air Compressor Equipment

Air Compressor Categories	Compressed Air Equipment
Positive Displacement	Air dryers
Reciprocating	Air coolers
Single acting	Air purifiers
Double acting	Pneumatic tools
Diaphragm	Pneumatic conveyor systems
Rotary, Sliding Vane	Pneumatic controls and instruments
Oil free	Processing equipment
Oil flooded	Special installations
Helical Screen	
Dry	
Oil injected	
Single screw	
Liquid Ring	
Dynamic	
Centrifugal	
Single stage	
Multistage	
Axial	
Multistage	

conveyor systems, controls and instruments are usually operated in a system supplied by a relatively low-capacity positive displacement compressor. Processing equipment and special installations utilize the huge volumes of air supplied by the dynamic compressors.

MAJOR INDUSTRIAL APPLICATIONS

The variety and scope of industrial applications serviced by compressed air systems are demonstrated in Table 11-2 showing some typical specific applications associated with the equipment types of Table 11-1. Some representative applications of compressed air associated with those industrial sectors of the U.S. economy that are the heaviest users of compressors are shown in Table 11-3 along with the type of air compressor being used. These industries account for more than half the U.S. market for compressed air equipment. The applications listed represent a core of usage that has been established for a long time and for which there is no indication of any dramatic changes in technical characteristics or magnitude relative to overall U.S. industrial activity.

Table 11-2. Functions of Compressed Air Equipment[1]

Tools	Control Devices	Materials Processing
Generally Operated from Central Storage Supplied by Positive Displacement Compressors	**Small Positive Displacement Compressors**	**Large Masses of Air Supplied by Dynamic Compressors**
Pneumatic Tools	Pneumatic logic	Oil refineries
Air Motors	Fluidic control	Chemical preparation
Abrasive Tools	Small part conveyors for automatic assembly	Sewage aeration
Chippers, Scalers, Rammers		Coke Ovens
Drilling and Tapping Tools		Pipeline boosters
Assembly Tools	Conveyor systems	Wind tunnels
Screw drivers	Air hoists	Blast furnace air supply
Right-angle nutsetters	Air flotation devices	
Impact wrenches	Pneumatic conveyors	
Multiple spindle tools		
Rivet squeezers	Miscellaneous equipment	
Medical and Dental Equipment	Presses	
Fixtured Machine Tooling	Pneumatic shears	
Air-operated Drills	Paint sprayers	
Drifter drills for mining	Atomizers	
Stope drill	Humidifiers	
Paving breakers	Bending machines	
Tampers	Forge machinery	
	Programmed automatic wire wrappers	

Table 11-3. Largest U.S. Industrial Compressed Air Use[2]

Industry	Representative Application of Compressed Air	Compressor Type
Chemicals	Reactant in butadiene, ammonia etc. Conveying, agitating, aeration	Centrifugal, reciprocal, rotary
Construction	Submarine, tunnelling, corrosion work	Centrifugal
Aircraft	Wind tunnels for aerodynamic test	Centrifugal
Iron and Steel Manufacturing	Oxygen enrichment	Centrifugal
Crude Petroleum and Natural Gas	Air pumping Repressuring oil and gas fields	Large centrifugal
Farm Machinery	Blowing oil furnacer Sand blasting	Centrifugal
Engines and Turbines	Chipping, sand blasting, forging hammers	Reciprocating
Paper	Air flotation, molding and drying	Centrifugal
Mining	Rock and hammer drills, stoper drills	Large reciprocating
Petroleum Refining	Cracking, alkylation	Centrifugal
Heating and Plumbing	Rollers, expanders, reamers, pneumatic boiler controls	Reciprocating

MANUFACTURERS OF COMPRESSED AIR EQUIPMENT AND TYPICAL PRICES

The leading manufacturers of compressed air equipment are members of the Compressed Air and Gas Association of Cleveland, Ohio, which conducts a program to promote the use of compressed air in U.S. industries. Its activities are organized under sections associated with the equipment being sold and following roughly the categories of Table 11-1. A list of some of these manufacturers with the equipment they sell is shown in Table 11-4. The prices of some of the equipment, which does not include dynamic compressors, are given in Table 11-5. Included in the price of the compressors, in addition to the compressor itself, are costs for necessary equipment to treat the air, such as interstage coolers and dyers, which comprise an integrated package. The dynamic compressors are assembled to special order out of standard modules, and there is usually no set price. They are usually part of a comprehensive plant addition and the subject of a bid competition.

Table 11-4. Leading Equipment Manufacturers[3]

Manufacturer	Equipment Categories
Ingersoll-Rand	Compressors (all to highest capacities), reciprocating, rotary helical, centrifugal, axial Large variety pneumatic tools and air hoists
Clark Div. of Dresser Industries	Compressors Centrifugal, axial (to highest capacities)
Nash Engineering	Liquid ring compressors
Garden-Denver Co.	Compressors and other heavy industrial equipment
Chicago Pneumatic Tool Co.	Pneumatic tools (largest variety)
Aro Corp.	Pneumatic control devices
Deltech Engineering	Air purifiers and dryers
Zurn Industries	Aftercoolers, dryers, oil separators
Fuller of GATX Corp.	Compressors and other heavy industrial equipment

MARKET FOR COMPRESSED AIR EQUIPMENT

Recent market trends for compressed air equipment are compared to overall U.S. economic activity and the more closely relevant portion of its, expenditures on fixed investment producer's durable equipment (FIPDE), in Table 11-6. There is a recent, strong but short-term upward trend in shipments of compressors and vacuum pumps, which is matched to some extent by the trend in FIPDE. A breakdown by type and capacity of compressor shipments for 1975 is shown in Table 11-7. This breakdown shows that there is no clearly dominant category of compressors so that one must always be cognizant of the wide variety of equipment in the market.

TECHNICAL TRENDS IN COMPRESSED AIR APPLICATIONS

From a standpoint of fundamental changes in the technical relation of compressed air and manufacturing practice, the increasing use of pneumatic control, especially by fluidic elements, has excited the most widespread interest among engineers. However, it has yet to make any noticeable impact on the market for compressed air equipment. The influence of greater cost for all forms of fuel has placed a great premium on the

Table 11-5. Representative Equipment Costs[3]

Compressors

	Max Press (psig)	Flow (cfm)	Required Power (hp)	Price ($)
	125	80-90	20	7,000
Reciprocating	125	1,000	200	41,000
Compressors	125	4,000	700	1 2,000
	230	320	100	33,000
Rotary Oil-Free	150	660	150	50,000
Compressors	150	3,450	800	105,000
Oil-Injected Rotary	110	120	25	7,000
Screw	125	320	75	13,000

Tools

	Price ($)
Belt Sander 1 1/8 x 21 in.	230
Pad Sander 4 x 4 in.	70
Air Filer and Sander	100
Disc Grinder, 4 in.	160
Jet Chisel Needle Scaler	100
Air Drill, 1/4 in.	100
Fiber Glass Cutter	260

efficient use of energy, and the use of compressed air as energy storage medium for the short term has attracted attention of both electric utilities and industry. A West German utility near Hanover has a power plant under construction in which compressed air will be stored in a salt cavern to assist in satisfying peak power electrical requirements. The spread of this practice will undoubtedly expand the compressed air market, particularly in the categories of large dynamic compressors, which are used to compress the enormous required volumes of air to high pressure. An outstanding example of local plant storage is a facility built for the Ames Research Center, National Aeronautics and Space Administration (NASA) to supply air for a hypersonic wind tunnel. A very recent experimental development is the use of air to partially burn coal underground and create a combustible gas of high heating valve. This is now under active consideration by the ERDA and the energy industry.

Table 11-6. U.S. Economic Activity and the Recent Market for Compressed Air Equipment[4,5]

	($ billions)						
Year	A	B	C	D	E	F	G
1968	1050	70.0					
1969	1080	73.0					
1970	1076	68.8					
1971	1110	67.5		.425			
1972	1171	74.3	.723	.437	.148	.94	.218
1973	1232	82.1	.853	.504	.193	.99	.259
1974	1216	81.6	1.030	.657	.303	.89	.226
1975	1190	74.6		.614			
1976	1271	77.6					

A. Gross national product, in 1972 dollars.
B. Fixed investment producer's durable equipment (FIPDE), in 1972 dollars.
C. Air and gas compressors and related equipment, in 1972 dollars.
D. Compressors and vacuum pumps, excluding refrigeration compressors, in 1972 dollars.
E. Parts and attachments for compressors, in 1972 dollars.
F. Industrial spraying equipment, in 1972 dollars.
G. Power-driven hand tools, in 1972 dollars.

Table 11-7. Quantity and Value of Compressor Shipments in 1975[5]

Products	No. of Companies	Quantity	Value ($ millions)
Compressors and Vacuum Pumps, Total	x	1440	802.5
Air compressors, total (value of drivers excluded) (10%, % imp. val.)	x	867	405.5
Stationary, total	x	249	258.5
Reciprocating, Single Action:			
Under 1 1/2 hp	26	153	17.4
1 1/2 - 5 hp	18	49.3	21.0
6 - 25 hp	17	23.3	27.0
26 hp and over	11	4.63	18.9
Reciprocating, Double Action:			
150 hp and under	14	8.501	21.16
Over 150 hp	6	.383	21.76
Rotary, Positive:			
Discharge press, 50 psig and under	6	1.54	9.44
>50 psig, 150 hp and under	11	6.16	48.8
>50 psig, >150 hp	9	1.28	14.28
Centrifugal and Axial:			
50 psig and below	3	.50	58.85
51 psig and over	4		
Portable, Total	x	618.1	146.9
under 75 cfm	18	605.6	31.9
75-124 cfm	13	2.34	8.55
125-249 cfm	14	5.96	27.49
250-599 cfm	12	1.57	14.50
600-899 cfm	12	1.96	41.70
900 cfm and over	10	0.020	2282
Pneumatic (air) Power Compressors and Motors (includes air brake systems)	8	433	34.39

GENERAL PRINCIPLES

Compressed air can be generated, stored and distributed for use very much like electricity. It differs from electricity in being inherently suited to a storage and discharge system rather than to a system in continuous constant operation. Transmission of compressed air is much more limited in distance due to both high costs of pipe relative to electric conductor and greater loss of available energy. In addition, the thermodynamic and mechanical processes of generating and using compressed air introduce unavoidable losses of energy so that the basic efficiency of the compressed air system is difficult to keep above 80%.

Nevertheless, compressed air in its application offers overriding benefits in economy and versatility.[1,6] Economy is achieved by relatively low capital costs of initially air-driven equipment and lower operating and maintenance costs. Versatility is demonstrated by the ever-growing array of unique applications that are hard to accomplish by any other means. In addition, a properly maintained compressed air system is relatively safe for industrial applications. It is extremely stable under widely varying ambient temperature, cannot shock or burn a human operator, and requires no cleanup in the event of a leak. Compressed air at a distant point of use does not entail the fire hazard of sparks or excessive overheating of electrically driven equipment. Hence, compressed air equipment is preferred in hazardous environments such as mines or where intermittent bursts of high energy are required for high-impact applications. In addition, compressed air tools are generally much lighter than comparable electric equipment. For example, the electric motor includes the same mass of magnetic circuit used in the electric generator. Equivalent air-driven motors are much lighter.

Pneumatic systems are not the best in all cases. Sometimes, the compressibility of air allows a softness that cannot be tolerated, and the rigidity and greater power of hydraulic or air/hydraulic systems should be utilized. Pneumatic signals propagated at the speed of sound may be too slow over great distances and there is need for the speed advantage of electronic and electronic circuitry. Finally, one must also recognize that fire and explosion hazards are not completely eliminated in compressed air systems[7] and that in such environments as mines, using long air lines, extensive measures must be taken to minimize these dangers.

Physical Fundamentals of Compressed Air Generation

Compressed air is generated prototypically in the reciprocal piston and valve arrangement shown in Figure 11-1a. The valves at A and B are elastically attached to the compressor outside wall. Initially, the piston is in contact with the plate in which the valves are placed. As it moves away, valve B remains constrained by its spring to keep the discharge port sealed while valve A is pulled toward the piston by the unbalanced pressure of air being drawn through the suction port. After the piston reaches full downward displacement with chamber pressure equal to that in the suction port, valve A is pulled by its spring to close the suction port. On the return of the piston, chamber pressure is built up until it is high enough to move valve B against its spring and pass air at this high pressure through the discharge port into a storage reservoir. If the storage reservoir is arranged as a constant-pressure variable–

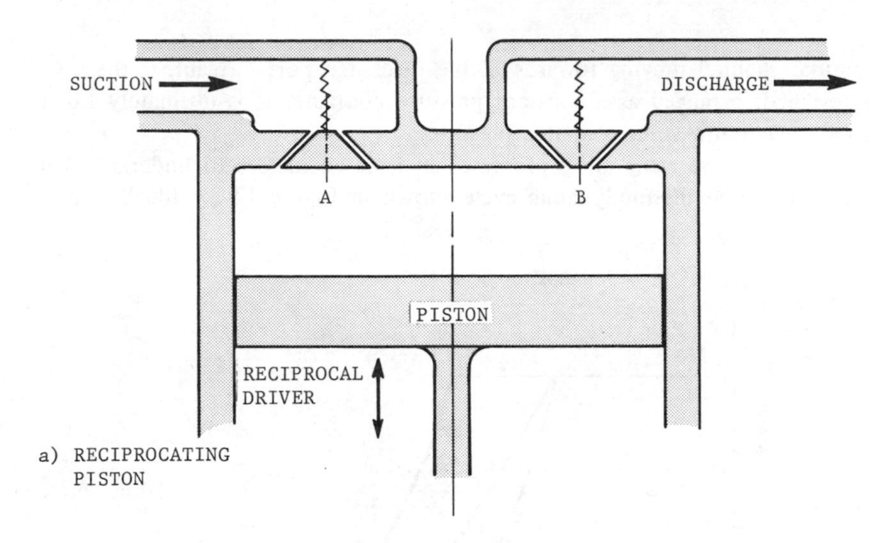

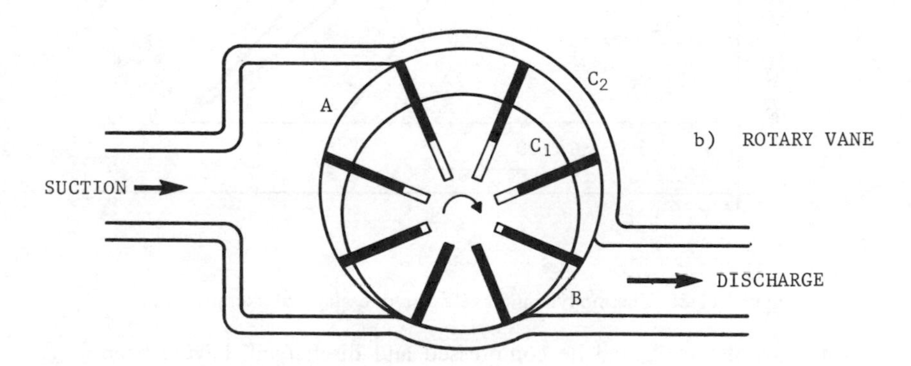

Figure 11-1. Compressed air generation.[1]

volume container, then repetition of the cycle will eventually fill the reservoir to its maximum volume at this constant pressure.

Figure 11-1b illustrates the same compression principle using a set of radial sliding vanes, mounted in a drum C_1, and undergoing reciprocating radial motion to maintain contact with the eccentric cylinder C_2. The vanes act as pistons, which collectively draw in air through the suction port at A_1 entrapping a mass of air between each successive pair of vanes, which then compress it as the vanes rotate through the narrowing annular channel moving towards B, the discharge port. Again, if the reservoir is arranged as a constant pressure container it is ultimately filled to its maximum volume at this pressure.

In both these cases air is processed in isolated masses to undergo essentially the same thermodynamic cycle shown in Figure 11-2. Ideally, the

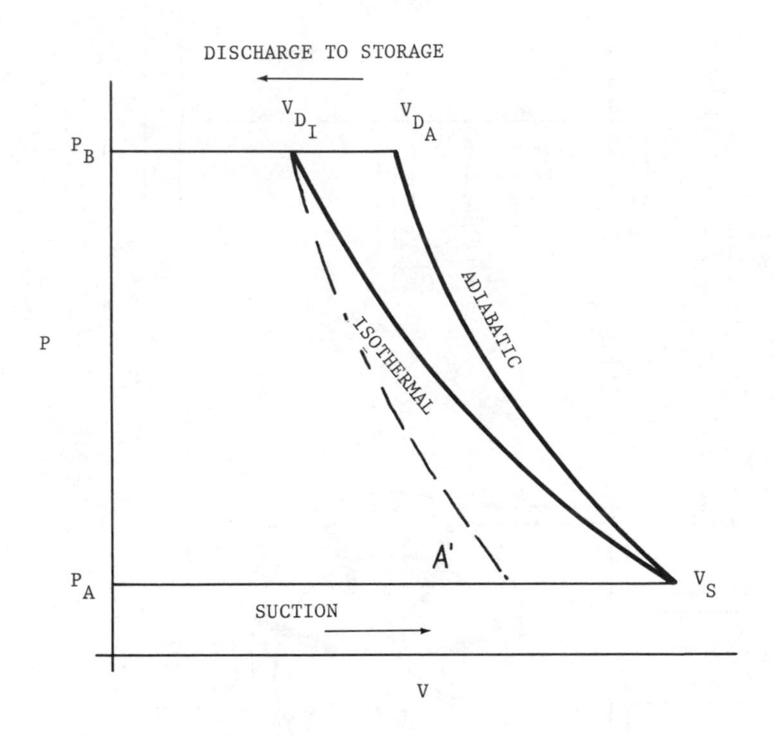

Figure 11-2. Thermodynamic cycle for one stage of compression.

entire mass drawn in will be compressed and discharged, having been raised in pressure from the suction P_A to the discharge P_B. For a relatively low ratio, P_B/P_A, with good thermal transfer between suction and discharge environments, the compression may take place isothermally. The air in volume V_s is then compressed to V_{D_1} before the discharge port opens. We then have air at pressure P_B stored at its original temperature. Generally, this is the cycle aimed at in practice with various measures taken to eliminate residual piston clearances, leakage across rotary sections, and to keep the temperature of the compressed air low. However, for high ratios P_B/P_A, compression normally takes place along the adiabatic segment and the air temperature has been appreciably elevated. In practice, this air is cooled by external means and results not only in the loss of the extra work done for the adiabatic compression but requires even more energy to operate the cooling equipment.

This is necessary because the air is normally either used in equipment, which cannot operate at elevated temperature or passed on to higher compression stage equipment, which has the same problem. Used to perform work, the stored compressed air will be expanded in a reverse cycle. In most cases the expansion is adiabatic and the temperature drop short-lived, so that it has little effect on the equipment. However, in some applications, such as using the air to operate turbines, the air must be re-heated to avoid too low a temperature on the final stages. In most practical cases, one probably cannot obtain for useful work more than 50-60% of the energy originally used to generate the compressed air.

Large compressed air systems in industrial plants are usually the result of a gradual plant development, and it is advisable to conduct rather comprehensive reviews of its energy transactions for cost reduction and power conservation.[8] In particular, the most prolific source of lack of capacity and unnecessarily reduced efficiency is air leakage, so an "audit" of the air system should be conducted at least on every occasion for its modification or extension.[9]

Major Categories of Compressed Air Equipment

Compressed air equipment is readily divided into the following groups, which can be considered independently:

- Air compressors;
- Compressed air dryers;
- Compressed air purifiers;
- Pneumatic tools;
- Pneumatic conveyor systems;
- Pneumatic controls and instruments;
- Processing equipment;
- Special purpose installations;

Air compressors probably account for about half the dollar volume of compressed air equipment sales. This is judged by the annual corporate reports concerning several major U.S. manufacturers of compressed air equipment, which account for most of the sales. Official U.S. Department of Commerce statistics are not broken down to sufficiently refined categories to provide an overall determination.

Along with compressors, the next four categories of equipment comprise items that are manufactured and sold, generally according to common specifications and standards, with additions from time to time of new general usage items that have been developed. The remaining categories comprise equipment that is usually designed for special purposes or is coordinated into special purpose systems such as large-scale assembly or chemical processing plants.

Air Compressors

Air compressors are generally useful for other gases as well and particularly in higher capacities have an extensive market in the chemical and petroleum process industries. Table 11-8 lists the air compressor categories as compiled by the Compressed Air and Gas Institute and Baniak.[1,10] In all these categories, the air being compressed undergoes a thermodynamic compression cycle similar to that described above. For positive displacement compressors, successive volumes of air are confined to a closed space which is then contracted to compress the air to a high pressure. Dynamic compressors accomplish compression by the mechanical

Table 11-8. Air Compressor Categories and Operational Characteristics[2]

Category	Ranges of Operational Characteristics: Volume of Free Air (ft^3/min)	Maximum Pressure ($lb/in.^2$)
Positive Displacement		
Reciprocating		
Single acting	100-900	40-100 (single stage)
Double acting	100-10,000	0-60,000
Diaphragm	.25	20-50
Rotary, Sliding Vane		
Oil free	30-3,000	5-250 (3-stage)
Oil flooded	120-600	80-125
Helical Screw		
Dry	-26,000	15-550 (booster)
Oil injected	100-1,800	-150 (2-stage)
Single screw	100-700	-125
Liquid Ring	2-16,000	5-100
Dynamic		
Centrifugal		
Single stage	-187,000	-150
Multistage	-20,000	-6,000
Axial		
Multistage	40,000-1,000,000	-500

action of rotating vanes which impart high velocity to the air being drawn in and through successive stages channel the air toward more confined space. The momentum gained from the rotor is converted to useful pressure energy by slowing down the gas in the stationary vanes of the stator. For light-weight applications in aircraft and satellites, dynamic compressors have operated at speeds in the range of 50,000-100,000 rpm.[10]

Reciprocating Compressors. The costs of reciprocating compressors vary in direct proportion to their throughput capacity and are generally the choice for low-volume requirements. Costs for dynamic compressors along with rotary screw do not rise directly with volume[11] so that they are used for high volumes.

The principles of reciprocating compressors and rotary sliding vane compressors have been described. Single-acting reciprocating compressors compress only on one stroke of the piston exactly as described. Double-acting compressors add another plate and set of valves opposing the one shown in Figure 11-1a so that when the piston is drawing air in on one side, it is compressing air on the other side. Diaphragm pumps operate on the same mechanical principle illustrated in Figure 11-1a except that the piston does not slide on a lubricated surface. To eliminate oil, as in medical or special scientific use, the piston is replaced by a flexible diaphragm or bellows which accomplishes, through limited reciprocating motion, sufficient pressure elevation to be useful. The reciprocating compressor is still regarded as the most efficient in terms of compressed air volume compared to input horsepower, and it is still a subject of active technical development. Matching capacity requirements precisely can now be done with methods of continuous capacity control.[12]

Rotary Sliding Vane Compressors. Operation of the oil-free rotary sliding vane compressor has been described (Figure 11-1b). In this form, it is extremely versatile with respect to the kinds of gases, and their physical states, it can handle. McGregor has described modifications of a conventional unit in which only stainless steel and graphite are exposed to the gas.[13] The unit has recirculated at a temperature of 350°C (design allows for 425°C) gaseous streams of H_2S, SO_2, COS, CO_2, N_2, H_2O and S_x.

The oil-flooded variety commonly used for air has liberal amounts of cooling oil injected directly into the compression chambers. Three purposes are served:

1. The oil continuously absorbs the adiabatic heat of compression to shift the path toward the isothermal;
2. Air and oil temperatures are maintained under 100°C; and
3. Rotor, vanes, bearings and shaft real are lubricated and appropriately sealed.

Air and water-cooled heat exchangers are used for cooling the oil. The main losses in rotary positive displacement compressors are in charging, leakage and warming-up period. Methods of managing these losses and determining empirically the pump operating efficiencies have been thoroughly discussed by Scheel.[14]

Liquid Ring Compressors. Liquid ring compressors operate on a principle similar to the rotary sliding vane. Compression takes place in a chamber formed in part by a captive ring of liquid, as shown in Figure 11-3a, which creates two sections of varying sequential air pockets, each operating like the similar sequence in the rotary vane. The curved blades of a rotor drive the ring of liquid within an elliptical casing. The inner surface of the liquid ring trapped between each pair of blades serves as the face of a liquid piston operating within each rotor chamber. At the inner diameter, these rotor chambers have openings that are sealed by, and revolve about, a stationary central plug. The ports are permanently open, an inlet-discharge pair for each of the two halves of the annulus formed by the liquid and casing operating like the inlet-discharge pair of the rotary vane. For every revolution, two compression cycles are completed in each rotor chamber, which result in a discharge flow that is practically free of pulsations.

A single-lobe design using eccentric axial placement of the rotor inside a circular cylinder is shown in Figure 11-3b.[15] It is designed for lower output pressures and higher flow rates.

Cooling of the liquid ring compressor is direct with required additional cooling fed into the ring, and excess liquid discharged with the gas. The discharged mixture is passed through a conventional bubble or centrifugal-type separator to remove the free liquid. Such compressors are useful in processes that require a low limit to the temperature rise throughout the compression cycle, approximating isothermal expansion.

Helical Screw Compressors. The helical screw compressor is shown in Figure 11-4.[16] Two mating rotating screws are the basic parts–one helical lobed male and one helical grooved female in the close fitting housing with helical timing gears to maintain clearances. Compression is accomplished by decreasing the volume of the inlet air pockets as the shafts rotate, and each pocket is moved from the suction to the discharge port.

The dry type has the following desirable features:[17]

- oil-free delivery;
- compact construction at moderate speeds;
- dynamic balance permitting a small simple foundation;
- smooth air flow;
- simple, low–cost installation; and
- low maintenance (simple construction, no rotor-to-rotor to housing contact).

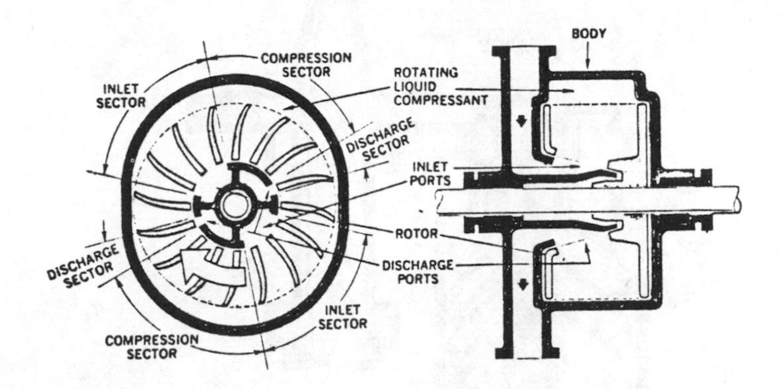

a) Double Lobe Design

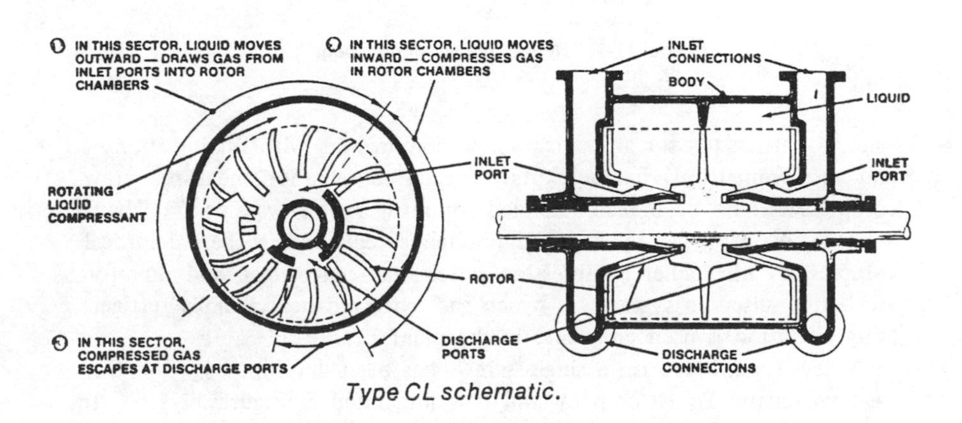

b) Single Lobe Design

Figure 11-3. Liquid ring compressor.[15]

It is also built with oil injection for the same advantage as in the rotary vane case when oil freedom is not essential in the delivered air. Both oil-free and oil-flooded are available in water-cooled or air-cooled versions. As a group, screw compressors comprise a relatively young generation,[18] with major new developments probably still forthcoming. Taking into account capital and operating costs, Weissenger has shown that screw compressors achieve lower compressed air costs than either the piston or rotary

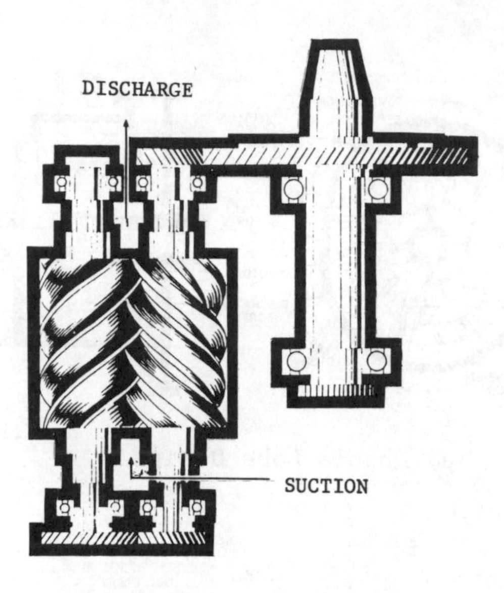

Figure 11-4. Helical screw compression.[16]

vane,[19] particularly for air-cooled screw compressors with oil injection and an asymmetrical chamber. As a special conservation measure, screw compressors have been recommended for refrigeration systems.[20] They are able to operate at much higher discharge temperature than reciprocal compressors and higher compression ratios than centrifugal, and therefore are better suited to systems in which the extracted heat from a refrigerating operation is used elsewhere in the plant.

A new type, based on a single screw, has been developed by the Chicago Pneumatic Tool Company and it is illustrated in Figure 11-5.[21] In the first model (11-5a), a central rotating screw mates with rotating seals (Roto-Seal) on either side. Each Roto-Seal is a series of vanes arranged like the spokes of a wheel. Compression starts when one tooth of the Roto-Seal enters a groove on the screw. A volume of air is trapped in the chamber formed by the groove, tooth and casing. As the screw rotates, the tooth follows along the groove's path, constantly reducing the volume available for the trapped air. Compression continues to increase until the tooth reaches the discharge port in the casing. The unit provides 575 cfm at 125 psig. The second model, shown in Figure 11-5, which replaces the central screw by spiral grooves, provides 700 cfm at 125 psig. Major advantages for these are due to their complete balance of pressures.[21] Their volumetric efficiency is 90-95% compared to 85-90% for twin-screw design. In addition, they promise greater durability in bearings and other components.

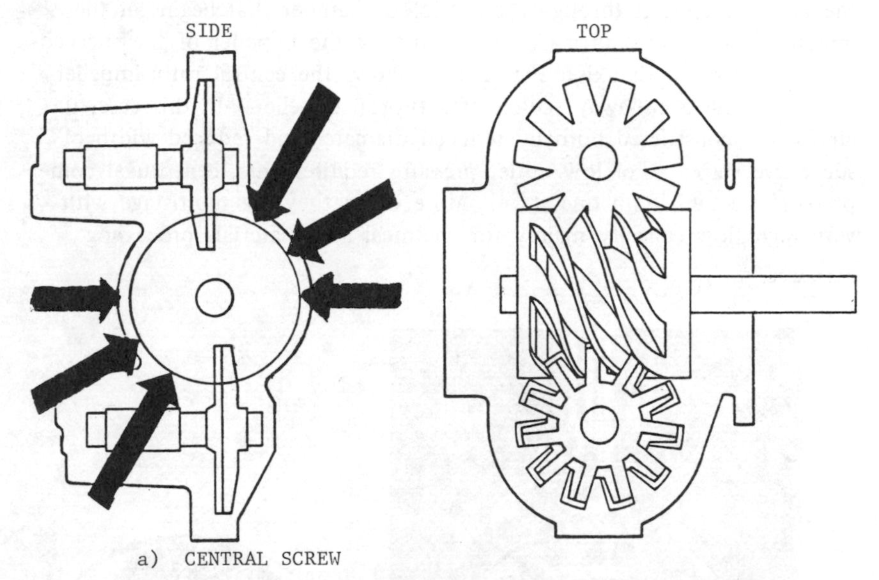

SIMULTANEOUS TWO-SIDED COMPRESSION PROVIDES BALANCED PRESSURES, NEUTRALIZING RADIAL AND AXIAL FORCES

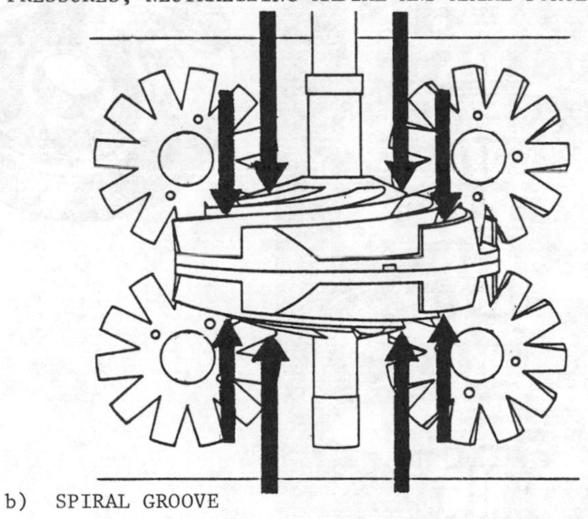

Figure 11-5. Single-screw compressors.

Dynamic Compressors. Dynamic compressors are divided into two types. Centrifugal compressors use a series of impellers rotating on a single shaft, which projects air radially outward. Each stage compresses the air by forcing it through a constricted chamber that begins at the impeller's outer rim and leads to an entry at the impeller of the succeeding stage close to its axis. Figure 11-6 shows the central rotor-impeller assembly with a cut-away sketch of a typical impeller. In this case, the air flow is constricted both by reduced diameter and reduced width of successive stages. For low outlet pressure requirements, centrifugal compressors are built with one stage. More often they are multistage with very high flow capacity, mainly for chemical and materials processing.

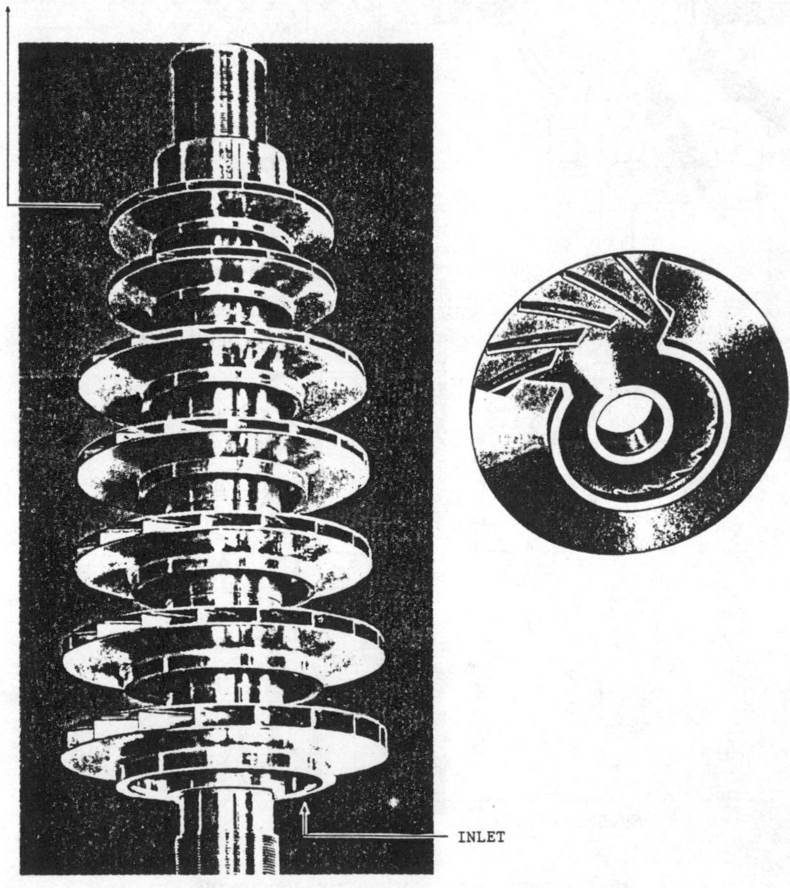

SOURCE: Ingersoll-Rand Centrifugal Compressor

Figure 11-6. Centrifugal compressor rotor-impeller assembly.

Axial compressors, the second category of dynamic compressors, operate on a similar principle of impelling air through a gradually constricted volume, except that the flow is constrained to follow the axis. This provides a much higher flow capacity with limited output pressure. Figure 11-7 shows the rotor-blade assembly of an axial compressor inside the casing. The blades of the rotor project the inlet air across their surface to travel in one direction through the gradually constricting annular space between the rotor hub and the casing. Between each pair of successive rows of rotor blades is a set of stator vanes projecting from the casing toward the axis and canted in the opposite direction to the rotor blades. The air flow vector across the circumference of the rotor is thereby reversed, slowing it and translating kinetic energy to a rise in pressure. In the centrifugal compressor, the same role is played by diffusers channeling the air flow from the impeller circumference to the axial entry into the succeeding impeller stage.

The centrifugal compressor has a limited stable operating range before surging occurs,[22] and therefore must be selected for the worst combination of circumstances and controlled to meet other requirements. Most centrifugal compressors operate around 20,000 rpm, which usually requires gearing up for electric motor drive,[23] but is ideal for steam and gas turbine drives. In addition, since the most economical and simplest way to control compressor output is speed variation; the steam turbine with its excellent speed characteristics is the ideal prime mover. A comprehensive review of field performance testing for centrifugal compressors has been provided by Pais of Clark-Dresser.[24] Size ranges vary from 20,000-650,000 cfm with horsepower requirements up to 35,000. Output pressures are from 10 to as high as 5,500 psig. Maintenance and attendant costs on centrifugals are low, and they frequently will operate continuously for two or three years without overhaul. In addition, they will tolerate dirty and corrosive gases better than any other compressor design, particularly those of single-stage design.

Axial flow compressors are also high-speed machines with normal operating ranges even more constrained. For equivalent size units, axials operate at higher speeds than centrifugals. They have less pressure rise per stage than centrifugals, requiring more stages for a given pressure rise. Efficiency generally will be better for an axial, depending on size and conditions. The axial will usually be smaller and lighter, requiring a smaller foundation with easier erection. On-line availability is about the same as for the centrifugal but is more sensitive to erosion and corrosion. Capacities range from 25,000 cfm to above 1,000,000 cfm. Pressures are usually well below 100 psig but may be as high as 500 psig. Pressure ratios vary from 2 to a limit of 7.

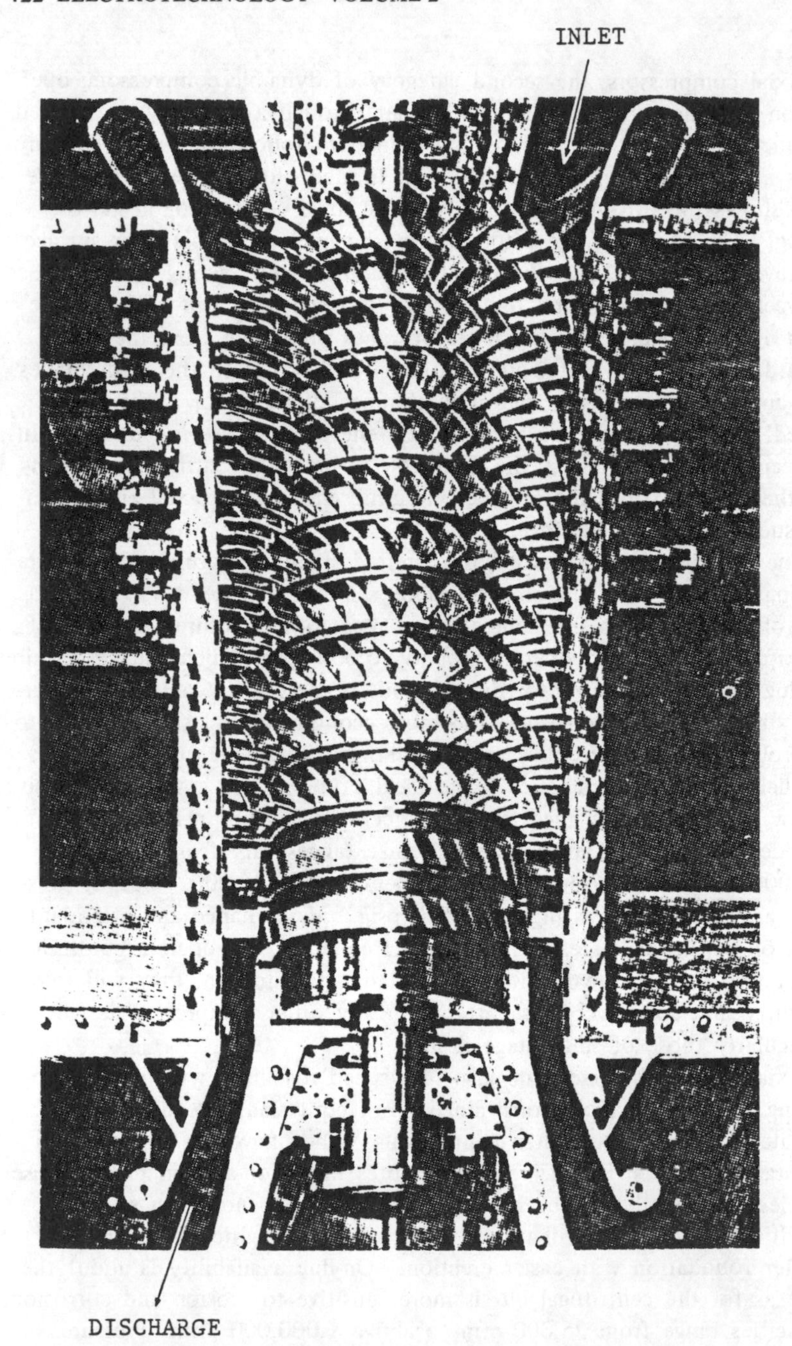

SOURCE: Ingersoll-Rand Axial Compressor

Figure 11-7. Axial compressor rotor and vane assembly.

Compressed Air Equipment

In addition to the compressor, compressed air systems are comprised of equipment to purify and otherwise condition the air, air mains to distribute it to points of use and the tools, devices and other facilities making use of the air. Ostrowski has discussed the typical air system deficiencies and maintenance problems involved in providing the enormous air requirements of a steel mill.[25] The layout and sizing of the air mains is a complicated problem,[26] and the demands of the equipment and flow rate limitations[27] must be scrupulously considered. Air leakage must be avoided and monitored after installation. More subtle effects, such as harmful air pressure pulsations, must be avoided in the original design and subsequent development of the system.[28] The following subsections discuss some of the characteristics of the equipment being served by the network of air mains, which must be considered in designing the entire system.

Equipment to Condition Compressed Air. Compressed air for industrial applications, which is generally hot when leaving the compressors, must be cooled before distribution, mainly to remove its adsorbed water content by condensation.[29-31] Dry air is particularly important for chemical plants,[32] and it has vastly improved the productivity of automobile assembly plants, 60 percent of whose tools are normally air operated.[33] After condensation in a chiller dryer unit, air is passed through a moisture separator and then reheated by thermal exchange with incoming air which is thereby precooled. A typical unit of this type is the Ingersoll-Rand R100 Model with a capacity of 25,000 cfm at 100 psig. A schematic for the unit is shown in Figure 11-8a. Other methods of drying include the use of dessicants such as silica gel,[34] which are recommended in unheated systems.[35]

Frequently combined with cooling and drying are various methods of filtration to remove dust particles, a growing requirement as air-driven equipment develops in complexity.[36,37] Benner[38] has reviewed the methods for removing oil aerosols from compressed air, a serious problem in handling paints,[39] and other easily contaminated materials. An example of compressed air filtration is the Deltech Filter (Deltech Engineering Co.), shown in Figure 11-8b, which is rated to eliminate 99.9% of solid and liquid contaminants normally found in compressed air.[40]

The major categories of equipment driven by compressed air and their functions are listed in Table 11-2. For materials processing, the system is usually constructed *in situ* and made to order either by or for the user industrial plant. Large-scale outside engineering contractors are involved, and the capital expenditures are major, running to hundreds

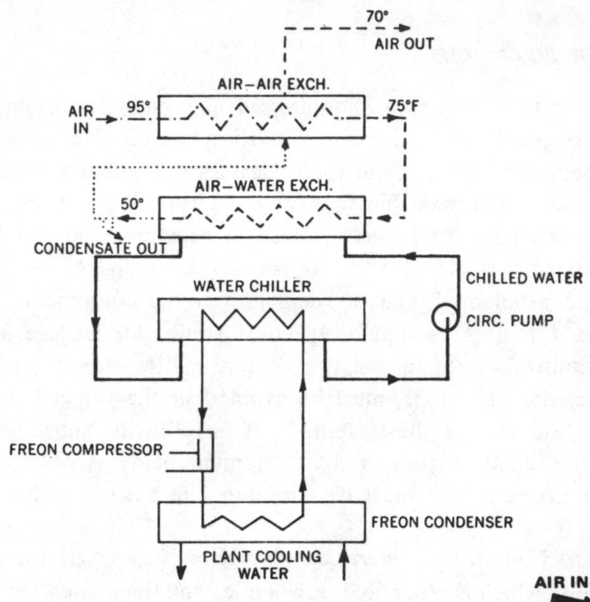

a) CHILLER-DRIER SYSTEM

b) FILTRATION UNIT

- LARGER PARTICLES SEPARATED INTO CONDENSED MOISTURE IN A & B AND DRAINED AT C. X = >U.C. "X"
- OIL TRAPPED IN D.
- FINE PARTICLES AND AEROSOL MISTS FILTERED IN E & F WHICH ARE REPLACEABLE AS COLOR INDICATES.

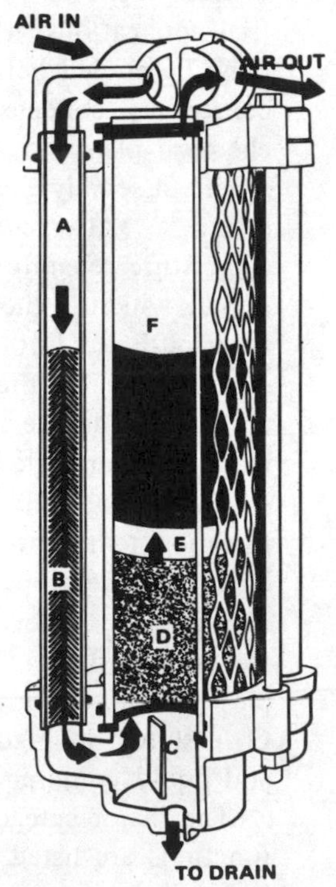

Figure 11-8. Air purification equipment.[1,40]

of millions of dollars for the entire installations. The other equipment is usually packaged and made by relatively specialized manufacturers.

Pneumatic Tools. Air-driven tools are characterized by the following general properties which contrast them with electrically driven tools:

- small size in relation to power used;
- freedom from burn out, shock hazard, or spark-ignited fires;
- simple variable speed control;
- simple construction; and
- low pilferage risk.

Rotary air motors in this category are prototypical as a direct substitute for the rotational drive supplied by the electric motor. Its operating characteristics are similar to those of a series-wound dc motor. For a constant inlet pressure, brake horsepower is zero at zero speed, increases to a maximum at the speed for most efficient operation and decreases again to zero at maximum speed. Torque is a maximum just above zero speed and decreases almost linearly as speed increases. This feature in fact provides a main advantage since no heating occurs when the rotor of a vane-type air motor is stalled.[41] One can lock the rotor, if required, by closing valves at the inlet and exhaust ports. The most prevalent construction principle is a reverse use of the rotary vane compressor (Figure 11-1b) with inlet in at the low–volume side and exhaust through the high–volume side. Zelenetskin has proposed a modification consisting of tangentially rather than radially arranged blades.[42] The increase in eccentricity that can be accommodated by them should provide more power.

Rotary air motors are used to drive most of the pneumatic tools such as the abrasive drills, screw drivers and most of the medical and dental equipment. Of five basic air motor designs, the rotary vane and piston versions are the most popular for industrial applications. Gear reduction and cost are usually the major selection considerations, and since gearing reduces efficiency and increases cost, the motor that demands the least is favored.[43,44] To drive tools that require a linear or sometimes rotary reciprocating motion, the percussion air motor is incorporated. This usually consists of an elongated cylinder and single piston. Using high-pressure inlet air, by suitable valving and porting, the piston is impelled to and fro at high speed to deliver a powerful forward blow through the operating tool. These are used wherever material must be broken up rapidly, chipped away, or impacted. Typical applications include breaking pavement, cleaning scale or sand from castings, removing rust, impacting sand in foundry work and driving rivets or taper pins.

Conveyance, Assembly and Control

Conveyor systems provide another important application of pneumatic drive. Air hoists with their freedom from shock hazard and easy control are ideal for most industrial environments where heavy workpieces must be handled under hazardous and sometimes awkward conditions. For example, the air hoist motor is always under positive internal pressure, which prevents the entrance of abrasive dust, corrosive vapors or other foreign material. Overhead air hoists are available with lifting capacities from 200 lb to 10 tons or more. Lifting speeds at rated loads and specified inlet pressures may range from 100 ft/min for small capacity, down to 5 ft/min for the largest capacity. Hoists can be fitted either with wire rope or chain links for coupling to the load.

In recent years, air flotation systems for extremely massive loads have been developed. They generally can be run from the existing high pressure air line of the plant and provide a way of moving and precise positioning for large systems where very slow or very small-scale movement is desired. Several examples are described among the applications in later sections.

Pneumatic conveyors using extended networks of tubes transfer materials either as a continuous stream or isolated items using relatively small unbalanced pressures. These are particularly important in automatic assembly systems. Either excess pressure or partial vacuums may be used. In the pressure system, the compressor precedes the system, while in the vacuum system it follows it. Generally, pressure differentials below 15 psi are sufficient to convey most materials. Differentials up to 30 psi are used for heavier products in mining operations such as iron and zinc ore, limestone and cement. Table 11-9 provides some idea of the variety of materials being conveyed pneumatically.

A great many devices are widely used for machining and other operations in manufacturing, and many of these devices are pneumatic. When they are combined into systems or machines, automatic controls are required for operations such as starting and stopping, sequencing, synchronization, feeding, and clamping. The pneumatic controls have in recent years been increasingly operated through pneumatic logic networks designed to operate on the same plant air pressure as the pneumatic devices they control. Some examples of pneumatic control valves operating as elements of a logic circuit are shown in Figure 11-9.

APPLICATIONS OF COMPRESSED AIR

The use of compressed air in U.S. industry has been established for about as long as mechanical motive power has been readily available. It

Table 11-9. Some Materials Handled by Pneumatic Conveyers[1]

Materials
Alum
Alumina (calcined)
Ammonium Sulfate
Arsenic (trioxide)
Asbestos Dust
Barium Sulfate
Barley
Beef Cracklings
Bentonite
Blood (dried)
Bone Char
Borax
Bran
Calcium Carbonate
Carbon Black
Catalysts
Cellulose Acetate
Cement
Clay (dried)
Clay (air-floated)
Cocoa Beans
Coffee Beans (green)
Copra
Corn
Corn Flakes (brewer's)
Cottonseed Hull Bran
Cycamide (pulverized)
Dolomite (crushed)
Feldspar (pulverized)
Ferrous Sulfate
Flax Seed
Flour
Fly Ash
Foundry Sand (green and core)
Fuller's Earth
Grain Dust
Grains (dry-spent)
Grits (corn)
Gypsum (calcined)
Gypsum (raw, pulverized)
Hops
Iron Oxide
Lime (hydrated)
Lime (pebble)
Limestone (pulverized)
Magnesium Chloride
Magnesium Oxide
Malt
Meat Scraps (dried)
Oats
Penolic Resin
Polyethylene (powdered)
Polystyrene Beads
Polyvinyl Chloride
Resin (synthetic)
Rice
Rubber Pellets
Rye
Salt
Salt Coke
Sawdust
Silex
Soap Chips
Soap Flakes
Soda Ash
Sodium Tetraphosphate
Soybean Meal
Starch
Steel Chips
Stucco (hydrocol)
Sugar
Titanium Dioxide
Vinylite
Volcanic Ash (pulverized)
Water-Conditioning Chemicals
Wood Chips
Wood Flour
Zinc Sulfide

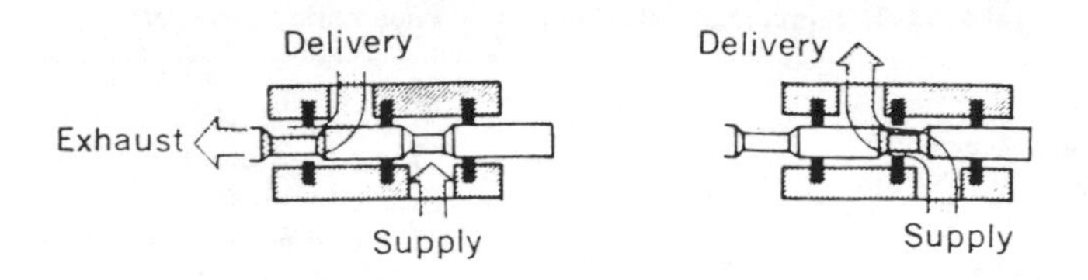

TWO POSITION

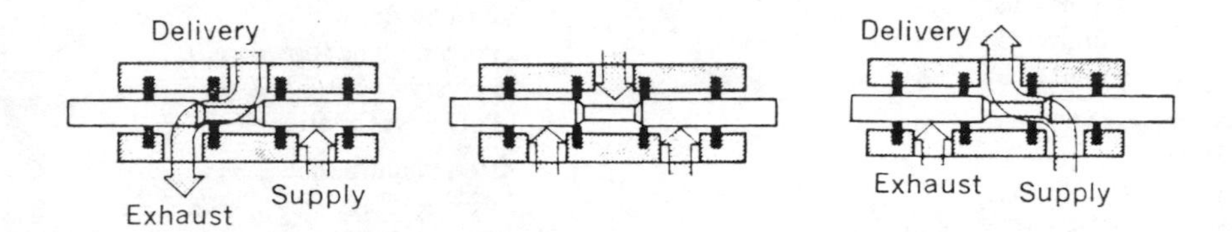

THREE POSITION

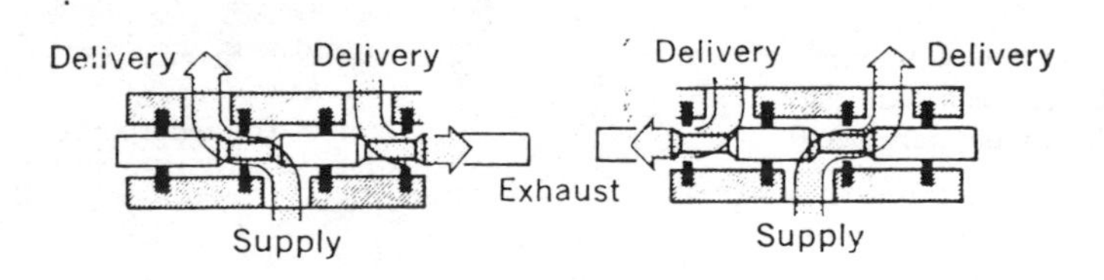

TWO POSITION 4-WAY

Figure 11-9. Pneumatic control valves.[45]

is generally installed in manufacturing plants as a widely distributed utility. However, the very high requirements of materials processing usually demand careful installation. The use of compressed air is also so well understood that new concepts find wide acceptance and soon result in established practice. For example, the United States Pavilion at Expo 70, Osaka, Japan was the first compressed air-supported cable roof structure of its size, an area of 270 by 462 ft. Most of the facilities used to provide tourist services in Washington, D.C. during the '76 Bicentennial celebrations were large temporary structures based on the same principle.

Semmerling[46] described the rapid spread of the use of fluidic control elements in hybrid electric-pneumatic and hydraulic-pneumatic systems while the relative merits of electronic and pneumatic control are still subjects of intense technical debate.[47]

Major Operational Categories of Industrial Compressed Air Applications

Established applications can be divided roughly into the following operational categories:[1,48]

1. driving force for an enormous variety of manufacturing tools and specialized equipment such as vibrators[49] and mining equipment;[50]
2. energy basis of assembly and control systems, consisting primarily of pneumatic devices;
3. conveyance medium, in such forms as (a) unbalanced pressure in pneumatic tube systems; (b) bubbles in fluid pumping systems; and (c) dynamic flotation to lift large masses for horizontal movement;
4. cleansing agent as a strong flow of purified air;
5. moisture and thermal transport medium for drying, humidifying and freezing; and
6. reactant in chemical, steel and other material preparation processes.

A complete list of the established applications of compressed air is presented in the Appendix, arranged according to the industrial sector categories of the U.S. national input-output model. Out of 79 sectors in the model used to describe productive sectors of the U.S. economy, compressed air applications are significantly represented in 48. Moreover, these industrial groups clearly account for the major position in the U.S. gross national product.

Recent Applications

These applications, which illustrate and define the trends in compressed air use, are discussed in two subsections. The first groups applications according to operational categories for which the application methods follow a well-established pattern that remains relatively unchanged. The second subsection includes applications specific to certain industries and for which there is not necessarily a standard pattern.

Applications by Operational Category

Of the six operational categories listed, all but the last are among the recent applications of sufficient interest to be subjects of articles and discussion, and are summarized in Table 11-10. The use of large-scale turbocompression to handle the large reactant air volumes required in materials preparation has been well developed,[51] and at present, most of the technical development in this area concerns refinements such as controls[52] and reduction of hazards.[10]

Pneumatic Manufacturing Tools and Specialized Equipment. There are a number of special air tools that have become increasingly popular recently for increasing productivity, improving quality, and in some cases for meeting government regulations.[53] These include newly developed tools for torque recovery, nut crimping and tensioning fasteners, all of which are used in automated assembly processes. In addition, special developments for computer controlling air tolls and tools developed specially for electronic component manufacture have been very prominent. Bizilia has pointed out the importance of retorquing to shrink the torque

Table 11-10. Recently Established Compressed Air Applications by Operational Category[2]

Operational Category	Application	Equipment Used
Pneumatic Manufacturing Tools	Bolt tightening, torque adjustment	Air motors, multiple nut rammer, torque recovery tools
Assembly and Control Systems	Automated assembly handling small irregularly shaped parts	Pneumatic control valves, fluidic elements
Continuous Conveyance	Transport of wide variety of materials in solid, slurry, or fluid form	Special control systems to prevent abrasion and clogging of pneumatic tubes. Air diffusion for liquid pumping
Large Manhandling	"Floating" large heavy structures for controlled movements	Controlled air pressure with flexible envelopes to contain compressed air
Sanitation	Cleansing air in hazardous working environment	Hoods and fresh, dry air supply, large air blowers
Moisture and Thermal Transport	Flash freezing with chilled air (-200°F)	Dynamic compressors for compression and cooling by adiabatic expansion

scatter in multifastener joints,[54] which traditionally has become an accepted method in high-speed assembly using tools such as multiple nut runners. The new breed of "retorquing air motors," such as those developed by the Chicago Pneumatic Tool Co., has also provided secondary advantages in tool maintenance cost and tool life.

Assembly and Control Systems. Pneumatic control elements and, increasingly, the use of fluidic amplification and logic, have become more prominent in the development of assembly systems and control of materials preparation processes. Buckley has developed the special analyses and principles to place the dynamic design of pneumatic control loops on a quantitative basis.[55] Alternate loop configurations, factors that affect frequency response and overall performance, and a modular approach are all essential elements in loop design. For example, it has been found that transmission distance has been overrated as a limiting factor in pneumatic control systems. Bailey has explained how process control requires a special balance between pneumatic and electronic elements.[52]

The decision of whether to use pneumatic or electronic control has been shown by Deuschle to depend relatively little on the cost factors involved.[56] One can arrive at the following criteria as most important in the choice between electronics and pneumatics:[47]

- performance–accuracy, resolution, sensitivity, repeatability and speed of response;
- flexibility;
- reliability;
- maintenance factors;
- safety considerations and environmental impact; and
- cost.

In general, electronic instrumentation suits applications having fast control loops, long transmission distances and digital logic requirements. Pneumatic instruments have the edge in corrosive, hazardous atmospheres, or where power supply disturbances are frequent.

For automated assembly systems requiring complicated distribution of irregularly shaped small parts, a new system, EDI-veyor from Engineering Designs Ltd (U.K.), uses elaborate sensing and control to keep parts moving.[57]

The basic section of the EDI-veyor system is a length of tubing, air–operated valves, two sensing heads, a decelerator and a pneumatic control panel. Pickup and delivery ends of the transport tube have electronic detectors to sense both entry and approach to the tubes' end point. Jamming is prevented by the system's design logic. Only one item is permitted in any specific segment with air flow stopped if the item fails to pass the second sensing head.

Continuous Conveyance. Compressed air for conveyance generally is managed in three basic forms. Discrete small masses are piped in pneumatic tube arrangements that have long been established as a workable practice. As the masses get larger, this technique is more difficult to manage economically. Fluids and fluid-like particulate material are managed in a similar way. Finally, very large masses have been assisted in small movements and special support by using air films as a bearing. These are discussed in the following subsection.

The limits of pneumatic conveying of particulate matter have been studied comprehensively by Capes and Nakamura using experiments on vertical conveyance.[58] Material was pumped through a 3"D riser, 30 ft in length using solids ranging in particle Reynolds number from 100-3500. Particle/gas mass flow ratio varied from 0-8, and pipeline conditions extended from dilute flow down to choking. Average particle hold-up was determined by means of a series of quick-closing valves, and particle slip velocity was often found to be greater than calculated terminal velocity. Frictional losses in the riser were also determined.

Most particulate conveyance systems have a serious problem handling fragile or abrasive products due to the high transport velocities normally required to keep the pneumatic lines from being plugged by the product. Trace air systems as developed by Semco of Houston, promise to solve this problem.[59] The systems use multiple sources of air to provide controlled boosts of pressure at short intervals along the transportation path. Items move slowly through the system with blockages relieved by a system of pressure differential sensing and supply. In a practical example, the Semco system is handling the catalysts, a very fragile material with a high unit cost, used in the manufacture of catalytic converters for automobiles.

The typical flow pattern of the system can be illustrated by considering a simple specific case: a short line with a pressure vessel, three booster stations and a use bin at the end of the line. The pilot air line, which controls the first booster, is tapped into the pressure vessel. When this vessel is empty or is being filled, no pressure is carried through the pilot line to the first booster, whose valve remains closed. This lack of pressure is sensed in sequence at the next two boosters, which also remain closed.

The transfer cycle is started by the buildup of pressure in the pressure vessel. The first booster responds to the pressure and opens to build the pressure over its segment of line and, similarly, for the successive segments controlled by their boosters. Product begins moving out of the vessel, and the conveying line quickly fills with plugs and spaces from the pressure vessel to the use bin. Suppose a lengthy plug were to bind the conveyor line because of its resistance to flow; the plug would stop moving, but the pressure behind it would continue to build. Pilot line pressure also would build, thereby opening the valve at the first booster to build a matching pressure. In succession, the next two boosters, sensing the rising pressure, would also build up pressure along their segments until the product downstream from the third is pushed out of the line into the use bin, producing a pressure drop. This clears the way for product movement between the second and third boosters and continues up the line until the inertia of the problem plug is overcome and the cause of the initial pressure buildup has been alleviated.

The net effect of pressure buildups is that in addition to clearing downstream resistance, they also break the movement of upstream product to prevent sudden high velocities when plug inertias are overcome. In operation, the plugs progress through the system in a series of stop-and-go movements with booster valves constantly operating to keep the system in balance.

A similar system has been developed by Sturtevant in England[60] and used with velocities as low as 10 ft/sec.

Fluids have been conveyed vertically using the principle of the bubble pump,[61] illustrated in Figure 11-10. Bubbles of air are released into a

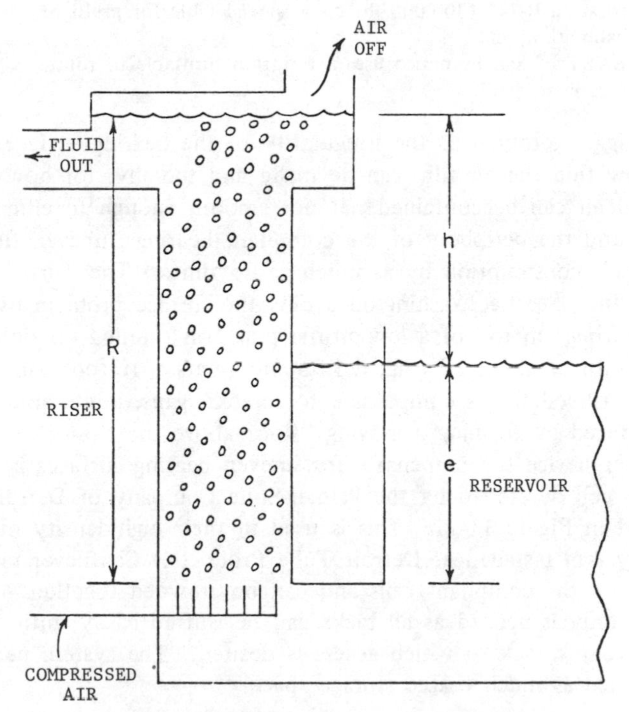

Figure 11-10. Bubble pump.[61]

confined liquid column to reduce the effective liquid density within the column below that of the surrounding reservoir. This creates, in effect, a pumping head that can be used for vertical conveyance. The most prevalent features of bubble pump applications are:

- suitability factor similar to those of an eductor;
- low efficiency and extra expense in initial cost when viewed as an independent system. Savings in cost realized when used as an adjunct to existing equipment;
- efficiency and relative low initial cost, realized at large low volume and low head; and
- aeration or a useful by-product, possibly saving an additional processing step.

Large Mass Handling. As early as 1912, an Australian, Alcock, demonstrated that for each pound of vertical pushing force exerted by a partially entrapped air pocket, approximately 1000 lb of mass can be moved horizontally. The most typical of such applications, usually managed under ideal bearing conditions, are similar to the following:

- an applied air film pressure of only 20 psig has been used to rotate a B-747 (300,000 lb) on a special table for preflight calibration; and
- BART of San Francisco uses a flotation turntable to rotate railroad cars in its yard.

The biggest problem is the irregularity of the bearing surface, which limits how thin the air film can be made and thereby, for how long the entrapped air can be contained. If not smooth enough to eliminate air leaks around the periphery of the compliant bearings, uneven floors could multiply air consumption by as much as 25 times. The Turtle from Aero-Go Inc., Seattle, Washington avoids the surface problem by mounting its air bearings on top of a low-profile platform running on ribbed tracks.[62] With 40 cfm of air at 18 psig, it lifts and floats a 10-ton load. The traversely ribbed tracks compensate for cracks, gaps, joints and rough floor surfaced by forming a moving "floor above the floor."

Another device to compensate for uneven bearing surfaces is the compliant air cell developed by the Palmer-Shile Company of Detroit[63] and illustrated in Figure 11-11. This is used in their high-density mobile storage system installed at Detroit Tube Products. Cantilever racks are mounted on the compliant cells and can be crowded together. Space for one aisle only is needed as all racks can be shifted easily until the aisle is adjacent to the rack to which access is desired. The system permits almost twice as much usable storage space.

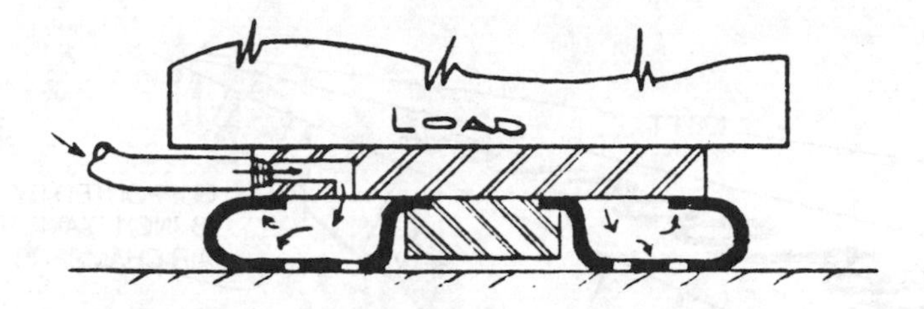

Air is pumped into inflatable cell to lift load of normal bearing pad.

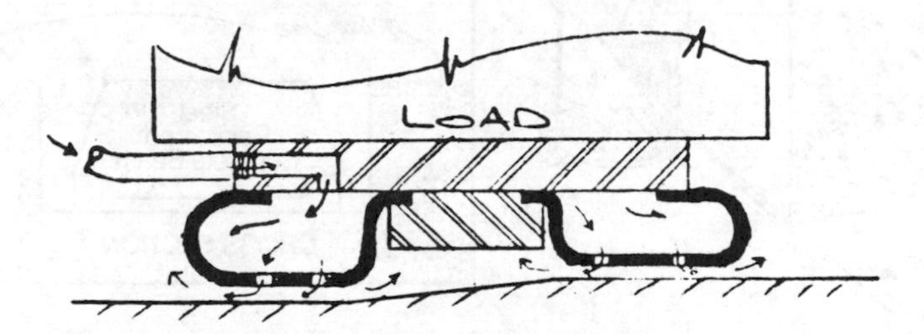

As air pressure is increased, air escapes into cell bearing area. Flexible inflated cells change shape to conform to uneven surface and maintain air film during load movement.

Figure 11-11. Air flotation cells.[63]

In delicate scientific experiments, one of the most difficult random disturbances to eliminate has been vibrations transmitted through the earth and floors to the equipment. They have contributed substantially to the residual uncertainty in many classical determinations of basic physical constants. Air flotation has been used to decouple these vibrations from equipment being used in laser measurements.[64] An optical bench of solid concrete, 106 ft long and weighing 132 tons has been floated on 12 interconnected pneumatic isolators. These support the entire weight of the bench, keep it level (within one quarter over its entire length) and free from ground vibration. Design and installation of the optical bench, shown in Figure 11-12, was directed by the Barry-Wright Corporation of Burbank, California and Watertown, Massachusetts, for the White Sands Missile Range. They make the 12 Serva-Levl units, which support the bench on 18-in. D compensated columns of compressed air. It is common to find normal building vibrations of 50-90 x 10^6 in. which is 4000 times the 5 Å resolution of a typical electron microscope. Compressed air has provided isolation from that disturbance in the White Sands laser measurements.

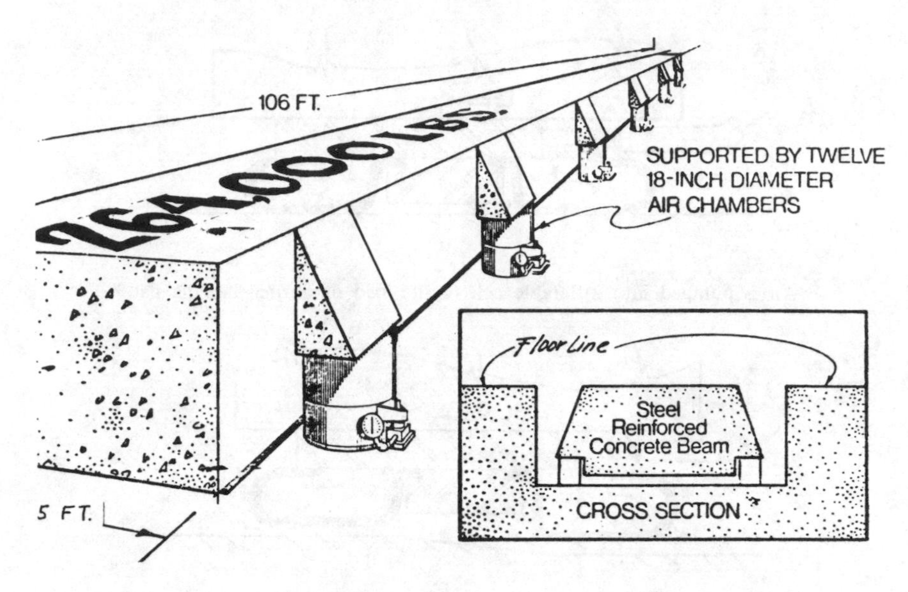

Figure 11-12. Floating concrete bench.[64]

Sanitation. One of the best ways to treat wastewater is to diffuse air through it as thoroughly as possible. The Water Pollution Control Corporation of Milwaukee, Wisconsin, has developed an effective air diffuser to overcome the chronic clogging problem of most others.[65] Their principal product is the Santaire aeration system, which incorporates an air diffuser of an inverted reservoir type. Air flows at rates varying from 8-50 cfm, with head losses controlled by an insert installed in a balancing nozzle, then passes from header into the diffuser. Exit ports are two rows of holes on each side of the diffuser's air reservoir. Below the diffuser's open base, a V-shaped deflector plate directs the flow of waste liquor up along its outer wall. As the air comes out, it is sheared into relatively small bubbles and then distributed well throughout the wastewater in the basin. The diffuser is fabricated of thin-walled stainless steel to minimize clogging due to formation of iron or calcium deposits. Water flow shearing of the bubbles has allowed relatively large ports without losing the high water interface contact area of small bubbles.

At the National Animal Disease Center (NADC) in Ames, Iowa, compressed air is used in a variety of ways to control the spread of infectious disease agents.[66] Laboratories are maintained at a negative air pressure of .03 in. of water, so that net air flow into them prevents the escape of microorganisms. Large blowers circulate 160,000 cfm of air through high-efficiency filters, which remove 99% of all particles down to 1 μm. Nonlubricated compressors using piston rings and packings of blended TFE fluorocarbon provide oil-free air. A primary use is breathing air for inflation of transplant hoods worn by experimenters. The air flows out positively from the bottom of the hood. Other air applications include the center's pneumatic controls, air agitation of plant effluent to speed oxidative breakdown of hazardous materials and daily spurts of 125 psig air to unclog a drain line carrying high-cellulose animal wastes that are highly viscous and hard to pump. Air is supplied by two 250-hp class XLE Ingersoll-Rand reciprocating compressor, each used on alternate days to provide 1300 cfm of oil-free air at 120 psig.

Moisture and Thermal Transport. The use of controlled precisely conditioned air to provide the moisture and temperature requirements of materials such a paper, textiles and food generally involves a combination of both since the saturated vapor content of water will be influenced by its temperature and pressure. Havighorst has described the Kryoaire system for flash freezing with air at -200°F.[67] The product moisture loss is minimized in an expansion turbine system that maintains a constant -80°F or below for direct contact freezing of foods. The system is mechanically simple–no coils, blowers or defrost cycles–and saves space. Typical refrigeration capacity achieved is 1.5 million Btu/hr. The turbine system continuously draws direct contact refrigerant air from the freezer. The air is compressed and cooled, then expanded to atmosphere to reduce its temperature to -200°F and returned to freezer. The small amount of make-up air, required due to leakage, is dried and purified through dessicant driers before entering the system.

New Applications by Major Industry Sectors

There are a large number and variety of compressed air applications that are primarily of specific interest to the operation of a particular industry. These are arranged by groups in the following subsections associated with those major industry sectors of the U.S. input-output model that seemed to derive the most immediate and unequivocal benefit. A summary of their descriptions is presented in Table 11-11. In most cases, tangible production or operating cost reductions were effected along with some dramatic benefits in safety, reliability and convenience.

Mining: Metals and Coals. Compressed air is being used more and more in one of the most important mining tasks–the transport of ore, rocks and other ponderous materials that are always difficult to handle. The pneumatic tube principle is being extended to the large ore carriers used

Table 11-11. Recently Established Compressed Air Applications for Major Industry Sectors[2]

Industry	Application	Equipment Used
Mining	Transport of ore, rocks, slurry	Large-diameter pneumatic tubes driven by 600-cfm compressor
	Thawing frozen coal masses	Compressed air in drilled holes
Crude Petroleum and Natural Gas	Air-powered drilling	Rotational drills powered by air used to expel cuttings
Primary Metals Manufacturing: Foundry Products	Automation of foundry procedures	Air-powered ladles and assembly line
Primary Metals: Metal Forming	Depth-controlled cold-working of sheet metal	Compressed air-powered ball peening
Special Industry Machines: Publishing	Automatic typesetting	
Food and Kindred Products	Cleaning, transport, heat transfer	Air substitute for water
State and Local Government Service	Fire-fighting, highway repair, sanitation	Air-powered chemical spray, pneumatic waste conveyors
Government Industry	Ammunition handling, deep sea operations	Pneumatic control circuits, self-contained compressor units

in mining,[68] both for increased safety and greater transport capacity. It has been found that relatively low pressures are sufficient to achieve practical speeds for trains of cargo ore.[69] Pneumatic ore conveyance, however, is still experimental and primarily being tried in surface operations[70] even though tunnel conditions are the major concern.

A recent experiment of considerable interest has been carried out by Martin and Faddick, who were aiming at an alternative to the slurry pipeline.[71] Western coal fields are not water rich and their experiment tested measures to minimize water requirements. Materials excavated by a tunnel borer were crushed and treated to form a muck, which was conveyed up a 500-ft 10-in. D pipe with vertical lift of 150 ft on a 27° slope. The test unit was considered full scale moving 100 tons/hr. Driving force was a 6000-cfm compressor powered by 350 hp. In similar work on a smaller scale, Konchesky has developed empirical relations describing the air and power requirements for the pneumatic transport of crushed coal in horizontal and vertical pipelines.[72] The test system included 2-, 4-, 6- and 8-in. D pipelines of 200 ft followed by a 90° bend and shorter sections. Air requirements are functions of coal rate and specific gravity, pipe diameter and length, ratio of cross-sectional area of pickup section to that of pipe. Although coal size was not a significant variable, maximum size cannot exceed .4 of the pipe diameter. Highest coal rate–55 tons/hr–was achieved in a pressure operation with 1.35 specific gravity coal transported through 8-in. D pipe. With pickup/pipe area ratio of 3:1, about 2200 cfm at pipeline entrance was required to keep the coal moving in full suspension with a 7.5 psi drop in pressure.

The use of pneumatic systems to handle coal has also been used at industrial plants that consume it. Hunter has described the experience with a prototype program-controlled system to convey coal and ash at a boiler plant.[73] A more recent application has been introduced by the Long-Airdox Company of Oak Hill, West Virginia in a system, marketed under the name Hopper-Popper, that uses air pressure to break up frozen coal masses.[74] Moisture on hopper cars tends to freeze coal contents. Previously, the only solution was to thaw the coal using vast amounts of energy. The Hopper-Popper was originally developed as a nonexplosive method for mining coal in gassy mines. High-pressure air tubes are placed into holes drilled in the mine face, and air is discharged at up to 10,000 psi. Compacted materials are broken up in storage bins and silos using pneumatic shells permanently installed in bin sides. Now frozen coal can be broken up in a hopper car in 6-8 min after car has been positioned under the main unit. Thawing took 4 hr.

Down-hole or in-the-hole drilling is usually limited by the height available in the tunnel for the driving equipment. The International Nickel

Company of Canada (INCO) has been using a new unit designed by the Ingersoll-Rand Company–the CMM/DHD jumbo, which can be "caged" from one mining level to another.[75] Air-driven, down-hole drills differ from conventional precision drills in that the hammer operates directly behind the drill bit at the bottom of the hole (Figure 11-13). Operating on compressed air at 85-100 psig, the piston strikes the drill bit; no power is lost in the transmission of driving energy through the drill string as in conventional long-hole drilling. Rotation of the drill rod and hammer assembly is supplied by the rotary head at 15-20 rpm. An air-water mixture is fed through the bit continuously to clean the hole and control the dust. Air consumption while drilling is about 400 cfm. Environmental conditions for the operations have been improved; physical effort is reduced since the driller has freedom to move about and perform other functions. There is a considerable reduction of fog, water, oil spray and dust. Sound levels have been reduced from 117 dBa for conventional drills to less than 100 dBa for down-hole drills. Several million tons of ore were mined by INCO in 1973 and 1974 using the new methods.

Crude Petroleum and Natural Gas. The rise in gas prices has provided a major incentive to develop once-marginal gas reserves with a high density of wells. Air drilling is suited to hard or abrasive formations.[76] The compressed air vented down hole carries the abrasive cuttings up and out of the hole, avoiding multiple drilling or cutting of the same layer of loose material, saving on drill bit wear and increasing drilling costs. A contractor with two air drilling rings can sink as many as 6 wells per month, averaging a completed well every 10 days. This compares with 30-90 days to drill a comparable 8000-foot well using fluid drilling techniques. Also, fluid drilling uses 25-30 drill bits per well. Some air-drilled wells use one 8.5-in. surface and two 6.5-in. production bits.

For offshore oil/gas operation, an air-powered winch has been developed.[77] The winch is model HA-155 of the Hydro Products Division of Tetra Technology. It holds 2500 ft of .88-in. D electromechanical cable. A piston air motor is coupled to a single-worm gear transmission sealed in oil. The complete winch can be purged for electrical safety. The model available is sized at 48 x 58 x 54 in., weighs 2000 lb, and its maximum retrieval rate is 150 ft/min.

Primary Metals Manufacturing: Foundry Products. The advantages of air-driven equipment for foundry operation have long been recognized. Temperature extremes of operating conditions in particular tend to favor total use of air for both operation and control.[78] The widespread use of air-operated devices has emphasized the requirements of air system

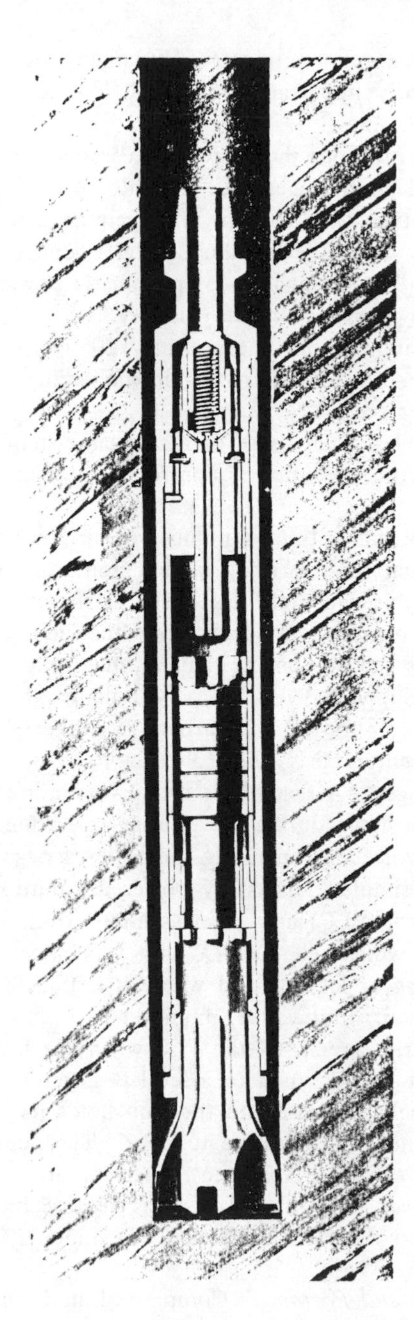

Figure 11-13. Down-hole drilling through rock strata.[75]

efficiency for both economy and maximum products. Huskonen, for example, has discussed the particularly important problem of low air pressure and contaminants in the air lines.[79]

A recently developed foundry continuous molding system is one approach to comprehensively designed air management. Foundry operations are notoriously difficult to systematize, with so many that require careful human attention and manipulation. Nevertheless, the Central Foundry Division of General Motors has used a combination of pneumatic power and logic to develop what amounts to an automated foundry.[80]

The foundry consists of a continuous mold line pouring 120 tons of iron into as many as 450 molds per hour. Different-sized molds are being handled on a programmed-mix basis.

The line's pace requirements are less than those of a manual system of comparable capacity, and the entire operation is controlled by a monitoring crew of two.

The system, known as Roto-pour, consists of a number of air-powered ladle carriages that move in a circular path carrying molten iron from a holding furnace to a continuous high-speed mold line. The carriage houses the air motors and drive elements required to rotate the ladle and propel the carriage itself. It carries the equipment that automatically sequences the pouring operation. Supported on two polyurethane wheels, only one of which is driven by the air motor, the carriage is connected to and rotates about a stationary center post, which also serves as the air supply column for all the carriages. A second air motor, operating through a speed reducer, rotates the ladle during the pouring operation.

No electrical components are involved in operating the ladle carriage systems. This allows the use of a simpler, less costly power distribution system. The pneumatic logic control (PLC) unit receives inputs from strategically placed air limit valves, which tell the PLC unit what area of the circle a given carriage is entering or leaving, and where it is in relation to other carriages. Two more air limit valves, mounted on opposite sides of the ladle shaft, actuating trip levers, tell the PLC when the ladle has completed pouring and when it has returned to a level position.

The pouring operation is extremely accurate and the iron holds its heat because it takes only six seconds from furnace to pour. The pouring rate is consistent with no false starts or stops, no overpours, and splash is negligible. Yield is improved over manual systems because the required amount of iron is placed in the ladle within plus or minus 2%.

Primary Metals Manufacturing: Metal Forming. Compressed air is the driving force of a particularly imaginative treatment of metal parts designed to protect them from the effects of fatigue. *Controlled shot peening* uses millions of tiny steel spheres or glass heads propelled by compressed air velocities of about 200 fps to bombard the surface of a critical part. Each piece of shot acts as a tiny peening hammer to indent the metal surface of the part being treated. A layer of residual compressive stress is thereby produced, equal to about 60% of the ultimate tensile strength of the metal being peened. The object of controlled shot peening is to produce a compressively stressed surface layer, in which the amount and uniformity of stresses and the depth of the layer can be held constant from piece to piece.

Fatigue failures are caused by tensile stresses that occur in the surface of a part, for instance as a result of a bending load or as the residual stresses present in a surface that has been ground, welded, machined, or plated. When these stresses are added to, by the operationally applied tensile loads under dynamic conditions, the combined stress could be great enough to cause premature failures. When such a part is shot peened, residual stresses become compressive, and since fatigue failures do not initiate in an area of compression, any applied loads must overcome the residual compression before fatigue failures can occur.

The variety of items that receive shot peening is impressive: gears of all sizes, shafts, staple drivers, wire terminal clinchers, shotgun cartridge extractors, steam and gas turbine parts and, of course, valves and blocks. "Peen forming," is undoutedly the most comprehensive application of this process.[81] It is used to add curvature to the wing panels of the DC-10 wide-body jet. Shot peening one side of the sheet of metal stretches that surface. Combined with the action of compressive stresses, this causes the metal sheet to develop a spherical curvature, convex on the peened side. By exercising rigid controls in this dieless process, panels that cover the wings of most of the larger aircraft can be formed to exacting aerodynamic curves.

One word that must be emphasized with respect to shot peening is "control." There is, at present, no nondestructive method of determining after treatment if a part has been correctly shot peened. Therefore, it is necessary to exercise stringent controls before the part is peened. This requires automated equipment, tight restrictions on shot quality, and careful determination of the relative kinetic energy delivered by the shot.

Trends in the use of steam and electricity as power sources are subject to a variety of contracting influences. As example we have the following two cases: the first, in which steam power has been replaced by electrically powered compressed air for environmental considerations, led to economies in work convenience and maintenance of associated equipment; and a second, in which a steam-driven compressor is being returned to full service for its energy efficiency and reliability.

In the first case, the LeFere Forge and Machine Company of Jackson, Mississippi switched from steam to compressed air to drive its large hammers.[82] The switch was originally forced by the U.S. EPA because the coal-fired boilers of the steam hammers were ruled too high in the emission of air pollutants. The cost of scrubbers was too high, and a cheaper way for the company to comply was to convert the steam hammers to be powered by electrically driven air compressors. The company, in fact, found that the conversion also made economic sense.

In 1973, the company installed electrically powered centrifugal compressors, each capable of supplying 7000 cfm at 125 psig. The old boiler headers were used as air receivers. The system uses no aftercoolers and insulated air pipes so as to utilize the full heat of compression to drive the hammers. As a result, they have reduced maintenance problems to a large extent. In place of two men full time, they use one man about two hours a week to check the equipment and perform routine care. Lubrication costs of the hammers dropped about 50% without the high-temperature tube loss. Hammer packings last one year as opposed to six weeks under steam. The cylinder-sleeve life span was lengthened from six to an estimated ten years. In addition, working conditions were improved with reduced visibility and added heat, due to steam removed in the air operation. There is added safety in that the centrifugal compressors supply oil-free air to eliminate any chance of down steam lubricant ignition. The hammers also work better and faster under the constant 125 psig pressure of the air line. The forge shop has become convinced of the advantages compressed air has over steam: low noise, high hammer efficiency, dependable operation, lower costs and safety.

In the second case, the Pepsi Cola Company Bottling Works of Long Island City in New York City is testing the relative economy of fossil-fueled steam and utility-supplied electricity for their compressors.[83] Currently it uses three compressors totalling 500 hp. One is a steam-driven reciprocating Ingersoll-Rand Class XPVR compressor rated at 200 hp, which has served the plant since 1937. The unit has two stages and generates 1000 cfm of air at 100 psig. Downtime has been negligible. Over the past 12 years, it has required only one major overhaul.

Air uses of the plant range from instrument and control, operating air agitators in the refinery, maintenance tools and general plant air to powering much of the automated equipment on the bottling line. As the facilities grew beyond the 1000-cfm capacity, new electrically powered compressors were added in 1972. In 1973 Pepsi Cola closed its sugar refinery, reducing requirements for steam, hot water and air. The steam-driven compressor was put on standby service. By 1975, however, electricity rates in New York City had risen so high, particularly for high-demand conditions, that use of the steam-driven compressor was resumed, especially for the possibility of using coal-fired boilers.

Special Industry Machine: Publishing. York Graphic Services of York, Pennsylvania has developed a very sophisticated application for compressed air in publishing.[84] One York Graphic patent, for an air-operated Automatic Pi Insertion (API) unit, has added unusual versatility to the firm's Fototronic typesetter, which photographs alphabetic characters at the rate of 30-40 characters per second. The machine then automatically selects individual letters from one of the five glass font disks, each containing 256 characters. After one of the font disks has moved into photographing position, it continues rotating at 3600 rpm while the correct character is being exposed. A movable prism, actuated by air pistons, allows the proper selection of one of two character rings on the disk. Exposure of the character is by a 6000-V spark discharged across tungsten electrodes. Compressed air is used to automatically:

- lock glass font disks in position;
- move the patented API mats into exposure position;
- move the font prism, which selects the character ring;
- clean the spark gap housing of accumulated dust;
- move the proper capacitor contacts in place to vary the intensity of the exposure spark;
- clean any dust buildup from the font disks as they change position;
- move the photocells that pick up spacing data from the font disk; and
- actuate a guillotine that automatically cuts an exposed sheet of film from the machine's leaf roll.

Clean, dry air is ensured for the operation by the installation of filters on the lines from the compressor, which supplies air at 125 psig pressure, regulated down to 40 psig at the machine.

Food and Kindred Products. Air has been described by Heldman as a useful substitute for water in certain food processing operations:[85] cleaning, transport and heat transfer. Available data in the literature are of limited use in evaluating the economic benefits. Examples of air use alternatives are a pneumatic transport system in preference to mechanical conveyance or evaporative vegetable cooling over flume or spray system.

While reduction in water use and hence effluent volume is an immediate benefit, savings are also possible in reduction of capital costs and the loss of solids through leaching. Optimum applications for air appear to be precleaning, transport over long distances and, in conjunction with a small quantity of water, evaporative cooling.

Pneumatic control in food processing can be of special value. Pure Lard of London has used a system supplied by Auxiliary Pneumatic to eliminate spillage of edible oil.[86] The control cabinet has an illuminated synoptic display with each node of the flume system controlled by two 5-port miniature values mounted on the synoptic display.

An outstanding example of an automated control system custom designed and built in the U.S. is the assembly process of Schrafft's factory in Boston.[87] The candies are transferred and allocated for a wide variety of assortments throughout the entire process up to final packaging. Each box is automatically filled almost to full capacity using air sensors, the last few pieces being added manually so that the box is then fully packed.

The system, made by Kaman Industries, consists primarily of two long tables. One holds feeder trays of vibrating chocolates, the other sprouts pneumatic pick up arms hovering over a moving line of receiving trays, each with pockets molded individually for separate pieces. The arm assembly, coordinated with the feeder trays movement, picks up a chocolate from them and places it squarely in the pocket of a receiving tray as it moves steadily forward. Each arm puts only one type of chocolate in its predetermined pocket. Any missing spaces at the end are filled by hand.

State and Local Government Services. There are a few new or growing users of compressed air in providing municipal services that are important in some of the equipment being used.

A chemical fire-fighting engine uses compressed air to propel both chemicals and water.[88] It is called Commando Mark III and was developed by the Fire Combat Corporation of Cape Canaveral, Florida. The apparatus works by ejecting the dry chemical in high volume and under high pressure (2400 psi) into the fire. The flow rate is adjustable to 10 lb/sec and storage capacity is 250 lb of dry chemical. Water is used primarily as a cooling agent and protective screen for the hose man.

For fast, efficient, quiet pavement–breaking, Ingersoll-Rand developed "Goblin," a hydraulic impulse breaker.[89] It delivers 135-600 blows/min with 250-500 foot-lb of energy per impact. The unit breaks up pavement quickly and silently by punching holes. The hydraulic system requirements are 25 gpm at 2000 psi. Impact energy is governed by the pressure of a nitrogen charge and frequency by rate of hydraulic flow. Oil from a hydraulic system enters the drive, raises the hammer to the top of its stroke and pushes the piston down, compressing a charge of nitrogen gas. At a preset hydraulic pressure, the stored gas is instantly released causing the hammer to strike the tool with a total energy output of from 33,800-300,000 foot-lb/min. The unit has also been much more effective in compacting earth after repair work with minimal subsequent settling.

Pneumatic methods of waste collection are gaining in popularity, as described by Dallaire.[90] For example, dozens of U.S. hospitals have installed vacuum networks to collect refuse and dirty linens. Initial capital costs are high with economics giving a break-even point in 6-8 years. Costs of pickup from handling and transport to disposal can be reduced to about 25% of conventional costs. A privately owned, low-income apartment complex in East Harlem of New York City has an ECI Airflyte-gravity-vacuum trash collection system. Trash falls to the bottom of each gravity chute where it is collected and compacted before being conveyed to a central collector. The system cost $400,000, about 1% of the entire cost of the complex.

Government Industry. Many government operations pose unusual requirements that are best met by the use of air-operated devices.

The hazards of handling ammunition in a depot are so high that locomotives driven by compressed air have been designed for their specific use.[1] Manufacturing and assembly operations for ammunition and other explosives pose similar hazards and have attracted an emphasis on air-operated systems. In particular, the use of pneumatic logic components for assembly operations is preferred over electronics. Eames and Sedergren describe the use of air logic in a packaging machine that organizes in proper sequence, 25 separate operations to fold, collate and box ammunition bandoleers.[91] To insure that each step in the operating sequence of the machine takes place only when called for, positive actuation is used on a series of pneumatic limit valves that signal each succeeding step, similar to the operation of a programmable controller. Hehn and Varacins describe the use of pneumatic logic in the automated assembly of explosive charges to make it a relatively safe operation.[92] Propellant charges, consisting of three or more cloth bags, ranging in weight from 0.5-16 lb are assembled for large-caliber howitzers. Pneumatic logic is used to count the stream of charges on the assembly network and to control the inspection-rejection operations. A conveyor transports the charges, arranged as increments, from loading booths to an assembly machine and then to a packer.

The schematic of pneumatic logic elements that operate the conveyor doors are shown in an example in Figure 11-14. It is a time delay circuit used to time the opening and closing of conveyor doors.

> "An inverter, A, supplies a signal to a spring-return, pilot-operated, 4-way valve which extends a 3/4-inch-bore air cylinder to close Door 1. When a bag reaches Door 1, it trips a pneumatic switch and generates a signal through or element, B, and inverter, C, to close Door 2.

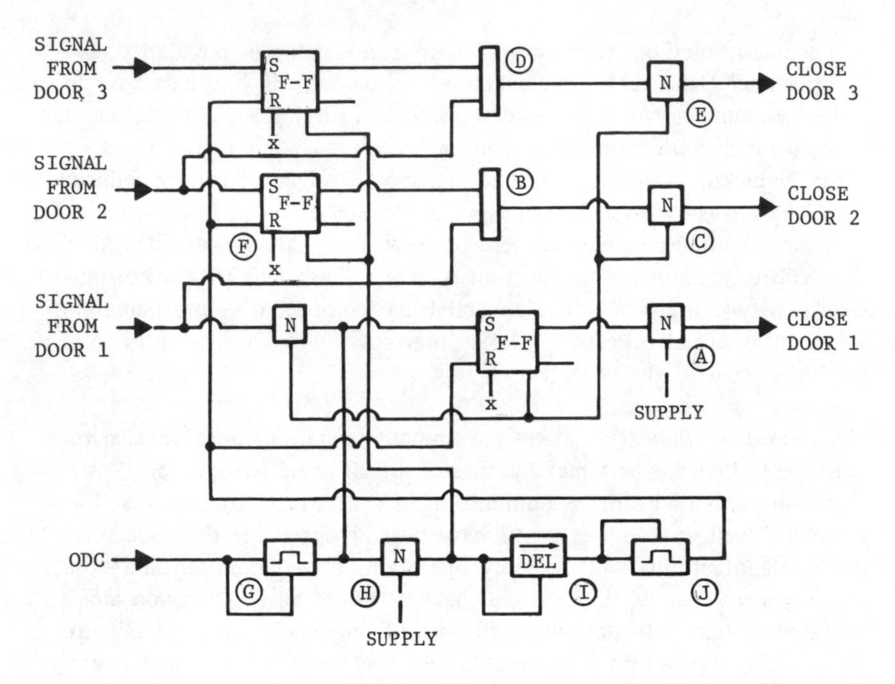

Figure 11-14. Time-delay circuit for control of conveyor doors.[92]

When a bag reaches Door 2, a signal through or element, D, and inverter, E, closes Door 3, and sets memory flip-flop, F, which remembers which door must close after a bag has been released from the buffer. When an open door command (ODC) arrives from the downstream machine, it triggers a fixed one-shot, or differentiator circuit, G, that controls the length of time the doors are open. Inverter, H, drives time delay, I, and one-shot, J. After the doors have opened and closed I, times-out and fires, J.

After completing the time-out period, which is long enough to insure that the bags are resting against the doors, J clears or resets the memory elements to prepare for the next cycle."

Many of the U.S. Navy's undersea operations, particularly emergency rescue-type take place in remote areas. Riegel describes a portable one-man recompression chamber, operating in a recirculating mode, which conserves air to enable operation in remote areas where a large air supply is not available.[93] The chamber permits rapid insertion and recompression of an injured diver and transportation to a suitable treatment facility where he can be removed to a large chamber and treated by medical personnel.

POTENTIAL APPLICATIONS

There are new industrial applications for compressed air continually being developed and tested. Possibilities of extending the use of compressed air, either to broader representation among all the industry sectors or to more intensive use in those sectors where it is already established, are indicated a good deal by the recent exploratory developments discussed below (Table 11-12). The collective implication, for future trends, of these and the other recent applications described above are summarized in a final section following a discussion of the market for compressed air equipment.

The developments described are usually the result of experiments and tests that are not necessarily economic or commercialized in any sense. There are few ready-made, manufactured products to help duplicate the applications. Nevertheless, they are generally aimed at and indicate a potential for sizeable economies, particularly in conserving and better managing energy use.

Fluidics in Pneumatic Logic and Control

Sensing and amplifying the air stream itself, or fluidics, to minimize as much as possible any electronic adjunct to pneumatic logic and control systems has attracted increasing interest since the early 1960s. Although the more subtle aerodynamic effects are used in devices not yet robust and reliable enough for practical industrial use, clear examples of the use of fundamental fluidic principles are becoming more frequent and, in fact, have stirred some technical controversy. Some of the most basic effects used in air jet sensors are illustrated in Figure 11-15,[94] which also includes a simple exemplary application.[95] Use of fluidics in logic devices typically rests on what may be considered less reliable second-order effects. A simple example,[96] the bistable wall reattachment amplifier, is shown in Figure 11-16. Wall reattachment of a fluid jet occurs when a boundary wall is placed in close proximity to it, causing the jet to bend and adhere to the wall. Normally, a jet of fluid, rising from a small cross section nozzle into a stationary fluid spreads outward if it moves downstream undeflected. The stationary fluid is entrained into the moving jet in a momentum exchange process. If the entrainment on one side of the jet is reduced by an adjacent boundary, however, then the local pressure on that side of the jet is also reduced, causing the jet to be deflected towards the boundary.

The wall reattachment amplifier has two boundary walls and a stream splitter positioned symmetrically downstream of the supply nozzle. Such devices are normally two dimensional, being formed by sandwiching the

Table 11-12. Potential Applications of Compressed Air

Typed Application	Procedure	Equipment Used
Fluidics for Pneumatic Logic and Control	Fluid dynamic air flow properties used to obtain application and discrete operational states	Fluidic elements operating in planar arrangement
Pneumatic Transport	Railroad-type operation. Capsules of one ton moved over closed loop. Unit train concept for petroleum	Special booster for pneumatic pipelines. Pneumatic valves for filling, emptying and load distribution
Energy Storage for Electricity Generation	Use of salt caverns and aquifers as large-scale storage of compressed air energy reserve to receive off-peak power generation	Power driver split into separated compressor and drive turbines
Energy Storage for Direct Industrial Plant	Deep-drilled holes at higher pressure for intermediate energy storage levels	High-pressure compressor, either centrifugal or double-acting reciprocating
Reactant	Underground *in situ* coal gasification	Low-pressure, high-volume compressors

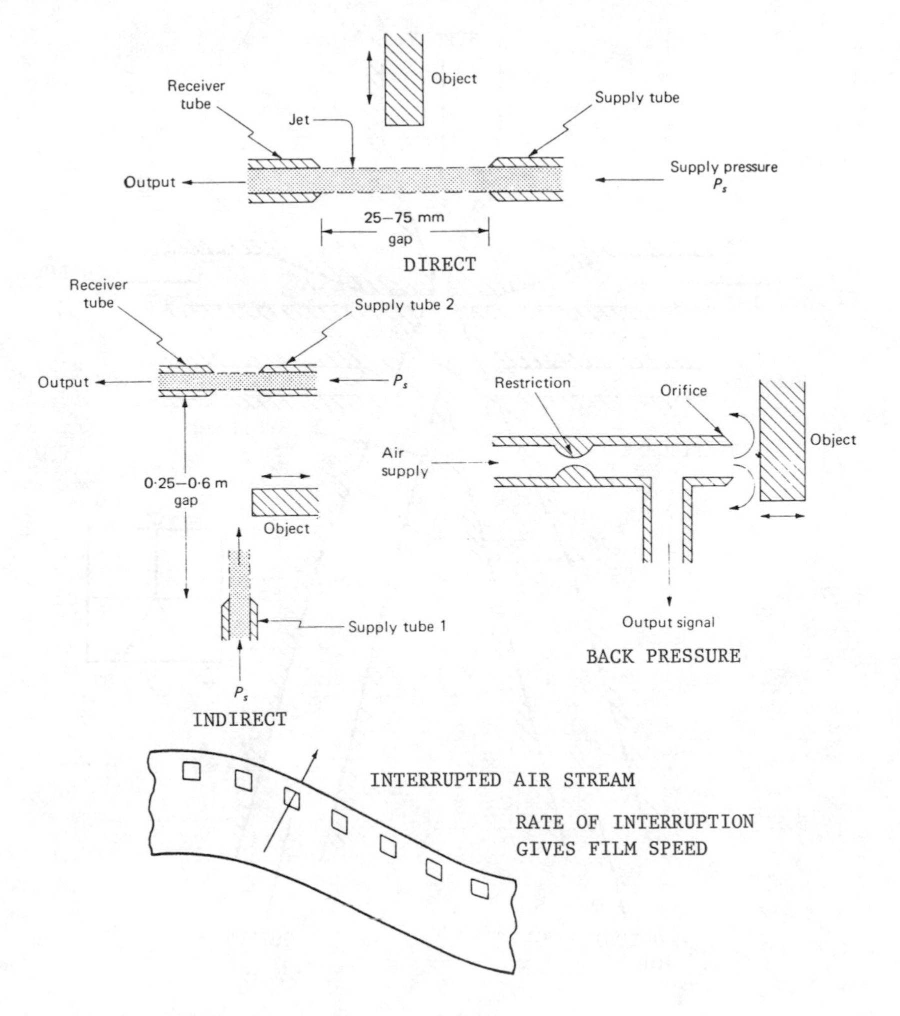

Figure 11-15. Fluidic air jet sensing arrangements.[94,95]

profile between top and bottom cover plates. Consider a jet that has been established down the left side. It is held there by a low-pressure region adjacent to the left wall and, not affected by the availability of output 02 downstream, issues primarily through 01. A small control flow through channel C1 is sufficient to raise the pressure in the wall attachment region so that the supply jet unlatches and deflects over to the other wall and issue through output 02. This holds even when C1 pressure is removed. The only way to return the jet to the original position is to supply similar control pressure to C2. The device then has the

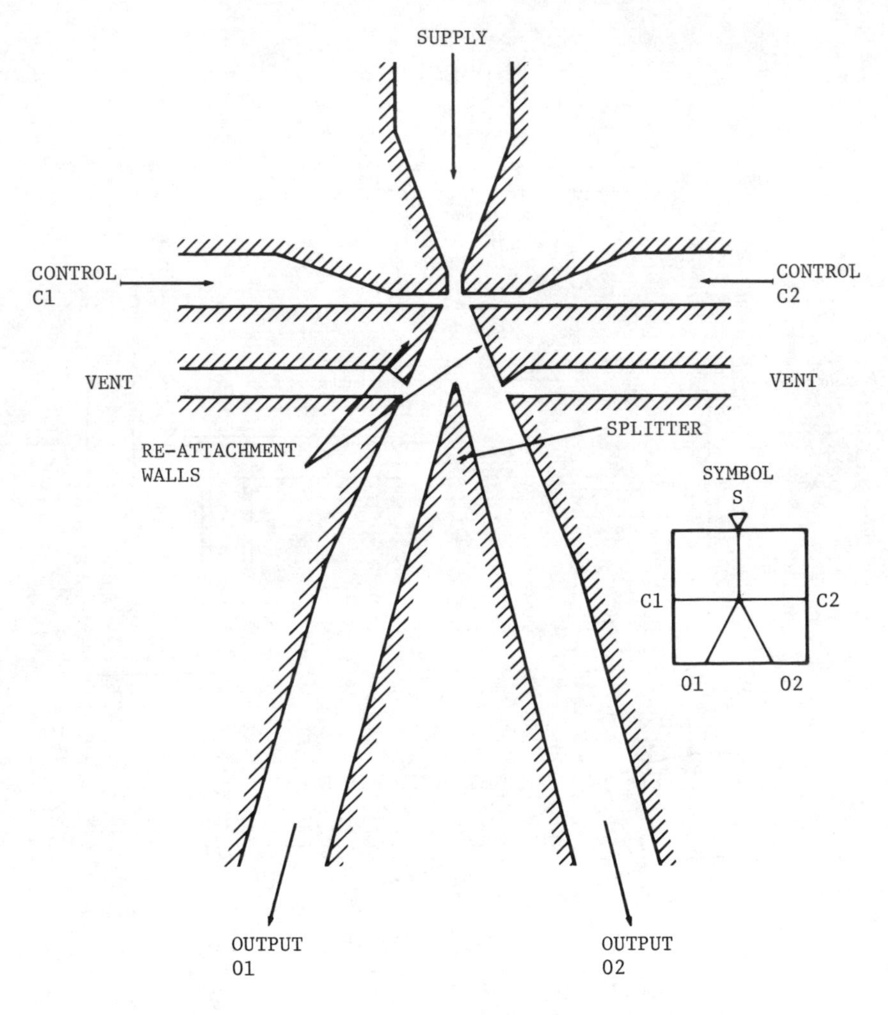

Figure 11-16. Bistable wall reattachment amplifier.[96]

property of a memory since it "remembers" the state of a previously applied signal in C1 and C2. It is also an amplifier of the control signal since the switching action is so sensitive to the control pressures. The vent channels allow the output channels to pass the required flow for the next element, any excess drawn off by them rather than through the opposite output channel.

Among the advantages that fluid logic has over other systems are the following:[94,97]

- Fluidic elements are physically robust with a wide operating temperature range;
- They are unaffected by many environmental conditions that affect the operation of other devices; *e.g.,* nuclear radiation, electromagnetic radiation, explosive atmospheres, etc.
- Their switching speed is adequate for most mechanical processes;
- Very simple methods of interconnection are used;
- They have an indefinite shelf life;
- The cost of fluidic systems promises to be low in the long term; and
- Sensors are simple and cheap to construct.

A major advantage of fluidic systems over conventional pneumatic elements is low air consumption which, in the latter, is often too great for relative economy.

Fluid logic devices are not appropriate where high operating speed is desired, as high flow conditions are being managed. Their most promising area of applications is in sequential systems with a relatively small number of elements involved. Some of the functions that can be carried by them are:

- machine tool control–inspection, sequential systems, numerical control;
- process control–simple computing functions, control of variables, *e.g.,* pressure, flow, temperature, feed rate, etc.;
- materials handling–manipulator control, conveyor sequence control, lift (elevator) control, position control, gauging; and
- medical engineering–artificial heart pump, respirators, cardiac compressors, artifical limb control.

Maas has suggested that fluidics are responsible for serious consideration of pneumatics as a control technology.[98] However, their frailty at some advanced design levels counters pneumatics' appeal in hazardous environments. New solid-state packages offer strong competition. They have relieved the major drawbacks of electronic instrumentation,[99] which are high voltage at low currents or high currents at low voltage, both adding risk of spark generation. Integrated circuits with zener barrier protection provide an intrinsically safe technology, which may interface better with computer control.

Mamzic, on the other hand, has described new developments in pneumatic interfacing equipment that make pneumatic analog controls and computers a desirable combination.[100] The mechanical multiplexer-transducer is accurate, practical and handles pneumatic input signal conversion at a cost comparable to that of connecting electronic analog signals to the computer. Furthermore, since control valves are almost invariably pneumatic, pneumatic controls require no extra cost for transducing valve signals. Pneumatic computer control stations respond as precisely to computer commands as do electronic stations, and the same easy switching is available. Kumar has also favored fluidics pointing out that their appeal remains the same as when they were first introduced.[101]

As an example of fluidics application in a rigorous environment, Chalmers describes the use of an air-powered and controlled welder to fabricate supports for the Alaska pipeline.[102] Fluidics are used to control the automated welding rig, which produces a smoother stronger weld in one half the time of a manual system. The air powered welder has a narrow 47-ft -long weld table, straddled by a gantry mounted on V-tracks. Two continuous-feed weld heads are mounted on each side of the gantry to simultaneously apply a bead down each side of the pipe assembly. Pneumatic sensors and fluidic interlock systems assure that positioning and welding operations take place according to a prescribed program.

The use of fluidic air logic in a complicated automated machining system for medical instruments has been described by Spinale.[103] The constituent parts of medical instruments were segregated into families with common denominators in terms of drilling and tapping operations. Each family could then be machined on fully or semiautomatic equipment capable of handling the entire range with slight set-up variations. Air cylinders clamp the parts and are activated from the logic module through operator command. Unclamping occurs when there is an automated END signal from the logic. Another machine application is the use of fluidics to control the speed of air motors.[104] Fluidic techniques are used to sense, amplify and compare the speed signal with a preselected reference. The system has been applied to industrial air motors from 1/25 to 5 hp.

Pneumatic Transport Experiments

Use of pneumatics for large-scale long distance transport is being tried in England. A 2-year project, funded at $250,000, calls for high-speed capsules of 1-ton capacity to move at 90 kmh on a 540-meter loop in Molten Keynes.[105] Three or four capsules are being designed and built to a size of 1.8 x 2.4 x 0.6 meters. Involved are two private engineering firms and three state interests. The private firms are General Descaling Co. of Worksop, Nottinghamshire and Sturtevant Engineering Products of London. A crucially important device being tried is the patented TRRL/BHRA "flap-gate booster" designed to cut pressure loss. Until this project, pneumatic pipelines were limited to short lengths due to the large pressure losses. The booster may permit pipelines of "unlimited length."

A somewhat different example of pneumatically assisted transport with greater likelihood of immediate practicability has been proposed by Ryan to facilitate petroleum transport where pipelines are not yet available.[106] This is the unit train concept for tank cars first used by the U.S. to move desperately needed petroleum to east coast ports during World War II. The tank cars are all filled hydraulically from one point using flexible tubing interconnecting the cars to each other. A system of pneumatic valves operated from a separate compressed air system is used to control the fluid weight in each car so that it is appropriately distributed among all the cars of the train. The entire string of cars may be unloaded from a single point by either a hydraulic pump or pressured air introduced at the loading end. This forces the liquid through the string and, at the same time, purges oil or chemical vapors that may be present. The system, as illustrated in Figure 11-17, has been implemented by the GATX Corporation in the U.S. Economically, the cost of such transport is somewhat greater than by an existing pipeline. However, if the pipeline is not available, the system essentially enables one to create the near equivalent of an instant pipeline for transport economy.

For material conveyance under water, compressed air is being used in a system for laying cables and tubes along the ocean's bottom.[107] The air is expelled from nozzles arranged on the frame of a sled built by J. Ray McDermott & Co. for work in the Ekofisk field of the North Sea, to carve out a channel for the cable as the sled is being towed. Towed along the path of a sea-floor pipeline, the sled is supplied with high-pressure air and water from the surface barge to perform its trenching operation. The sled straddles the pipe (previously set in place by a lay barge) and digs into the sea bottom 8 ft on either side as it is pulled along by a towline from the front of the surface barge.

Hose lines from the rear of the barge pump air and water to the sled at high pressures. As horizontal water jets cut and emulsify the ocean floor material (see front-view illustration), air is fed into an eductor, creating differential specific gravities to carry the loose mud and sand up through the pipes for side dispersal.

Compressed air has been used to efficiently convert double glass windows to insulating walls, and vice versa.[108] The *Beadwall* is a double-glazed window that uses air power to fill or empty its 3-in.-wide internal void with thousands of opaque plastic beads, on command. It was invented by David Harrison who says the bead-filled void is thermally comparable to a standard insulated wall. A New Mexico design firm, Zomeworks Corp. sells plans and licenses to builders. Normal installation requires blowers that move air at 100 cfm.

The system takes the insulating property of a normal double-glazed window one step further, by offering dual use of the dead-air space

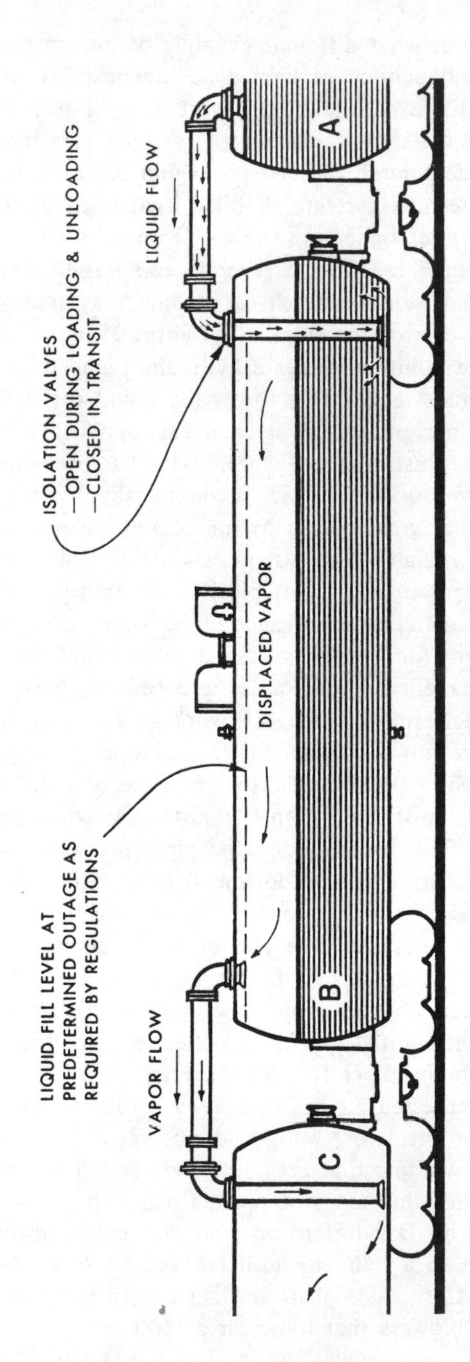

Figure 11-17. Basic flow diagram during loading phase. Interconnected tank cars.

between two panes of glass. Additional equipment required includes a special venting arrangement, storage tanks for granular material, and either simple vacuum cleaner-type blowers for small windows, or the projected use of a standard compressed air supply for blowing beads in and out of skyscraper-scale installations. Positive air pressure moves the beads from the storage tank into the window when transparency is no longer desired, or vacuum draws the beads out when the user wants transparency returned.

For energy conservation during winter, the window would face south and be transparent during the day to receive incident solar radiation and opaque at night to block loss through the window of the interior heat energy. During summer, the reverse procedure would be followed. Experimentation has shown that, as beads, expanded polystyrene or Styrofoam work best when treated with an anti-static agent.

This concept has been applied to a greenhouse and airport in Colorado. The developers claim it would also be economical for large industrial or commercial buildings.

Compressed Air Energy Storage for Generation of Electricity

Shortages of premium fuel, natural gas and low-sulfur oil which present little environmental difficulty, are forcing U.S. electric utilities to look for ways of storing base-load capacity for peaking purposes.[109] The utilities have been studying and experimenting with large-scale compressed air storage systems, which are relatively inexpensive and which would store electrical energy generated overnight for resupply during peak daylight hours. Enormous well-sealed underground caverns have been proposed as air-storage vessels. Salt caverns, created by using fresh water to leach out underground chambers in salt domes or bedded salt deposits have been used since 1951 to store liquid hydrocarbons, such as propane, butane and ethylene. These are now being proposed for compressed air storage.[110] The method for their development is shown in Figure 11-18.

The use of such a cavern for overnight energy storage in an electric utility off-peak generating program faces some serious difficulties.[111] The caverns are severely limited in the compressed air temperatures they can withstand. Moreover, it is not clear how effectively high-temperature stored air can be thermally insulated by the cavern envelope. Hence, much of the adiabatic heat of compression must be deliberately extracted. To be used effectively for electricity generation, the stored air must be expanded through a turbine. The economic way is currently estimated to require using some fuel to combust the air entering the turbine so as to raise its temperature. The use of water injection, which contributes extra gas flow momentum to raise power output has been proposed by

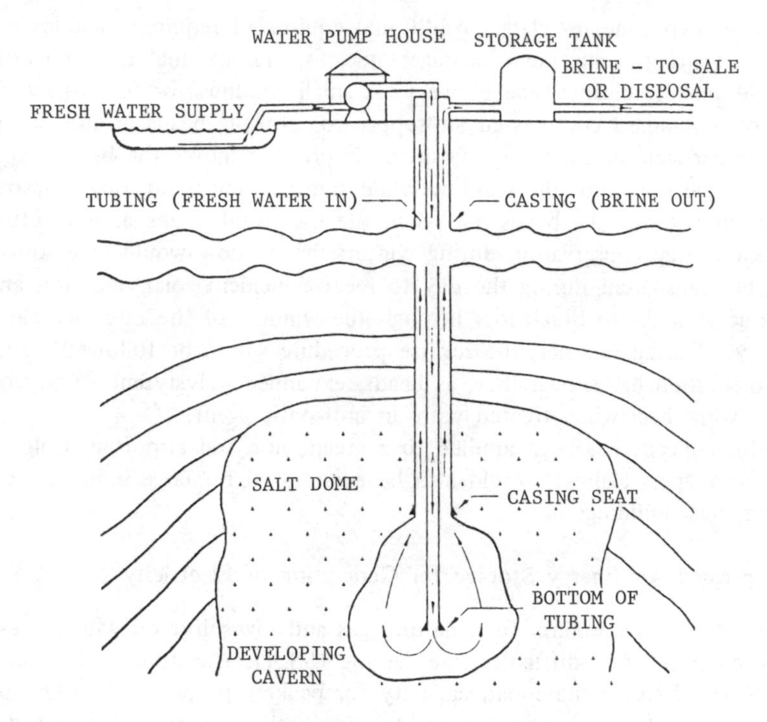

Figure 11-18. Salt cavern development.[110]

Gasparovic.[112] This has an advantageous side effect of reducing nitrogen oxide emission by 50%.

Such procedures still do not leave a comfortable margin for a positive economic gain, and Glendenning argues that it is insufficient for long-term requirements on a cycle of storing energy for 6-8 hr and generating it for 12-16 hr.[113] He proposes to use uncooled compression as a true energy storage scheme which would require no fuel other than the off-peak energy taken from the electrical grid.

A facility designed on the General Electric concept[111] was ordered by a West German utility[114] and is expected to be on-line in mid-1977.[115] It is a 290-MW peaking facility, which uses underground compressed air storage cavities leached from salt domes. It is estimated that these caverns will take 490 years to lose half their volume. Brown Boveri, supplying the power units, has combined steam and gas turbine technologies. A conventional gas turbine/air compressor system was adopted with a clutch between turbine and generator. During off-peak hours, the generator, driven by electricity from a base load unit, acts as a motor to compress air to storage. Air is compressed in an axial-flow compressor followed by a centrifugal blower, which raises its pressure to 1000 psi. Cooling stages are one precooling, two intercooling, and one aftercooling. Heat from the cooling water is removed in a closed cycle cooling system so that no water is wasted. During peak demand, air is removed from storage, heated in the combustion chamber, and expanded in the turbine.

Nordwestdeutsche Kraftwerke A.G. of Hamburg ordered the plant from the Mannheim Division of Brown Boveri. It will be located at Huntorf near Oldenburg and will operate on a cycle of eight off-peak hours for storage, two peak hours for generation.

Compressed Air Energy Storage for Direct Industrial Plant Use

A smaller-scale development using man-made underground holes has so far been successful in practice. Beam designed and constructed an overnight compressed air storage system for the National Aeronautics and Space Administration, Ames Research Center, to provide high-pressure air for the hypersonic velocity wind tunnel.[116] Compressed air storage was in a set of seven pipes set 1,200 ft deep and consisting of oil well casings as shown in Figure 11-19. The holes are each about 9 in. inside diameter and provide an aggregate volume of 30,000 ft^3. The maximum pressure used is 3,000 psi. Pressure failure was prevented by a careful cementing procedure. Cement was applied to surround the casings and was cured for 24 hr under 3,000 psi using water to fill the holes. Total cost of the installation, if new casing had been used, would be $1.57 million. The designers estimated that an aboveground installation of equivalent storage capacity following American Society of Mechanical Engineers (ASME) safety standards that govern such installations would be $4.85 million. With the used surplus casing actually employed, the construction cost was $.77 million.

As presently used, the gas is compressed to 3,000 psi with no provision for either retaining the adiabatic heat of compression or cooling it to avoid overheating. Some of the heat is conducted away through the ground and when the stored air is used in an expansion down to 1,500 psi, the temperature drops to -40° F. The designers estimated very roughly that as much as 75% of the original work energy used to compress the air may have been lost. The Ames hypersonic wind tunnel facility primarily needed the high capacity made available via prior storage.

An alternative, more conventional use for general manufacturing power would require that the compressive energy loss be alleviated. This would be accomplished in either of three ways. In one case, heat of compression

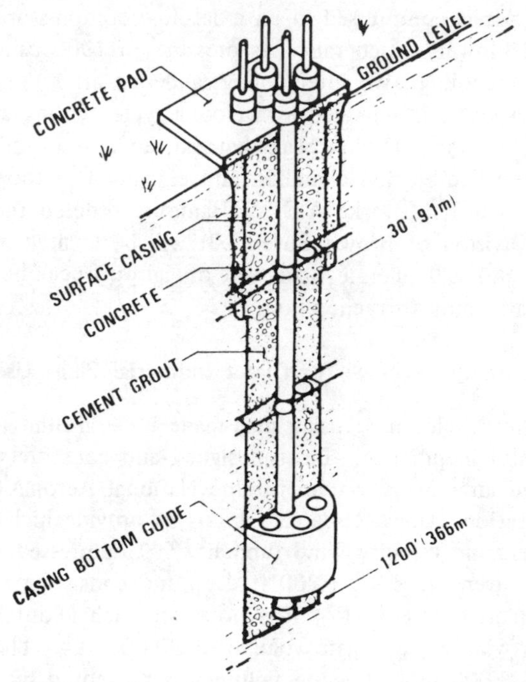

a) SKETCH OF SEVEN-CASING CLUSTER CONCEPT USED AT AMES RESEARCH

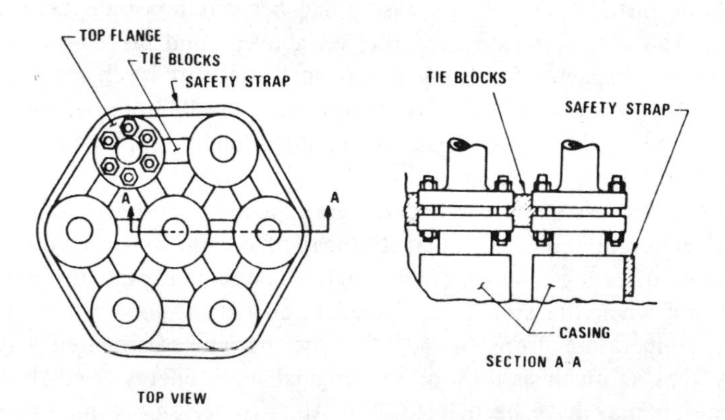

b) SKETCH OF SAFETY-RESTRAINT USED FOR THE TOP OF THE SEVEN-CASING CLUSTER

Figure 11-19. Seven-casing cluster of underground compressed air storage holes.[116]

could be retained by suitable thermal insulation of the storage holes and air pipelines. Tools and other manufacturing machinery would be equipped to handle high-temperature input air. Alternatively, the compressed air could be cooled, the extracted heat retained, and this heat resupplied to the air during the expansion process. In a third system, the very high storage pressure would be traded off for much higher volume and the reduced compression ratio would considerably reduce the potential energy sacrifice as well as the temperature excursion of the storage and work cycle.

Compressed Air as Reactant for *In Situ* Coal Gasification

A central role is played by compressed air in a new method of *in situ* coal gasification, which shows strong promise in its first tests.[117] The method essentially facilitates access to the coal seam by drilling wells, igniting the seam, sustaining gasification by injection of air or oxygen, and withdrawing the product gas from neighboring wells. The gas can be combusted to generate electricity on-site, be used as a chemical feedstock, or possibly be upgraded to synthetic natural gas by further reaction at high pressure. There are significant environmental advantages because most ash and sulfur contaminants remain underground. The process may permit exploitation of some deposits that cannot be mined with conventional techniques. In other cases, it offers greater energy extraction because conventional mining leaves much of the coal in place.

The experiment has been conducted at ERDA's Laramie Energy Research Center near Hanna, Wyoming, on a coal seam 30 ft thick, 300 ft below the surface. Four process wells were located at the corners of a 60-foot square. Coal was ignited at the bottom of one well, while air was injected into an adjacent well. Sufficient air percolated through the coal seam to sustain fire in the ignition well and caused it to burn toward the air-injected well. The coal was carbonized by the flame forming a highly permeable pathway between the two wells. When fire reached the injection well, high volumes of air could be injected through it into the seam at low pressure. The reaction zone then expanded back toward the ignition well to encompass the full 30-foot seam thickness, comsuming all of the coal for a distance of 20-30 ft on either side pathway. The process was repeated with the two remaining wells.

A total of 6600 tons of coal were gasified even though the prime 60-foot square bounded by the wells was estimated to contain only 4600 tons. Gas production averaged 8.5×10^6 scf per day with heating values in the range 175-185 Btu/scf. This is at the high end of the usual producer gas values. No previous underground gasification tests have given sustained yields of such high-quality gas.

MARKET

The U.S. market for compressed air equipment is portrayed below, treating separately the leading manufacturers of the equipment, examples of current costs for typical categories of the equipment and, finally, an economic review of the recent trends in the market for compressed air equipment both in its relation to other sectors of U.S. industry and as a distinct factor in U.S. national industrial growth.

Leading Manufacturers of Compressed Air Equipment

The leading manufacturers of compressed air equipment, (Table 11-4) are those companies whose annual sales volume is more than $100 million in compressed air equipment. Liquid rings compressors form an example of a specialty item available primarily from Nash Engineering. These and many of the other firms making compressed air equipment are members of the Compressed Air and Gas Association. The Association, currently numbering 38 members, conducts a program of activities designed to promote the use of compressed air in U.S. industry and subdivided according to separate sections for the following categories of compressed air equipment:

- rock drills;
- pneumatic tools;
- pneumatic hoists;
- air dryers;
- centrifugal compressors;
- reciprocating compressors, single-acting and unit;
- portable compressors;
- rotary drills;
- rotary positive compressors; and
- stationary reciprocating double-acting compressors.

Of the companies listed, Ingersoll-Rand, with annual sales revenue over $1 billion, is dominant. At least half its business is in compressed air equipment, and as a result, the company's market largely determines the kind of equipment and cost ranges offered on the market. Dresser Industries is as large, but its market is in oil field equipment such as drills and drill bits. Oil field operations are a major application area of pumps and compressors. GATX Corporation is not as large as Dresser or Ingersoll-Rand. Its position is similar to Dresser in having substantial compressor equipment sales, while its other activities are in heavy industrial operations that have already developed a great many compressed air applications.

Range of Approximate Costs for Equipment

For almost any compressed air system, the most expensive single unit will be the compressor. Table 11-5 shows the costs for those compressors, which are sold in packaged units, and some costs for a few pneumatic tools. Individual tools are generally fairly cheap due to their simple construction and the high level of competition among vendors. The capital barrier for new entrepreneurs is relatively low. The large compressors, particularly the dynamic type used in process industries, range into millions of dollars, are highly individualized for specific plants and frequently are the subject of bid competitions.

Economic Position and Current Trends of Compressed Air in U.S. Industry

The relationship of compressed air use to the major industrial section of the U.S. economy must first be placed in the context of recent patterns of U.S. economic growth and categories of activity. The long-term directions of U.S. national product are presented in Table 11-13 in terms of both actual values and these values normalized to the U.S. price level of 1972. Long-term trends over the 21-year span from 1955-1976 are almost continuously positive for both the gross national product and the part of it which is fixed investment producers' durable equipment (FIPDE). This latter portion, now the major portion of all capital spending, includes compressed air equipment. It exhibits, particularly in the normalized values, much greater sensitivity to short-term recessions such as those of 1958, 1961, 1970-71 and 1974-75. At the same time, FIPDE has maintained roughly a constant ratio to gross national product over the entire 21-year period, staying within the range, .055-.069.

The use of compressed air equipment by other sectors of U.S. industry in aggregated economic terms is most comprehensively presented in the input-output (I/O) structure of the U.S. economy. The two tables of this structure most recently published by the U.S. Commerce Department are for the years 1967 and 1963. The elements of the table relevant to compressors are shown in Table 11-7 for those sections of U.S. industry which bought the largest amounts of related equipment.

For the input-output model industry is divided into establishments which primarily produce the indicated categories of equipment. It is the sales of these establishments among each other by category and to other sectors of the U.S. economy such as government and consumers which are traced in the entire I/O table. The entries of Table 11-14 concern the purchases by establishments (whose primary activity is the indicated industry) of products sold by establishments whose primary activity is manufacturing pumps and compressors. For I/O analysis, this category is not

Table 11-13. U.S. National Product[4,118]

Year	Gross ($ billions)	Fixed Investments Producers' Durable Equipment ($ billions)	GNP Price Deflator for 1972 $	Gross in 1972 $ ($ billions)	FIPDE in 1972 $ ($ billions)	Ratio: FIPDE/GNP
1955	399.3	23.9	60.78	656	39.2	.060
1956	420.7	26.3	62.90	670	41.9	.063
1957	442.8	28.6	65.02	680	44.0	.065
1958	448.9	24.9	66.06	679	37.7	.056
1959	486.5	28.3	67.52	711	41.9	.059
1960	506.0	29.5	68.67	736	43.0	.058
1961	523.3	28.7	69.28	756	41.4	.055
1962	563.8	31.8	70.55	798	45.1	.057
1963	544.7	34.0	71.59	830	47.5	.057
1964	635.7	38.2	72.71	876	52.6	.060
1965	688.1	45.1	74.32	927	60.7	.066
1966	753.0	52.2	76.76	981	68.1	.069
1967	796.3	52.6	79.02	1007	66.6	.065
1968	868.5	57.7	82.57	1050	70.0	.067
1969	935.5	63.3	86.72	1080	73.0	.068
1970	982.4	62.8	91.36	1076	68.8	.064
1971	1063.4	64.7	96.02	1110	67.5	.061
1972	1171.1	74.3	100.00	1171	74.3	.063
1973	1306.6	87.0	105.92	1232	82.1	.067
1974	1413.2	95.1	116.4	1216	81.6	.067
1975	1516.3	95.1	127.3	1190	74.6	.063
1976	1695.0	103.4	133.3	1271	77.6	.061

Table 11-14. Input/Output Relation of Pumps and Compressors (P & C) to Other U.S. Industry Sections (in 1967 $)[119-121]

	1967				1963		
Consuming Industry (CI)	Supply by P & C Industry ($ millions)	CI Sales ($ billions)	P & C Supply / CI Sales	CI Code	Supply by P & C Industry ($ millions)	CI Sales ($ billions)	P & C Supply / CI Sales x 10^3
Crude Petrol and Natural Gas	87.0	15.03	5.78	8	79.2	13.5	5.86
New Construction	121.7	79.89	1.52	11	77.0	72.4	1.06
Maintenance and Repair Construction	35.4	23.40	1.51	12	49.6	21.9	2.27
Paper, except Containers and Boxes	71.1	17.82	3.99	24	60.6	15.6	3.88
Chemicals and Related Chemical Products	139.8	23.82	5.86	27	61.6	19.2	3.21
Petroleum Refining and Related Products	69.3	26.99	2.56	31	9.3	24.2	.38
Primary Iron and Steel Manufacturing	120.7	32.21	3.74	37	79.8	27.5	2.90
Heat, Plumbing and Fabricated Metal Structures	60.4	13.17	4.58	40	25.5	10.0	2.55
Other Fabricated Metal Products	54.1	12.65	4.28	42	46.1	10.0	4.61
Engines and Turbines	73.2	3.82	19.18	43	42.5	2.7	15.71
Farm Machinery	82.3	4.83	16.70	44	27.0	3.4	7.94
Construction Mining Oil Field Equip.	70.4	6.06	11.60	45	85.9	4.5	19.10
Metalworking Machinery and Equip.	56.7	9.02	6.29	47	42.9	7.0	6.12
Special Industrial Mach. & Equip.	26.9	5.84	4.61	48	16.9	4.2	4.02
Aircraft and Parts	121.4	23.86	5.09	60	51.4	17.3	2.97
Other Transportation Equipment	23.4	7.88	2.97	61	57.1	5.4	10.59
Transportation and Warehouse	38.2	43.17	.88	65	67.6	43.6	1.55
Total Related Industries	1252.0	349.5	3.54		880.0	302.4	2.91
Total of All Industries	1957.9	796.3 (GNP)			1527.4	646.0 (GNP)	

broken down further by the Commerce Department, even for the still anticipated 1972 Table. However, in subsequent tables drawn from the Census of Manufactures, we can separate out compressors, and in fact, its subcategories. However, the breakdown by sales to consuming industries is not provided in those tables.

From Table 11-14 we see that in general there was somewhat more intensive use in 1967 of pumps and compressors by the listed industry groups in that their cost represented a somewhat larger portion of the consuming industry sales. These portions do not show any dramatic increase, with the exception of the petroleum refining industry. That change is most likely related to specific fluctuations in petroleum refining capital expansion. Hence, Table 11-14 does not indicate any far-reaching new technological direction in the industrial applications of compressed air.

We can gain some idea of the proportion that compressors have to pumps in the I/O tables by Table 11-15, taken from the 1972 Census of Manufactures.[122] These entries show that in 1972, roughly one third of the shipments by the combined category of pumps and compressor establishments originated with compressor establishments. Another point is that the shipments of these establishments as primary product consisted mostly of compressors. The Census of Manufactures also has data broken

Table 11-15. Pump and Compressor Industry Breakdown and Relation to 1967 Data[122]

	Value of Shipments ($ millions)			
Industry and Census Year	Total	Primary Product	Secondary Product	Miscellaneous Receipts
Pumps and Compressors, 1967	2207.5	1704.9	265.4	237.2
Pumps and Compressors in 1972 $	2800.0	2160.0	336.0	300.0
Pumps and Compressors, 1972	2775.4	2057.8	430.8	286.8
Pumps and Pump Equipment, 1972	1917.3	1408.1	296.1	213.1
Air and Gas Compressors, 1972, and Related Equipment	858.5	649.7	134.7	73.7

down according to the equipment itself. These show, for example, that in 1972, while compressor establishments shipped $649.7 million of compressors, other establishments also shipped compressors to raise the total to $722.7 million. Table 11-16, in terms of this latter analysis, shows the trends in shipments of compressors and compressor equipment over the span 1971-1975. A definite upward trend for compressors, compressor parts and attachments, which is much stronger than the trends in both GNP and fixed investment equipment of Table 11-13 is clearly shown. The estimated statistical errors of 1974 figures are in agreement

Table 11-16. Value of Shipments of Air and Gas Compressor Equipment by Product Category for Years 1971-1975[5,122-124]

Class of Product	1972 $ (millions)					% Standard Error for 1974
	1971	1972	1973	1974	1975	
Air and Gas Compressors and Related Equipment		723	853	1030		3
Compressors and Vacuum Pumps, Excluding Refrigeration Compressors	425	437	504	657	614	3
Parts and Attachments for Compressors		148	193	303		4
Industrial Spraying Equipment		94	99	89		1
Power-driven Hand Tools: Pneumatic, Hydraulic and Power Actuated		218	259	226		4

with what one would judge from the statistical revision in census bureau statistics repeated in their monthly sequential reports. These errors are too small to be responsible for the upward trend. From Table 11-16 we see from the 1972 dollar values that in contract in this same period, 1971-75, both gross national product and fixed investment producers' durable equipment rose and fell with a small net advance.

We can extract from the Census of Manufactures continuing reports, data on the quantity and value of shipments by category of compressor for the years 1967, 1972, 1974 and 1975. Table 11-17 shows data for the years 1967 and 1972. From the 1967 columns, we see that in 1967 the compressor portion of pumps and compressors bore approximately the same relation to the whole as shown for 1972 in Table 11-15. While the total values were similar for both years, 1972 had a much greater portion in related equipment. Every compressor category, except gas, was reduced in value from 1967 to 1972. These values should be counted as producer's prices for the bare unit alone, which does not count distribution and certain other costs such as drivers and necessary extra equipment included in the packages of Table 11-5. We therefore find fairly

Table 11-17. Quantity and Valued Categories of Compressor Shipments of all Producers, 1967 and 1972[125]

	1967		1972	
Product	**Quantity (1000s)**	**Value ($ millions)**	**Quantity (1000s)**	**Value in 1972 $ ($ millions)**
Air and Air Compressors and Related Equipment	x	724	x	722.7
Air and Gas Compressors and Vacuum Pumps,	x	634	x	480.8
Excluding Refrigeration Compressors	494.7	592	962.5	437.0
Air Compressor Stationary	282.8	275	604.0	212.0
Stationary	203.8	197	323.1	148.7
1.5 hp and under	121.3	13.3	218.5	18.5
1.5 → 5 hp	48.0	19.4	58.6	18.4
5 → 15 hp	19.9	21.2	29.1	22.6
16 → 100 hp	12.1	68.8	14.4	49.6
101 → 250 hp	1.8	32.4	1.9	20.5
251 → 1000 hp	.6	26.3	.5	12.1
Over 1000 hp	.1	15.6	.1	7.1
Portable	79.0	78.0	281.0	63.3
Under 75 cfm	69.1	11.0	268.6	12.4
75-150 cfm	6.8	25.8	6.4	14.4
151-400 cfm	1.9	15.1	4.1	14.0
Over 400 cfm	1.2	26.1	1.9	22.6
Gas Compressors	4.2	24.2	2.7	113.4

low average values. For example, the category of stationary air compressors driven at 101 to 250 hp has an average producer's price of $10,800 in 1972 and $18,000 in 1967 in 1972 dollars. This indicates a shift toward smaller units in this category. Similar shifts can be discerned in the other categories and these probably reflect some general economizing on capital expenditures.

In Table 11-18 we have a similar data comparison for 1974 and 1975 according to new compressor equipment categorizations adopted by the Census Bureau. (Data for 1975 alone are presented in Table 11-7.) Here, the average values by category are shown in the table and they range from $114 from compressors under 1.5 hp to $117,300 for the dynamic compressors, centrifugal and axial. Again, these values are related to producer prices for the base unit alone. Including distribution costs, installation and necessary ancillary equipment would multiply the price at least two or three times.

The outstanding area of technologically propelled expansion is in pneumatic control asserted by fluidic elements. International trade competition has added to the already existing business pressure to constantly increase manufacturing productivity. As a result, wherever possible, automated assembly systems are placed in operation to eliminate manual labor requirements to a maximum degree. Fluidics is the cutting edge of their technological advance and has some inherent limitations that will be difficult to overcome. The response speed of fluidic systems is limited by the propagation velocity of sound so that a simple dc or step-function-signal suffers a transport lag that will always act to limit the spatial extent of the system. In addition, ac signals suffer a degradation due to the influence of fluid dynamics effects inside a pipe, which smear the wavefront shape of the acoustic signal. Finally, the devices are vulnerable to air leakage into their vents and control channels. This defeats air supply filtering and entrains contaminants which eventually accumulate to cause malfunctions.

Energy costs have been the driving force in the other major areas of new compressed air applications. These are the use of air to transport fluid and bulky solid masses, what may be a whole new area of high pressure air as massive stored energy, and the use of air as a process reactant in coal gasification. Each of these applications has its distinct limitations. Conveyance is still limited in distance by the pressure drop, which must be compensated for without losing the basic economy. Energy storage faces the difficulties in handling the thermodynamic cycle of pressuring and expanding stored air for conversion to mechanical work. Use of air as a reactant is probably the most completely developed technology measured against the requirements, which are primarily to maintain the appropriate quality with adequate preconditioning.

Table 11-18. Quantity and Value of Shipments of Compressors According to New Categories, 1974 and 1975.[124]

Products	No. of Companies	1974 Quantity	1974 Value ($ millions)	1974 Average Value ($ thousands)	1975 Quantity	1975 Value ($ millions)	1975 Average Value ($ thousands)
Compressors and Vacuum Pumps, Total	x	1460	766.3	0.525	1440	802.5	0.557
Air Compressors, Total	x	761	429.0	0.564	867	405.5	0.468
(Value of drivers excluded)							
(~ 10%, % imp. val.)							
Stationary, Total	x	344	301.9	0.878	249	258.5	1.039
Reciprocating, Single Action							
Under 1½ hp	26				153	17.4	0.114
1½-5 hp	18				49.3	21.0	0.426
6-25 hp	17				23.3	27.0	1.16
26 hp and over	11				4.63	18.9	4.07
Reciprocating, Double Action							
150 hp and under	14				8.501	21.16	2.37
Over 150 hp	6				0.383	21.76	56.8
Rotary, Positive							
Discharge Press, 50 psig and under	6				1.54	9.44	6.13
> 50 psig, 150 hp and under	11				6.16	48.8	7.91
> 50 psig, > 150 hp	9				1.28	14.28	11.13
Centrifugal and Axial							
50 psig and below	3				0.50	58.85	117.3
51 psig and over	4						
Portable, Total	x	417.2	127.1	0.305	618.1	146.9	0.238
Under 75 cfm	18				605.6	31.9	0.053
75-124 cfm	13				2.34	8.55	3.660
125-249 cfm	14				5.96	27.49	4.620
250-599 cfm	12				1.57	14.50	9.220
600-899 cfm	12				1.96	41.70	21.3
900 cfm and over	10				0.020	2282	36.5
Pneumatic (air) power compressors and Motors (includes air brake systems)	8				433	34.39	0.079

In other industrial work the special advantages of compressed air in handling and safety[125-127] must overcome energy inefficiency, and there will undoubtedly be cross currents as people devise ways, either to adopt it or dispense with it in favor of alternatives such as steam or direct electrical use. Its energy difficulties may well be worked out as methods are developed to avoid thermal losses in the storage and use cycle, and compressed air may continue to gain favor as a manageable compact medium of energy storage and transmission.

Growth in the use of pneumatic control entails so little increase in either compressed air generation or pneumatic toll use that we cannot expect it to influence the discernible market for compressed air equipment in these categories. On the other hand, growth of compressed air as an energy storage medium can have a measurable impact in both the areas. The small upward trend in the compressed air market for 1971-75 relative to other capital industrial equipment, may well have been the beginning of an economic trend associated with that change in U.S. industrial technology.[128] We must also anticipate that if the *in situ* coal gasification use for compressed air proves practicable (furthermore, recent reports indicate favorable appraisal by U.S. ERDA and industry observer) we would see a specific marked increase in sales to the mining sector of the U.S. economy.

REFERENCES

1. Compressed Air and Gas Institute (CAGI), 1973.
2. MITRE. Assessment from Compressor Equipment Catalogs.
3. MITRE. Survey
4. U.S. Commerce Department. "Business Statistics 1975: 20th Biennial Edition" (May 1976).
5. U.S. Commerce Department. "Annual Survey of Manufacturer, Value of Product Shipments (October 1976).
6. Ewing, W. "Compressed Air. Economical and Versatile Assembly Power - 1," *Assembly Eng.* 13(5):58-63 (May 1970).
7. Burgoyne, J. H. and A. D. Craven. "Fire and Explosion Hazards in Compressed Air Systems," *Am. Inst. Chem. Eng., Low Prevention Symposium 7,* New York, November 28-30, 1972.
8. Kauffman, W. H. "Audit Your Air System to Save Energy," *Power* 119(5):40 (May 1975).
9. Crossley, A. M. "How to Assess Your Compressed Air System," *Works Eng. Fact Serv.* 65(763):20-23 (January 1970).
10. Baniak, E. A. "Compressors: I - Principles and Types, II - Application and Lubrication," *Lubrication* 59:1-20, 21-48 (January - March 1973); (April-June 1973).
11. Rampfer, L. E. "100 lbs. Air - Centrifugal or Reciprocating," *Purdue Computer Tech. Conf.* July 10-12, 1974.

12. Tuymer, W. J. "Stepless Variable Capacity Control," *Purdue Compressor Tech. Conf.*, July 10-12, 1975.
13. McGregor, D. E. "Recirculation Compressor for Hot Gases," *Rev. Sci. Instru.* 44(2):236-238 (February 1973).
14. Scheel, L. F. "Technology for Rotary Compressors," *J. Eng. Power Trans. ASME* 92, Ser A 3:207-216 (July 1970).
15. Nash Engineering Co.
16. Atlas COPCO. Catalog.
17. Dunn, R. L. "Rotary Screw Air Compressors," *Plant Eng.* 28:102-107 (May 16, 1974).
18. Elsner, F. "Screw Compressors, a Young Compressor Generation," *Ind. Anzeig.* 65(27):520-521 (March 30, 1973).
19. Weissenger, W. "Economy of Screw Compressors," *Ind. Anzeig.* 96(84): 1879-1899 (October 9, 1974).
20. Barker, P. K. "Let's Reassess Screw Compressors," *Power* 118:76-78 (April 1974).
21. Chicago Pneumatic Tool Co. "A Revolution in Air Compressor Technology," CP-3884 (April 1976).
22. Holloway, C. and D. Pavlov. "Anti-Surge Control of Large Compressors," *Chemsa* 1(9):190-192 (July 1975).
23. Hendricks, J. F. "Understanding Centrifugal Air Compressors," *Plant Eng.* 29(23):161-164 (November 13, 1975).
24. Pais, R. B. "Field Performance Testing of Centrifugal Compressors," *Turbomach. Symp.* Texas A&M Univ., October 15-17, 1974.
25. Ostrowski, R. "Compressed Air System Deficiencies and Maintenance," *Iron Steel Eng.* 48(7):79-86 (July 1971).
26. Marx, F. M. "Layout and Sizing of Compressed Air Mains," *Hyd. Pneum. Power* 19(255):341-344 (September 1973).
27. Pousma, J. G. "Maximum Recommended Compressed Air Flows," *Hyd. Pneum.* 24(8) (August 1971).
28. Iocco, D. E. "Air Piping System Design for Reciprocating Compressors," *Purdue Compressors Tech. Conf.*, July 10-12, 1974.
29. Pueschel, I. "Compressed Air Cooling in Industrial Compressors," *Werkstatt Betr.* 105(9):683-686 (September 1972).
30. Weiss, W. H. "In Compressed Air Systems, Why an Aftercooler?" *Plant Eng.* 27(1):66-68 (January 11, 1973).
31. Cunningham, E. R. "Selecting a Compressed Air Dryer," *Plant Eng.* 28(25):74-78 (December 12, 1975).
32. "Dry Air for Chemical Plant," *Compressed Air Mag.* 12-13 (September 1974).
33. "Dry Air Ups Productivity," *Compressed Air Mag.* 8-11 (March 1974).
34. Rya, T. "Drying Compressed Air," *Plant Eng.* 28(5):197-201 (March 7, 1974).
35. Hurst, L. G. "Compressed Air Drying," *Fluid Power Internat.* 36(416): 31-34 (February 1971).
36. Akerman, I. "Compressed Air Must Come Clean and Dry," *Fluid Power Internat.* 36(427):30-32 (October 1971).
37. Bruckshaw, N. B. "Removal of Contamination from Compressed Air," *Filt. Sep.* 10(3):296 (May-June 1973).
38. Benner, D. W. "Removing Oil Aerosols from Compressed Air," *Automation* 18(10):42-44 (October 1971).

39. Harwood, J. "Why Dry Air," *Prod. Finish (Lond)* 27(4): 27-28 (April 1974).
40. Deltech Engineering Co. "Compressed Air Filters," *Bull.* 101k.
41. Allen, E. "Stalled or Locked Rotors. A Breeze for Air Motors," *Power Trans. Design* 13(6):36-38 (June 1971).
42. Zelenetski, R. "Design of Rotary Pneumatic Motors with Tangentially Arranged Blades," *Eng. J.* 54(8):10-14 (1974).
43. Brooks, E. "Application of Air Motors in Industry," *Aust. Mach. Prod. Eng.* 27(3):21 (March 1975).
44. Webb, J.R. "Air Motors - Putting Them to Work," *Plant Eng.* 27(8): 117-120 (April 19, 1973).
45. Compressed Air and Gas Institute (CAGI). "Compressed Air Handbook."
46. Semmerling, W. "Compressed Air - Economic and Versatile Power-3," *Assembly Eng.* 13(7):44-48 (July 1970).
47. Hordeski, M. "When Should You Use Pneumatics, When Electricity," *Instr. Control System* 51-55 (November 1976).
48. Ericksson, H.B. "Industrial Applications of Compressed Air," *Chart, Mech. Eng.* 21(11):57 (December 1974).
49. Fischer, P.A. "Pneumatic Industrial Vibrators," *Plant Eng.* 29:77-79 (June 26, 1975); 89-92 (August 21, 1975).
50. Compressed Air and Gas Institute (CAGI), 1975.
51. Bauermeister, K.J. "Turbocompressors in Process Plants," *Chem. Process Eng.* 50(9):79-81 (September 1969).
52. Bailey, S.J. "Process Controllers; a Case of Pneumatic-Electronic Coexistence," *Control Eng.* 20:46-54 (December 1973).
53. Knoche, J.R. "What's New in Automated Assembly," *Proc. Society of Mechanical Engineers,* November 12-14, 1974, SME Tech. Paper AD 74-424.
54. Bizilia, P.K. "Re-torquing Air Motors–Minimize Torque Scatter," *Proc. of Society of Mechanical Engineers,* November 12-14, 1974. SME Tech. Paper AD 74-421 (1974), pp. 1-8.
55. Buckley, P.S. *Instr. Technol.* 22(4):33-40 (April 6, 1975); 39-42 (January 1975).
56. Deuschle, R. "Pneumatic? Electronic?–Instrument Cost Alone is Less than Half the Tale," *Instr. Control System* 48(2):51-54 (February 1975).
57. "Parts Transfer by Pneumatic Tube," *Compressed Air Mag.* 12-14 (April 1974).
58. Capes, C.E. and K. Nakamura. "Vertical Pneumatic Conveying: an Experiment with Particles in the Intermediate and Turbulent Flow Regimes," *J. Chem. Eng.* 51:31-46 (February 1973); 391-392 (June 1973).
59. "Trace Air Conveying," *Compressed Air Mag.* 81(12):16-18 (December 1976).
60. Sumner, J. "Makers look Forward to Likely Growth in Pneumatic Handling," *Engineer* 240:32-33 (February 27, 1975).
61. Andeen, A.B. "Bubble Pumps," *Compressed Air Mag.* 16 (January 1974).
62. "Air Bearing Improvement," *Compressed Air Mag.* 18 (March 1973).
63. "High-Density Mobile Storage," *Compressed Air Mag.* 18 (March 1973).

64. "Floating Concrete Bench," *Compressed Air Mag.* 8 (December 1973).
65. "They're Forever Blowing Bubbles," *Compressed Air Mag.* 79(8):10-11 (August 1974).
66. "Waging War on Animal Diseases," *Compressed Air Mag.* 6-9 (July 1975).
67. Havighorst, C.R. "Freezer with -200F Air; Kryoaire," *Food Eng.* 44: 54-56 (November 1972).
68. "Shipping Cargo in Pneumatic Tubes," Automation 21:52-53 (December 1974).
69. "Airborne in a Tube; Loaded Trains of Cargo are Whisked by Low-Pressure Air," *Compressed Air Mag.* 79:18-19 (August 1974).
70. "Mechanical, Hydraulic Pneumatic Options in Deep-Mine Transport," *Coal Age* 79:61-77 (July 1974).
71. Martin, J.W. and R.R. Faddick. "Experimental Verification of a Pneumatic Transport System for the Rapid Excavation of Tunnels," Report DOT-TST-76-63 to U.S. DOT Contr. DOT-OS-50100 (March 1976).
72. Konchesky, J.L. "Air and Power Requirements for the Pneumatic Transport of Coal in Horizontal and Vertical Pipelines," *J. Eng. Ind.* 97:94-106 (February 1975).
73. Hunter, J. "Pneumatic Transportation of Coal and Ash in Industrial Boiler Plants," *Mech. Eng. Proc.* 189(10):29-43 (1975).
74. "Hopper-Popper Cures Frozen Coal Problem Using Air Compressors," *Eng. Min. J.* 177:99 (January 1976).
75. "Bench-Blast Drilling Goes Underground," *Compressed Air Mag.* 14-15 (July 1975).
76. "Gas Wells by the Dozens," *Compressed Air Mag.* 14-15 (Mary 1975).
77. "Air-Powered Winch is Developed for Offshore Use," *Marine Eng/Log* 79:108 (December 1974).
78. Andreiev, N. "Temperature Extremes Favor an All-Pneumatic Foundry Control," *Control Eng.* 23:44-45 (February 1976).
79. Huskonen, W.D. "Maintaining Air System Efficiency," *Foundry* 99 (8):92-94 (August 1971).
80. "Pneumatic Logic Controls Iron Pouring System," *Hyd. Pneum.* 29:20 (June 1976).
81. Eckersley, J.S. "Controlled Shot Peening," *Compressed Air Mag.* 12-14 (August 1973).
82. "From Steam to Air Power," *Compressed Air Mag.* 16 (July 1975).
83. "Reevaluating a Workhorse," *Compressed Air Mag.* 16 (June 1975).
84. "Putting Words on Paper," *Compressed Air Mag.* 6-7 (October 1975).
85. Heldman, D.R. "Air as a Substitute for Water in Food Processing," *Food Technol.* 28:40 (February 1974).
86. "Pneumatic Control Eliminates Spillage of Edible Oil," *Process Eng.* 15 (March 1976).
87. "Schrafft Candy Company," *Compressed Air Mag.* 14-15 (November 1974).
88. "Commando Mark III," *Compressed Air Mag.* 11 (February 1976).
89. "Goblins All Year Round," *Compressed Air Mag.* 18 (March 1973).
90. Dallaire, A. "Pneumatic Waste Collection on the Rise," *Civil Eng.* 44:82-85 (August 1974).
91. Eames, J.O. and E.D. Sedergen. "Air Logic Insures Safe Handling of Ammo," *Hyd. Pneum.* 28:74-75 (July 1975).

92. Hehn, A.H. and A. Varacins. "Pneumatic Logic Controls Conveyor Line, *Hyd. Pneum.* 29:70-71 (August 1976).
93. Riegel, P.S. "Portable One-Man Recompression Chamber," *Mech. Eng.* 97:26-31 (December 1975).
94. Morris, N.M. *Introduction to Fluid Logic* (London: McGraw-Hill Book Co., 1973).
95. "Indexing Film with a Fluidic Sensor," *Compressed Air Mag.* 20 (October 1973).
96. Foster, K. and G.A. Parker, *Fluidic Components and Circuits* (London: Wiley-Interscience, 1970).
97. Walle, L.D. "Understanding Air Jet Sensors," *Hyd. Pneum.* 27:66-70 (June 1974).
98. Maas, M.A. "Pneumatics; it's Not Dead Yet," *Instr. Control System* 47:47-52 (November 1974).
99. Millichamp, D. "Pneumatic Technology is Just Hot Air When Electronics are Around," *Engineer* 238:34-35 (February 14, 1974).
100. Mamzic, C.L. "Pneumatic Controls Interface with Computers," *Instr. Control System* 48:31-36 (October 1975).
101. Kumar, V.S. "Fluidics–a Practical Solution for Impractical Environments," *Instr. Control System* 35-37 (December 1976).
102. Chalmers, D. "A Revolution in Air Compressor Technology," *Hyd. Pneum.* 29:(April 1976).
103. Spinale, J.R. "Air Logic Automates Machining of Medical Instruments," *Hyd. Pneum.* 69-71 (November 1976).
104. Woodson, C.W. "Fluidics Control Speed of Air Motors," *Design Eng.* 19(7):52-55 (July 1973).
105. McLain, L. "Few Takers for Pneumatic Transport Plan," *Engineer* 241:11 (December 4, 1975).
106. Ryan, R. "Application of the Unit Tank Train Concept to Petroleum Transportation Requirements of the French National Railways," MITRE WP-12106, (February 22, 1977).
107. "Pipe Burying Barge," *Compressed Air Mag.* 12-13 (November 1974).
108. "Beadwalls," *Compressed Air Mag.* 10-11 (June 1976).
109. Ayers, D.L. and R.E. Strong. "Compressed Air Storage; Another Answer to the Peaking Problem," *Power Eng.* 79:36-39 (August 1975).
110. Slater, G.F. "Salt Caverns–Multipurpose Storage Vessels," *Compressed Air Mag.* 16-17 (November 1975).
111. General Electric Co. "Economic and Technical Feasibility Study of Compressed Air Storage," Report NC-946/SRD-76-037 to Office of Conservation U.S. ERDA contr. E (11-1)-2559 (March 1976).
112. Gasparovic, N. and D. Stupersma. "Gas Turbines with Heat Exchanger and Water Injection into Compressed Air," *Combustion* 45:6-16 (December 1973).
113. Glendenning, I. "Long-term Prospects for Compressed Air Storage," *Appl. Energy* 2(1):39-56 (January 1976).
114. Stys, Z.S. "New Energy-Storage Concept Sold," *Elect. World* 183:46-47 (June 15, 1975).
115. "World's First Air Storage Plant Nearing Completion," *Gas Turb. World* 16-20 (November 1976).

116. Beam, B.H. and A. Giovanetti. "Underground Storage Systems for High-Pressure Air and Gases," *National Congress on Pressure Vessels & Piping,* San Francisco, California 23-7 June 1975 ASME Paper #75-PVP-5 (1975).
117. "*In Situ* Coal Gasification," *Compressed Air Mag.* 82(1):14-15 (January 1977).
118. U.S. Commerce Department. "Survey of Current Business," (September 1976).
119. U.S. Commerce Department. "Input-Output Structure of the U.S. Economy: 1963" (1969).
120. U.S. Commerce Department. "Input-Output Structure of the U.S. Economy: 1967" (1974).
121. U.S. Commerce Department. "Inter-industry Transactions in New Structures and Equipment, 1963 and 1967 (1975).
122. U.S. Commerce Department. "1972 Census of Manufacturers, SIC 3561" (October 1976).
123. U.S. Commerce Department, March 1975.
124. U.S. Commerce Department. "Pumps and Compressors, 1975" (October 1976).
125. U.S. Commerce Department. "1972 Census of Manufacturers, SIC, 356" (March 1975).
126. Watson, K.P.A. "Trends in Compressed Air Applications," *Hyd. Pneum. Power* 17(194):66-72 (February 1971).
127. Watson, K.P.A. "Trends in Compressed Air Applications," *Hyd. Pneum. Power* 18(207):114-120 (March 1972).
128. Watson, K.P.A. "Trend in Compressed Air," *Hyd. Pneum. Power* 19 (218):42-45 (Februrary 1975).

Appendix. Established Compressed Air Use

Industry	Representative Applications	Compressor Equipment
Agriculture	Applying fertilizer & pesticides	Small portable rotary
	Leveling sand dunes	
	Pneumatic digging for harvest	
Mining: Metals and Coal	Sand conveying	Large units–reciprocating
	Filling cracks and seams with cement	
	Rock and hammer drills, pile driving	
Crude Petroleum and Natural Gas	Air pumping for lifting oil and gas	Large centrifugal
	Repressuring fields	
	Loading nitroglycerin in oil wells	
Chemicals and Fertilizer Mining	Plug drills	Large centrifugal
	Brine pumping	
Construction	Subacqueous tunneling	Large centrifugal
	Submarine drills	
	Caisson work	
	Sand blasting, pneumatic tools, etc.	Small reciprocating and rotary
Food	Conveying and agitating bulk liquids, granular material	Centrifugal
	Canning punch presses	Reciprocation and rotary
	Drying grain	
	Sausage stuffing	
Tobacco Manufacturing	Tobacco dust cleaning	Centrifugal
	Moistening tobacco for rolling	

Appendix, Continued

Industry	Representative Applications	Compressor Equipment
Textiles and Apparel	Conveying, agitating, dyes and other solutions	Unit reciprocating and rotary
	Baling presser	
	Humidifying and moistening systems	
	Cleaning flaxmill boiler flues	
Lumber and Wood Products	Unloading logs	Unit reciprocating and rotary
	Cutting panels and friezes with sand blasts	
	Automatic sprinkler system	
Furniture	Air clamps and vises	Unit reciprocating and rotary
	Wood carving	
	Air lift tables	
	Pneumatic buffing, enameling and varnishing	
Paper and Paperboard	Fabricating newsprint	Centrifugal
	Air flotation	
	Moling, drying and sterilizing pulp-paper containers	
Chemicals	Reactant air in butadiene, ammonia, ethylene, etc.	Centrifugal
	Air separation	
	Conveying, agitating, aeration of acids and other liquids	Reciprocating and rotary
	Testing and caulking tanks and pipelines	
Drugs	Mixing ingredients	
	Fermentation process air	Centrifugal
	Evaporation systems	

Paints and Allied Products	Filling and sealing cans	Reciprocating and rotary
Petroleum Refining	Catalyst circulation for cracking process	Centrifugal
	Alkylation process pressurizing	
	In situ combustion for secondary recovery of high-viscosity oil	
	Gasoline extraction from natural gas	
Rubber and Miscellaneous Plastics	Vulcanizing	Centrifugal
	Drip sprays	
	Cleaning rubber goods	
	Cutting tires and tubes in molds	
Glass and Glass Products	Blowing and etching glass	Reciprocating and rotary
	Molds and presser	
	Pumping glass-sand by return-air systems	
Stone and Clay Products	Aeration of cement bins and silos	Centrifugal
	Bag cleaning and baling	
	Slurry agitation	
	Sand sifters	
Primary Metals Manufacturing	Aerating metal to remove impurities	Centrifugal (estimate 10^5 ft^3 of compressed air for 1 ton of steel
	Oxygen enrichment	
	Applying core paste	Reciprocating
	Chipping, sand blasting, wirebrushing castings and billets	
	Air cylinders for bending and straightening	
	Forge equipment	
	Furnace flame curtain	
Heating and Plumbing	Pneumatic boiler controls	Reciprocating
	Flue rollers, expanders, reamers, etc.	

Appendix, Continued

Industry	Representative Applications	Compressor Equipment
Farm Machinery	Blowing oil furnaces Pneumatic painting and whitewashing Sandblasting for cleaning castings and removing paint	Centrifugal
Special Industry Machinery	Grain conveyance for distilleries Automatic temperature control in dairies Bending wood Air vises and clamps	All types
Machine Shop Products	Air chucks and vises Air motors for grinding, buffing, etc. Belt shifters for heavy machine tools Workpiece feeding and ejection devices	Reciprocating and rotary
Service Industry Machines	Handling dangerous solutions Laundry equipment Air-operated presser	Reciprocating and rotary
Motor Vehicles and Equipment	Testing radiators and gasoline tanks Pneumatic tool machine work Air hoist of heavy assemblies Driving shackle bolts Compressing and fastening cushion springs	Unit reciprocating and rotary
Aircraft and Parts	Refueling airplanes Pneumatic toolwork of all varieties Compressed air jet engine starting systems Air-operated safety devices	Reciprocating and rotary
	Wind tunnels for testing airplanes	Centrifugal and axial

Ships and other Transportation Equipment	Air hoists, lifts and motors	Reciprocating and rotary
	Riveting, caulking, sanding drills	
Miscellaneous Manufacturing	Compressed air locomotive	Centrifugal
	Explosive farming using air-operated pulse	Reciprocating
	Air instrument control	
	Die casting	
	Precision measuring devices	
	Testing containers	
Transportation	Air brakes	All types
	Rail sandblasting prior to welding	
	Third-rail ice scraper	
	Starting and maneuvering motor chips	
	Switch and signal systems	
Electric Gas and Water Service Utilities	Caulking boiler and pipelines	All types
	High-pressure gas transmission	
	Plug drills and rock drills for trenching	
	Water filtration	
	Pumping sewage	
Wholesale and Retail Trade	Cleaning by air jet	Reciprocating
	Pneumatic mail system	
Hotels and Lodging	Accumulator system	Reciprocating
	Sandblasting buildings	
Auto Repair and Service	Pneumatic jacks, lifts and hoists	Reciprocating and rotary
	Fender and bumper straightening	
Government Enterprises	Salvaging sunken ships	Centrifugal
	Torpedo charging	
	Highway construction and repair	
	Naval ship use	

INDEX

INDEX